Think Straight

An Owner's Manual for the Mind

JON GUY

Prometheus Books

Essex, Connecticut

Prometheus Books

An imprint of Globe Pequot, the trade division of
The Rowman & Littlefield Publishing Group, Inc.
4501 Forbes Blvd., Ste. 200
Lanham, MD 20706
www.rowman.com

Distributed by NATIONAL BOOK NETWORK

British Library Cataloguing in Publication Information Available

Library of Congress Cataloging-in-Publication Data

Names: Guy, Jon (Skeptic), author.
Title: Think straight : an owner's manual for the mind / Jon Guy.
Description: Lanham, MD : Prometheus, [2022] | Includes bibliographical
   references and index. | Summary: "Our world is awash in mis- and
   disinformation, baseless conspiracy theories, New Age ideology,
   antiscience propaganda, and all manner of magical thinking. In careful
   detail, author Jon Guy investigates the art of thinking critically,
   offering readers the ability to empower themselves and our society at
   large"—Provided by publisher.
Identifiers: LCCN 2022001247 (print) | LCCN 2022001248 (ebook) | ISBN
   9781633887978 (paperback) | ISBN 9781633887985 (epub)
Subjects: LCSH: Critical thinking. | Magical thinking. | Conspiracy
   theories.
Classification: LCC BF442 .G89 2022 (print) | LCC BF442 (ebook) | DDC
   153.4/3—dc23/eng/20220118
LC record available at https://lccn.loc.gov/2022001247
LC ebook record available at https://lccn.loc.gov/2022001248

∞™ The paper used in this publication meets the minimum requirements of American
National Standard for Information Sciences—Permanence of Paper
for Printed Library Materials, ANSI/NISO Z39.48-1992.

Contents

Foreword

As a critical thinking teacher, during the past several years, I have looked at the news and often thought to myself, "Why do I bother?" Conspiracy theories and pseudoscientific practices like the ones discussed in this book, some of which you may believe in, abound. As I write, despite all evidence to the contrary, the overwhelming majority of a major political party in the United States thinks that the presidential election of 2020 was stolen.[1] Even though a remarkably effective, safe, and unquestionably lifesaving vaccine has been available for more than a year for free, 35 percent of American adults have not yet been fully vaccinated against SARS-CoV-2.[2] Despite experiencing extreme weather as a direct result of climate change, society's production and consumption of oil is only expected to increase.[3] To someone who values evidence and reason and the scientific mindset, such lapses in critical thinking on such a vast scale are certainly disheartening.

Nonetheless, while critical thinking is not something that is practiced at the national or international levels, it certainly can be nurtured within individuals, though it takes work. I'm not sure it's possible to tell someone how to think critically any more than you can tell them how to play a piano. Critical thinking is a skill that is developed through practice. It takes effort, as well as a willingness to be wrong and admit it; it often demands that you suspend your judgment and be comfortable with uncertainty; it may well require you to go against the crowd when the evidence is just not there; hopefully, it will also compel you to suspend your own stubborn individualism when the crowd has a point. It requires you to be humble enough to admit your ignorance, brave enough to trust experts, and discerning enough to recognize genuine expertise. It might also require you to be imaginative enough to realize that sincere conviction is just a feeling that we attach to both true and untrue propositions.

Unfortunately, not everyone is receptive to the idea of suspending judgment until the evidence comes in. Someone whose personal, emotional, or social needs are satisfied by a flawed understanding of the universe will often prefer to remain comfortable and fight to maintain their beliefs. Asking them to give up that comfort is a scary ask. Furthermore, strong emotion is contagious. Sadly, reasoning is not. And even the most motivated critical thinking enthusiasts come up against the limitations of their own brains. Learning to recognize the mental shortcuts, biases, assumptions, and logical fallacies by which we all mislead ourselves is crucial for developing a skeptical, open mindset. As has been said many times before, the skepticism implicit to critical thinking is rooted in doubt, but it is not cynicism, denialism, or knee-jerk nihilism. Skepticism is productive doubt, a means by which we test our ideas against the reality of the world and allow the accumulation of evidence to improve our understanding by progressively excluding errors. *Think Straight* is both a product of this process and a prompt to encourage others to practice critical thinking.

Think Straight was conceived, researched, and written almost entirely in the most adverse circumstances, and it is a genuine marvel that it was written at all. The number of gatekeepers and intermediaries between Jon Guy and the research he was incorporating in his book, to someone unfamiliar with the corrections system in this country, was staggering. Clearly, such obstacles make it hard for those in the system to cultivate their curiosity about the world. Even though this book by all rights should probably not exist, you are nonetheless holding it in your hands. *Think Straight* is a testament to the power of critical thinking to transform lives, and I think that's why one bothers to teach critical thinking in dark times: the fact you are holding this book means that there is hope.

Robert Blaskiewicz
Atlantic City
February 2022

CHAPTER 1

Meet Skepticism

An introduction to any topic should begin with an understanding of what that topic is. Unfortunately, society is currently structured so that most of us begin learning long before we learn how to think. This makes about as much sense as teaching math before introducing numbers or teaching writing before introducing letters. Accordingly, this book begins by endeavoring to explain critical thinking (CT)—an amazingly powerful subject that can revolutionize your worldview.

Building Vocabulary

Objective:	Veridical:
Not influenced by emotion or prejudice; lacking any presupposition.	The face value acceptance of information; belief that lacks thought or questioning.

Theory:	Skepticism:
A coherent statement or set of ideas that explains observed facts or phenomena or that sets out the laws and principles of something known or observed.	An approach to truth claims that encourages doubt and demands reliable evidence.

Toward an Understanding of Critical Thinking

Famed theoretical physicist Albert Einstein (1879–1955) is often credited with saying that you don't understand something unless you can teach it to children. Another famous popularizer of science, Carl Sagan, refined this adage, arguing that you don't understand something until you can explain it to someone else so that they understand it as well. Critical thinking is a big topic, and my goal is that by the end of this chapter, you'll be able to explain it to children so that they understand it as well as you do.

With real-world examples, and drawing from the latest science, I'll demonstrate what it means to think critically and explain how it's done. Put simply, CT is the process of formulating effective methods of evaluating the validity of claims and propositions and applying those methods to claims made by ourselves and others. CT has two aims:

1. *demonstrating* that a claim or proposition is true (or false); or
2. *determining* whether a claim is true (or false) (Schick & Vaughn, 2005, p. 154).

> Our whole dignity consists in thought. Let us endeavor, then, to think well: this is the principle of ethics.
> —Blaise Pascal

Either position can be undermined by how we think and by what we think. That is, misconstruing reality is sometimes a matter of how we believe, and sometimes it's a matter of what we believe. Critical thinkers care about what's true and have built arsenals to blast through massive amounts of nonsense to discover the wonders of reality.

Throughout this book, we'll meet some of the biggest names in science and **skepticism**, psychology and philosophy, and we'll learn about their discoveries and ideas, where they come from, and how that knowledge helps us Think Straight. On this journey, we'll discuss famous experiments and experimenters, while familiarizing ourselves with these amazingly powerful tools we keep inside of our heads that almost none of us have ever been taught how to use. If you think about it, that's pretty scary: the human brain is easily the most powerful and potentially dangerous tool in the known universe, and nearly every one of us goes our entire lives without an operator's manual! CT is just that: an operator's manual for learning how to use a brain effectively.

By contrast, *un*critical thinking has been called "perhaps the most unrecognized and underreported global crisis of all" (Harrison, 2013, p. 18), because it leads to poor decisions, wasted time and money, and numerous health complications, including death. Sloppy thinking stifles progress, promotes nonsense, and cripples our independence. Sure, saving money beats wasting it; believing what's true beats believing falsehoods; being healthy beats being sickly. Nevertheless, far too many people harbor beliefs that critically undermine these basic truths, which is one reason CT education should begin in grade school (Lilienfeld, 2017).[1] Since the best remedy for bad thinking is good thinking, you cannot go wrong with improving the quality of your thinking. Of course, simply knowing how to Think Straight doesn't mean that you'll never be wrong or that you'll never be tricked again. Far from it. The act of knowing is a cognitive process, and cognitive processes are the very things that lead us astray. Accordingly, we must do more than simply think: we must learn *how* to think.

Throughout this book, we'll explore many of the successes of science and the relevance of science to our everyday lives. And, on top of providing the tools needed to intellectually defend science, this book will introduce readers to the enduring conflict between science and non-science, between what we believe and what we know, and between how we see the world and how the world really is. Society is driven largely by science, yet frighteningly few of us know anything about it. Moreover, science and scientific advancement are constantly undermined by an onslaught of mis- and disinformation. Some of that information presents itself in the form of impossible allegations that I call *incredible claims*, which I define *as an assertion or proposition that contradicts well-established scientific facts and/or generally accepted standards of proof (GASPs)* (see chapter 10). Others have called them weird things (Schick & Vaughn, 2005; Shermer, 1997)[2] or extraordinary claims,[3] both being valid descriptors of incredible claims. I prefer *incredible claims* because the claims we encounter are often incredible in both senses of the word: that is, they're incredible in that they're *not* credible, and they're incredible in the sense of being astonishing. However, skeptics understand that initially, some incredible-sounding claims turn out to be true. Before the invention of the microscope, who would've thought that sickness could be explained by the existence of tiny creatures invading our bodies like invisible armies? Before the discovery of plate tectonics, who would've thought hidden forces could move continents and create mountains? Before Einstein's theory of special relativity, who would've thought that whether two things happen at the same time depends on who's looking and from where?[4]

Incredible claim are everywhere, but virtually none of them are supported with good evidence. As we meet the seemingly endless array of incredible claims, we'll learn about how toxic they can be to ourselves, to our culture, to our communities, and to our minds. If we're not careful, without realizing it, false beliefs become lead anchors, pulling us deeper and deeper into a sea of irrationality. Most of us wouldn't intentionally believe something we thought was false, yet billions of people not only maintain a host of false beliefs; many people tragically die for them.

Breaking Bonds

Born a slave, second-century Stoic philosopher Epictetus[5] once said, "Wherever anyone is against his will, that is to him a prison." We can be imprisoned in many ways, including by dis- or misinformation. CT is the key to unlocking us from the cognitive bondage that too many of us live our lives constrained by.

Over the course of thirteen chapters, we'll see that many perceptual and cognitive flaws go hand-in-hand with untrained, uncritical thinking, which, like that ever-growing snowball, prevents us from intelligently dealing with those flaws. Detailed explanations can be quite complicated, and most of us lack the training needed to deal with or understand the complexities of reality. Scientific facts and discoveries are sometimes unfamiliar, and for many, uncritical thinking offers simpler, easier explanations. The term *cognitive miser* is used by psychologists to explain how the mind tries to expend as little effort as possible; thus we make up our internal models of reality in simpler, rather than more sophisticated, thought processes. As readers will see, being cognitive misers restricts us from enjoying reality on its own terms.

> Critical thinking is the
> most important driver
> of the advancement of
> the human race.
> —Brian Dunning

Fortunately, many of us are already skeptics to some degree. When someone tells stories about dodging bullets, fist-fighting bears, or becoming millionaires, we demand proof. Having seen too many people spin a yarn, we adhere to the Royal Society's motto, *Nullius in Verba*: Take nobody's word for it. We need solid evidence to believe that the same guy who made millions selling drugs needs to borrow a Ramen until payday.

Astonishingly, when it comes to claims that are *even more* incredible, such as the ability to communicate with dead people or read people's minds, some people think even *less* critically. For instance, few of us understand how magicians perform their tricks, yet we're usually unwilling to believe that they're performing real magic. But psychics performing those same tricks convince many of their omniscient powers (Hines, 2003, p. 131). Such debunked claims fully contradict what we know about the laws of nature, yet people still find them compelling truths about the world. A claim that contradicts well-established scientific facts may not be wrong, but it should always be approached skeptically. *That* is CT, and since there will never be a shortage of incredible claims, it will never be wise to turn our brains off and believe something before we have evidence. If we want to avoid being fooled, conned, or taken advantage of, evidence must precede belief!

From time to time, I'll draw on some far-fetched examples such as aliens or ghosts to illustrate my points. My goal is to convey what causes people to believe incredible claims so that we might avoid the sort of thinking that leads us to false conclusions in the first place. And while we might laugh at those who believe they saw Bigfoot at a shopping mall or were abducted by aliens, uncritical thinking is no laughing matter. Research show that millions of Americans think Bigfoot (13.5 percent) and telekinesis (19.1 percent) are real, space aliens have visited Earth in modern times (24.7 percent), and global warming is a hoax (42.1 percent) (Poppy, 2017). In fact, about 75 percent of Americans hold at least one paranormal belief (Moore, 2005; Newport & Strausberg, 2001). These numbers are especially alarming since these are the same people who believe they're making rational decisions when voting to elect public officials or deciding who gets parole and who doesn't, for example. And what about those elected officials? Their own lack of CT leads to runaway pandemics, bizarre distortions of science, witch hunts, and more.

Thinkers . . . Start Your Brains!

Thinking Straight is an approach to examining the world objectively. It begins by asking two questions: "Is X the case?" and "How do we know?" This approach provides a basic level of skepticism, including a technique for challenging any claim or belief. And, when it comes to claims and beliefs, we can try to figure out the truth behind them, or we can try to reinforce what we already believe. If we choose the former, sometimes we'll discover that we believe what is true, and sometimes we'll realize what we believe is false. If we choose the latter, we can be right by accident or monumentally wrong and unwilling to admit it.

Facing a death sentence for supposedly corrupting the youth of Athens (he urged them to think for themselves), the famous Greek philosopher Socrates (470–399 BCE) reportedly said, "The unexamined life is not worth living" (Cooper, 2000, p. 39). Some have modified this quote, exclaiming that "the unexamined thought is not worth thinking." I agree with Socrates but doubt that every thought must be examined to have value. We needn't examine the thought "Boy, I just love my kids" or "My wife is beautiful" or "These roses smell amazing." Such thoughts are well worth thinking without any examination. Others have used the process of thinking to confirm their very existence, such as René Descartes (1596–1650), who famously said, "I think, therefore I am." Descartes decided that *everything* should be questioned, including his own existence. Socrates and Descartes shared an understanding of the process and importance of CT. They understood that CT is not about what you know or what you believe; it's something you do.

The name of this book was inspired by the failed "scared straight" programs (Lilienfeld, 2005; McCarthy, 2005), which were ineffective, costly, and likely caused more harm than they prevented (Lilienfeld, 2007; Petrosino, Turpin-Petrosino, & Buehler, 2004). We know that challenging prisoners' thinking and developing their thinking skills positively affects their ability to reason (Boghossian, 2006) and, hopefully, helps them make better decisions in the real world, and we relied on science to figure these things out. Since one of our goals here is learning how to think, meeting the challenge demands that we learn how to think clearly about how to clearly think.

The human rights activist Mahatma Gandhi once said, "If we could change ourselves, the tendencies in the world would also change." The world Gandhi knew is full of influence technicians (Taylor, 2017) and compliance professionals (Cialdini, 2007)—for example, con artists, marketers, media, advertisers, salespeople, cult/military/religious recruiters, authorities, governments, and more—who believe, as brain scientist Susan Greenfield does, that by changing the world they can change the people in it (Taylor, 2017, p. 353). By predicting our behavior, these "agents of influence" ingeniously manipulate facts to make us think and behave in specific ways (Pariser, 2012). This book explains how to identify and resist many of the triggers and traps that these agents of influence rely on. I take Greenfield's warning seriously but turn it inward because I believe that change begins with individuals.

CT is also largely about asking question (Wade & Tavris, 2017, pp. 7–8). The Socratic Method was first introduced by Socrates in Plato's dialogue *Theaetetus*. We'll learn how to apply the Socratic Method in chapter 10. For now it suffices to say that the Socratic Method is a way of presenting, comparing, and examining different points of view in an attempt to discover the truth. And "if it is important to know the truth," as Carl Sagan understood, "then nothing other than a committed, skeptical scrutiny is required" (Sagan, 2007, p. 145). Questions like "How do you know?" "What evidence supports your position?" "What evidence rejects it?" or "What do experts think?" help us decide whether a claim is worthy of belief. And skeptics don't allow beliefs to invade their heads for free. The price for entry is evidence and reason (Harrison, 2015, p. 222).

Critical Thinking Defined

CT covers a lot of ground, so no one definition fully explains this far-reaching subject. Psychologists define CT as "the ability and willingness to assess claims and make judgments on the basis of well-supported reasons and evidence rather than emotions or anecdote" (Wade & Tavris, 2017, p. 6). The National Council for Excellence in Critical Thinking defines CT as the "intellectually disciplined process of actively and skillfully conceptualizing, applying, analyzing, synthesizing, or evaluating information gathered from, or generated by, observation, experience, reflection, reasoning, or communication, as a guide to belief and action" (Scriven & Paul, 1987). That's a bit of a brainful. In simpler terms, CT is characterized by the following:

- clear, disciplined thinking that is rational, open-minded, and guided by evidence
- reasonable, reflective thinking focused on deciding what to believe or do
- a commitment to using reason in the formulation of beliefs
- using logic and applying systematic doubt to any claim or belief
- consistency in the standards and application of GASPs

Attractive traits, to be sure, and many people eagerly admit what great critical thinkers they are. However, one of my goals here is to show that everyone can be fooled, myself included. The psychologist James E. Alcock explains: "Possibly the most common pitfall with regard to critical thinking is the belief that one is already a good critical thinker" (Alcock, 2018, p. 531). Everyone harbors intuitive beliefs about psychology, physics, biology, and so on, because everyone has personal experience that supports their beliefs. Unfortunately, problems arise when intuitive beliefs get more credibility than established facts (Stanovich, 2013, pp. 199–200) because our experience is more compelling than scientific research.

Thinking Styles and Veridical Vending

Without an obvious reason to doubt our senses, we treat the information they bring us veridically, meaning we accept it as true prior to evaluating it. Since doubt is more cognitively taxing, we also judge statements as "true" significantly faster than we judge them as "false" (Gilbert, 1991; Harris, 2011, p. 120; Shermer, 2011a, pp. 133–37), supporting the hypothesis that veridicality is our default, or *unthinking*, mode (Alcock, 2018, pp. 151–53). Making matters worse, we're significantly less critical of information that's agreeable or aligns with our biases (Petty, Wegener, & White, 1998).[6]

The idea that we initially process sensory information as true traces back to the seventeenth-century philosopher Baruch Spinoza (1632–1677) (Gilbert, 1991). Modern research supports Spinoza's hypotheses that our brains use two distinct systems to process information (Evans, 2003): an intuitive system and a rational system. As we'll see, both systems have advantages and disadvantages. Moreover, whether people are intuitive or rational thinkers often has nothing to do with intelligence but rather is the result of their biology and life experience (Epstein, Pacini, Denes-Raj, & Heier, 1996).

Veridicality is a product of what the psychologist Daniel Kahneman calls System 1 thinking (Kahneman, 2013, pp. 20, 450; Stanovich & West, 2000). System 1 thinking is one of two ways our brains process information. According to Kahneman, "*System 1* operates automatically and quickly, with little or no effort and no sense of voluntary control" (Kahneman, 2013, p. 20). Some of what System 1 does includes the following:

- detecting that one object is more distant than another
- orienting to the source of a sudden sound
- reaching out to catch something falling out of your hands
- making a "disgust face" when shown something gross
- detecting hostility in a voice
- answering 2 + 2 = ?
- driving a car on an empty road
- understanding simple sentences (Kahneman, 2013, p. 21)

Such tasks are performed without conscious thought and sometimes without us realizing a task was performed. If a loud noise suddenly interrupts your consciousness, you'll direct your attention to it without thought.

Our other mode of information processing is System 2, which we use to focus attention, make tough choices, concentrate, calculate (Kahneman, 2013, p. 21), and perform other functions, including the following:

- focusing on the voice of a particular person in a crowded room
- searching memory to identify a surprising sound
- maintaining a faster walking speed than is normal for you
- monitoring the appropriateness of your behavior in a social situation
- counting the occurrences of the letter *a* in a page of text
- telling someone your phone number
- parking in a narrow space
- questioning whether something is true or accurate
- checking the validity of a logical argument (Kahneman, 2013, p. 21)

Research shows that the busier we get, the more rushed we feel, and the more we have on our minds, the more System 1 is in control (Chugh, 2004). We'll revisit the concepts of System 1 and System 2 throughout this book to help understand how we respond to our environment, how we evaluate information, and how much influence we have on our own thinking. Before moving forward, however, I want to tackle the term "**veridical**," which we'll also return to again and again. Understanding veridicality is so important that I want to belabor it with the following analogy.

Human brains process information like vending machines process bills. Anything fed into the machine resembling a bill will trigger a mechanism instructing the machine to accept the "bill" without deciding whether the bill is counterfeit.

According to its programming instructions, the machine veridically accepts anything resembling a bill. That is, the machine automatically accepts all "bills" as legal tender. Not until something is *obviously* wrong with the bill—when the machine's programming instructions are obviously violated—is it rejected. If legal tender is rejected, as we know, a bit of smoothing out and the machine usually accepts it. However, sometimes a perfectly good bill is rejected if the machine "thinks" something is amiss. Moreover, sometimes counterfeit bills are deceptive enough to trick the programming instructions. And so it is with our brains.

Core critical-thinking skills:

- Self-regulation
- Interpretation
- Analysis

- Inference
- Evaluation
- Explanation

These core elements allow us to think critically by using logic and evidence as guides to reality.

Sensory information (bills) entering the brain is processed veridically by System 1. If we see a birth following a pregnancy, we have no reason to suspect anything is wrong with the chain of events: births follow pregnancies, not the other way around. Unless something is obviously wrong with the information we receive—and what's "obviously wrong" depends on one's internal model of reality—our brains accept the information as true. "Obviously wrong" information goes to System 2 for further processing. If we saw a chicken give birth to a frog, System 2 would activate to analyze why the information doesn't match our internal model of reality.

As often as this system works, it isn't flawless, yet without CT skills, our brains operate like vending machines without any programming instructions. Nobody knows everything, and even false information can be "smoothed out," fed back to System 1, and accepted as valid. The more sophisticated the vending machine's programming instructions, the less susceptible it is to good counterfeit bills. Likewise, the better thinkers we are, the less susceptible we are to counterfeit information.

Critical Thinking for Everyone

CT is the backbone of science, and both are highly effective tools for examining reality. Science and CT are the tools we've honed over centuries to help us learn from our mistakes and compensate for the many flaws that make us human.

Skepticism is a set of skills—a process, if you like—for analyzing and thinking about information. All this analyzing and thinking happens in the brain, so it makes sense to understand how the brain works, including how it doesn't. Since learning about the brain can be challenging, I've devised a metaphor that will help visualize some of the processes we'll be discussing.

Imagine the brain not as a three-pound lump of meat but as a planet, covered inside and out by a vast network of interconnected highways and byways. I call it Street World. On Street World, every imaginable type of roadway exists, from dirt roads and city streets to freeways, interstates, highways, and superhighways. In the brain, information takes the form of electrochemical signals that travel across vast networks of interconnected neurons called neural networks. These neural networks make up our *worldview*, a philosophy of life incorporating many beliefs that help us make sense of a wide range of issues (Vaughn, 2019, p. 418). On Street World, information takes the form of travelers using this vast system of roadways. Neural networks are strengthened or weakened based on how frequently they're used (Burton, 2008, pp. 42–43), such as by the strength of a belief and how often that belief is reinforced. Likewise, roadways grow in size and connectivity based on their popularity. Strongly held beliefs or deeply ingrained memories, for example, might be represented by an expressway or a superhighway with lots of travel, whereas lesser beliefs might be represented by a walking trail or an alley. The bigger the roads, the more connectivity. But if a belief goes out of fashion or gets replaced, the road slowly deteriorates due to the lack of upkeep and travel. Such experiences are manifested as memory loss, new opinions, and even epiphanies.

> The world is changed by your example, not by your opinion.—Paulo Coelho

Changing a belief is like demolishing a road and replacing it. It follows, then, that destroying a superhighway (abandoning a strongly held belief) requires substantially more construction than paving a dirt road (changing a weakly held belief). However, the destruction of a motorway doesn't mean travel is no longer necessary; it means another route is needed. Moreover, if we're replacing existing beliefs, that is, if we're building new roadways, we want to ensure we're doing it right. We want to use the best material, recruit the best engineers, and build our structures with the latest technologies. Thinking Straight is like the blueprint for how to build reality-based beliefs that are founded on the best way to examine information.

Traits of the Skeptic

In announcing a consensus statement on CT, the American Philosophical Association (APA) stated that "the ideal critical thinker is habitually inquisitive, well-informed, trustful of reason, open-minded, flexible, fair minded in evaluation, honest in facing personal biases, prudent in making judgments, willing to reconsider, clear about issues, orderly in complex matters, diligent in seeking relevant information, reasonable in the selection of criteria, focused in inquiry, and persistent in seeking results which are as precise as the subject and the circumstances on inquiry permit" (American Philosophical Association, 1990, p. 2). The APA also detailed what they believed are six core elements of CT: interpretation, analysis, evaluation, inference, explanation, and self-regulation. While no one could personify all these traits at all times, CT demands a sincere attempt to incorporate these skills into our lives.

Critical thinkers are inquisitive, analytical, systematic, and judicious truth-seekers, valuing curiosity, skepticism, independent reasoning, and empirical judgment. They appreciate the importance and value of logic, argument, evidence, and truth. Accordingly, critical thinkers apply these skills when evaluating information. CT is clear thinking, informed by evidence, guided by reason, bound by logic, committed to truth, and aware of biases and thinking errors.

Skeptics place emphasis on *clearly* and *rationally* analyzing and critiquing everything—and I do mean *everything*. In 1941, Edward M. Glaser called CT "a persistent effort to examine any belief or supposed form of knowledge in the light of the evidence that supports or refutes it." Glaser's words are instructive and underscore the many goals of rational, critical thinking. And what's so special about being rational? For one, better reasoning abilities decrease one's chances of believing incredible claims (Hergovich & Arendasy, 2005). But more importantly, rationality curtails extremism and racism, inhibits women's inequality, preserves biodiversity, and keeps emotions like envy, greed, and anger in check (Zeigler, 2020). Moreover, rationality grounds our beliefs and allows us to appreciate life on its own terms. Being rational is important because basing conclusions on logic and evidence gets us much further than basing them on hopes and speculations.

Even more than these practical matters, however, CT skills are important tools for personal growth and maturity. CT involves problem solving, decision making, metacognition, rationality, rational thinking, reasoning, knowledge, and intelligence accompanied by moral components such as reflective thinking (Kompf & Bond, 2001). Skeptics, therefore, must reach a level of maturity in their development and possess a certain attitude along with a set of taught skills.

How many times have you done something that you immediately regretted because you hadn't thought it all the way through? How many times have knee-jerk reactions gotten you into trouble? Have you ever defended a position you really didn't know how to defend—or even why you were defending it at all? CT slows down our instinctual reactions to information and stimuli, buying us those crucial moments we may need before responding emotionally to inevitabilities such as insults, deception, or misinformation.

In this age of technology and instant information, our lives are constantly besieged with incredible claims. News media misreport stories while putting partisan slants on the information they spread. We blind ourselves to reality by selectively choosing what news to watch, what books to read, what authorities to listen to, and even what religion to follow. Doing so ensures that we encounter much in the way of arguments and evidence that support our beliefs and little in the way of arguments and evidence that oppose them.

And we aren't alone. Megacorporations such as Google, YouTube, Amazon, and Facebook use sophisticated algorithms that mine our personal information by monitoring our digital activity to maximize advertising returns (Harrison, 2017). These algorithms create what internet activist Eli Pariser calls "filter bubbles" (Pariser, 2012), barriers that insulate us from conflicting viewpoints and information. Filter bubbles expose us to information we may want and agree with but filter out information that may be important or true, effectively distorting reality and diminishing our ability to act as informed citizens.

Nearly every moment presents an opportunity to make a decision, from small decisions like whether to have extra dessert, to much bigger decisions like whether to steal to feed our families. For skeptics, it's important to use reasoned judgment and think critically before making decisions or taking a stance. Doing so requires awareness of both good thinking and bad, and motivation to practice our skills as critical thinkers (Vaughn, 2019, p. 31). To this end, Glaser proposed that Thinking Straight requires the following:

1. an attitude of being disposed to consider in a thoughtful way the problems and subjects that come within the range of one's experiences
2. knowledge of the methods of logical inquiry and reasoning
3. some skill in applying those methods (Glaser, 1941)

> We cannot solve the problems we have created with the same thinking that created them.
> —Albert Einstein

Nevertheless, better thinking and decision making don't get us far if we still can't make good decisions. If CT helps you identify a scam and you respond by pummeling the scammer, well, you're not much better off than you were without CT. If, on the other hand, you simply save money by not falling for the scam, then CT paid off, literally. Most of us try to make rational decisions, and many of us naïvely (and wrongly) believe our decisions are guided by reason. Truthfully, our decisions are "bounded" by certain limitations, such as our intelligence, access to information, cost restraints, timing, or who the decision affects (Bazerman, 2018, pp. 243, 248–49).

So while CT itself won't fix our lives, Thinking Straight plays a vital role in freeing us from the confines of bad thinking and poor decision making. Like CT, good decision making requires motivation, System 2, and the ability to use System 2 well—a mirror of Glaser's CT criteria. Business psychologist and Harvard professor of business administration Max H. Bazerman provides the know-how for good decision making by suggesting that we take the following steps before making a decision:

1. Define the problem.
2. Identify the criteria necessary to judge the options.
3. Weigh the criteria.
4. Generate alternatives.
5. Rate each alternative on each criterion.
6. Compute the optimal decision. (Bazerman, 2018, p. 244)

One of my goals is to put Glaser's and Bazerman's advice to good use by developing skills that help us make the best decisions we're capable of making. The first step toward better decision making is better thinking. Thus, skeptics should know about the confirmation bias, be able to identify logical fallacies, understand the purpose and structure of logical arguments, spot both hucksters and experts, and understand the many problems with memory and perception. Adopting this strategy reduces the effects of biases and improves our ability to objectively evaluate information and make good decisions by transitioning from reliance on intuitive System 1 thinking, to engaging in more meaningful and deliberative System 2 thought.

None of this is to say that thinking is valid only if it's devoid of emotion. Emotions evolved to help us deal with the challenges of life, and they drive our motivations and goals (Nesse & Ellsworth, 2009). Without emotions, we wouldn't experience humor, joy, awe, or love. Music would sound like any other noise, and we'd be unmoved by the sound of a crying baby. Thinking critically doesn't attempt to replace emotions with cold facts or rigid logic. Instead, CT helps us understand that taking a moment of pause to examine our emotions might be a good—or even necessary—idea. We wouldn't use CT to determine whether we should love

> It is far better to grasp the Universe as it really is than to persist in delusion, however satisfying and reassuring.—Carl Sagan

our children, though thinking critically when marketers are vying to treat their illnesses should be second nature. Emotions can lead us astray and we shouldn't rely on them when evidence and analyses provide better answers or solutions.

Getting Serious

In perhaps the most famous psychology experiment ever conducted, psychologist Philip Zimbardo created a mock prison, randomly assigning college students to play the role of guard or prisoner. After being "incarcerated" for less than three days, prisoner 8612 reported that the frustrating prison conditions brought on "a mental dullness like a fog surrounding everything" (Zimbardo, 2008, p. 109). Elaborating, he went on to say that "the arbitrariness and idiot work of the guards grates on you" (Zimbardo, 2008, p. 109). Reflecting on his experience, in his final evaluation he noted that "the psychological part . . . was the worst. . . . It's not having the choice that's the tearing apart thing" (Zimbardo, 2008, p. 186). Anyone who has ever done time knows exactly what he's feeling. Prisoners rarely get a break from the mental monotony of incarceration. While CT cannot solve this problem, I'll show that when put to good use, it can help stimulate the mind and clear up some of that fog in those otherwise dull and dragging moments.

Let's get one thing clear: it's absolutely critical to think critically! Cognitive scientist Susan Greenfield (2000) tells us, "You are your brain." These same brains are inherently flawed, and, by extension, all of our thoughts, beliefs, and opinions are far from perfect. Developing CT skills can empower and liberate us from the shortcomings of these three-pound lumps of meat we call our brains. Thinking Straight can improve nearly every aspect of our lives. As one skeptic puts it, "Good thinking won't necessarily make you more intelligent, but it can make you less prone to making dumb decisions. It probably won't make you rich, but it could prevent you from wasting many thousands of dollars over a lifetime. It can't guarantee you a long and healthy life, but it will make you safer" (Harrison, 2015, p. 11).

The skills you'll learn will also prepare you to thrive in this unpredictable world. Without knowing it, we're constantly being deceived, whether by our own brains or by the world around us, whether by accident or by design. Our first line of defense are the tools of CT. These tools can free us from the confusion of the false and mutually exclusive beliefs we cling to because of little more than our cognitive makeup. Our priorities and desire for self-worth motivate us to believe we are smarter; are more logical, attractive, fair-minded, consistent, ethical, competent, honorable, companionate, and generous; are less prejudiced; are better drivers; and have a better sense of humor than the average person

(Balcetis, Dunning, & Miller, 2008; Brenner & Molander, 1977; Brown, 2012; Chabris & Simons, 2010, pp. 91–92; Epley & Dunning, 2000; Gilovich, 1991; Van Lange, Taris, & Vonk, 1997; Wiseman, 2011, p. 26). As if drunk on our own ideas of ourselves, almost everyone believes they're superior to everyone else on practically any desirable trait (Williams & Gilovich, 2008).

With such *self-serving biases*, we falsely convince ourselves that we're special (Harris, 2011, p. 89). As we weave our way through the wonders and complications of our lives, our brains are constantly constructing an extremely narrow picture of reality—our internal models of reality—which we then interpret according to our personal likes and dislikes, culture, environment, past experiences, religious and/or political affiliations, and a laundry list of cognitive flaws and biases. These quirks and traits are seamlessly woven together with our memories, beliefs, passions, convictions, abilities, interests, and a slew of other characteristics to form what we call a self.

Psychologists recognize that self-serving biases boost our self-esteem and "enable us to explain away our failures, take credit for our successes, and disown responsibility for bad decisions, perceiving our subjective world through rainbow prisms" (Zimbardo, 2008, p. 261). These biases are also "maladaptive . . . blinding us to our similarity to others and distancing us from the reality that people just like us behave badly in certain toxic situations" (Zimbardo, 2008, p. 261). Identifying our biases is one step in the right direction, and loosening their hold on us takes knowledge and effort. Unfortunately, this is an uphill battle as research shows that most people think they aren't as vulnerable to self-serving biases as others, even after being shown they are (Pronin, Kruger, Savitsky, & Ross, 2011).

CT is about applying logic and reason in the pursuit of taking a stance or formulating an opinion. Since opposing logic and reason is illogical and unreasonable by definition, if you don't value reason and logic, perhaps it's time you start! Imagine taking a serious stance *against* evidence, reason, and logic: "Sir, your logic is sound, your evidence is reliable, and the reasons you offer are convincing. But why should I believe you?" Through the lens of a skilled skeptic, witnessing others' sloppy thinking can be pretty disappointing. As the mathematician John Allen Paulos laments, "In a society where genetic engineering, laser technology, and microchip circuits are daily adding to our understanding of the world, it's especially sad that a significant portion of our adult population still believes in Tarot cards, channeling mediums, and crystal power" (Paulos, 2001, p. 5).

The science popularizer and research scientist Kathleen Taylor says, "Sometimes, when we are sufficiently motivated, we stop and think about the influences we experience. When we do not we are open to exploitation" (Taylor, 2017, p. 76). The (in)ability to pause and think is often situational and out of our control, another cognitive flaw known as the fundamental attribution error (see chapter 7). To make matters worse, what psychologists and neuroscientists have learned over the last several decades shows that we can be exploited by ourselves as easily as by others. While avoiding cunning con artists and mental pitfalls requires effort, completely overcoming the many cognitive illusions and other foibles of our brains is impossible, and living in a constant state of mental hypervigilance is far from practical. What CT helps with is to "learn to recognize situations in which mistakes are likely and try harder to avoid significant mistakes when the stakes are high" (Kahneman, 2013, p. 28). Reason, evidence, and logic are useful tools but only if you know how to use them.

Dangerous Thinking

Many incredible claims have been relegated to the past. Nevertheless, we should still take CT seriously for a number of practical reasons (Lindsay, 2017; Pigliucci, 2017b). Harrison stresses that "critical thinking is the indispensable skill for smart living in modern society, and skepticism is the essential posture for the fully awake twenty-first-century human being. Unfortunately, most people fail to appreciate and utilize either" (Harrison, 2015, p. 11). Why? Because we live in a society that is largely driven by science and technology. As a society, we have to make decisions about issues like climate change, vaccine safety, stem cell research, genetic modification, and water fluoridation, and the uncritical promotion of incredible claims that undermine scientific progress and technological advancement can compromise the facts on these contentious issues (Lobato & Zimmerman, 2018). Moreover, in capitalistic societies like ours, we'll always be a target of claims made by others pursuing their own agendas by misrepresenting facts or narrowly construing reality to influence us for personal gain. Such claims come from marketers, lawyers, politicians, salespeople, lobbyists, media, friends, family, social groups, special interest groups—basically everyone.

Instead of accepting the claims people make veridically, skeptics ask, "What evidence supports your claim?" and "Where does it come from?" "How personally invested are those making the claim?" "What are the motivations behind making them?" "Are those making the claim reliable sources?" In other words, skeptics are from Missouri, the "Show Me" State.

Morality is also a part of discovering the truth. Edmund Way Teale asks, "Is it morally as bad not to care whether a thing is true or not, so long as it makes you feel good, as it is not to care how you got your money as long as you have got it?" (quoted in Sagan, 1996, p. 12). The morality behind Teale's inquiry is reflected in deeds like mandating ethical standards for scientific research or prohibiting blood diamonds from entering our markets. We understand that it's morally wrong to financially advance by robbing another. But when it comes to knowledge, many of us miss the point that ignoring the truth is morally unacceptable, and being unable to discover what's true is far from harmless.

Baseless superstitions support illicit black markets for folk "medicines" that drive beloved animals into premature extinction (Ladendorf & Ladendorf, 2018; Stromberg & Zielinski, 2011). The so-called alternative medicine (SCAM) industry makes provably false claims about the efficacy of their products, leading people to forgo science-based medicine,[7] risking horrors from permanent paralysis and viral infections to brain swelling and even death. AIDS denialists and COVID-19 hoaxers circulate myths that lead vulnerable people to forgo behaviors and treatments that could save their lives (Kalichman, 2018; Novella, 2018, pp. 419–24). People calling themselves "breatharians" die because they believe they can survive on air and "universal energy" (Polidoro, 2020a), whatever that is. In his book *Do You Believe in Magic?* pediatrician and virologist Paul A. Offit shows how the (completely unregulated) vitamin and supplement industry callously puts profits before people, recounting thousands of cases of sickness and death caused by dietary supplements and vitamins, including studies demonstrating that vitamins and multivitamins can increase one's risk of heart disease, cancer, and death (Offit, 2014, pp. 47–62; Tavel, 2018).[8] Against this evidence, the vitamin and supplement industry pulls in $28 billion a year (Offit, 2014, p. 61).

Whether it's drinking unfluoridated water, suffering the effects of climate change, or risking exposure to deadly viruses, you don't even have to be the one believing nonsense to suffer the consequences. During the coronavirus pandemic, many people wrongly believed wearing masks was a political statement, a belief that contributed to the needless spread of a deadly virus. Although the success of vaccines has saved hundreds of millions of lives (Offit, 2015; Pinker, 2018, p. 64), mis-, dis-, and uninformed parents deny their usefulness, putting the health of their children—and the health of their communities—at serious risk. Belief in repressed memories (see chapter 7) and facilitated communication (FC) (see chapter 3), often perpetuated by professionals who should know better (Jacobson, Mulick, & Schwartz, 1995), has resulted in numerous prison sentences (Hagen, 2018) and ripped families apart (Sobel, 2018; Vyse, 2019b) with completely evidence-free accusations. For almost two hundred years, the *Malleus Maleficurum*—the infamous "Hammer of Witches," written by fifteenth-century inquisitors as a guide to hunt and torture "witches"—was the second-best-selling book in Europe next to the Bible (Guiley, 2008, p. 223). Anyone claiming that superstition and misinformation are harmless need only look at the number of critically endangered species or the resurgence of once eradicated diseases to see otherwise.

And the dangers don't end there. Psychics and mediums cause untold amounts of mental anguish and financial losses. Using shameful techniques like cold reading (chapter 9) and other dubious tactics, so-called psychics exploit the naïve, the innocent, and the aggrieved to satisfy their egos and greed. Outrageous medical claims harm the gullible and desperate who rely on "alternative" treatments instead of mainstream medicine. What's wrong with such fantasies is obvious: the truth matters because our actions are based on our beliefs, and, although we need to believe many things to survive, it stands to reason that ethical actions and good decisions are an unlikely consequence of faulty beliefs. As biologist and anatomist T. H. Huxley put it, "The foundation of morality is to . . . give up pretending to believe that for which there is no evidence, and repeating unintelligible propositions about things beyond the possibilities of knowledge" (quoted in Sagan, 1996, p. 207). In the case of SCAM, which Americans spend over $34 billion a year on, this couldn't be any more obvious (Harrison, 2015, p. 196; Offit, 2014, p. 93). CT can be deadly serious, and irrational beliefs are not faraway things that cannot harm us: they're a part of life that can cost us dearly. As attorney John W. Miner—who investigated the deaths of Marilyn Monroe, Robert F. Kennedy, and the Manson victims—reveals, "Quackery kills more people than . . . all crimes of violence put together" (Schick & Vaughn, 2005, p. 12).

Falling for incredible claims not only endangers our lives; it can also deeply encroach on our livelihoods (Dobelli, 2014). For instance, the internet has no shortage of Tarot card readers and psychic mediums who charge exorbitant prices to reveal spiritual secrets only they can reveal. Psychic hotlines—a multi-million-dollar-per-year industry (Schick & Vaughn, 2005, p. 12), though not because psychics can foretell the next card at the blackjack table or where the roulette ball will land—charge more than most therapists and lawyers do. In fact, most psychic hotlines are staffed by unemployed housewives, not enlightened spiritual gurus (Woodruff, 1998).

Psychics never accurately anticipate important events. No psychic predicted 9/11, 7/7, Sandy Hook, the Bay of Pigs, the rise of Hitler, the coronavirus pandemic, or any other large-scale event. Government secrets are exposed by spies and whistleblowers. Pandemics are explained by epidemiologists and virologists. And, despite the false claim that psychics assist police in solving crimes, crimes are solved by ordinary police work and investigations (Alcock, 2018, pp. 506–7; Hines, 2003, pp. 73–77; Radford, 2009b; Radford, 2010b; Radford, 2011; Reiser & Klyver, 1982; Reiser, Ludwig, Saxe, & Wagner, 1979; Wiseman, West, & Stemman, 1996a; Wiseman, West, & Stemman, 1996b). If psychics could really do what they claim, you wouldn't contact them on 900 numbers spending $50 for a fifteen-minute call. They'd be working for the FBI and CIA, wearing suits to work and behaving like normal professionals. On second thought, they wouldn't because there'd be no such thing as the CIA because there'd be no such thing as a state secret![9]

As you can see, CT is about protecting ourselves from schemes and scams rather than debunking nonsense. Therefore, skepticism is a plea to Think Straight, rather than a stance against something. Exposing incredible claims is a consequence of CT, not one of its aims. Despite its negative reputation, skepticism is a positive endeavor that's exercised with good intentions. Perhaps skepticism's bad rap is a vestige from the process of declaring saints. During the canonization process, the Catholic Church employed lawyers to take a skeptical view on a candidate's character, poke holes in the evidence, or argue how any miracles attributed to the candidate could be fraudulent. Because they were hired to argue against the Church, they were called Devil's advocates.

Skepticism isn't just about Thinking Straight: it's an effective means of discovering truth. Sure, discovering that Earth orbits the Sun debunked geocentrism, but we now know the truth. Learning that unicorns don't exist is negative in that it pops a cherished bubble from our childhood imaginations. We overcome our disappointment because we prefer seeing the world for what it is rather than living in a false reality. Skepticism is negative only in the sense that it's against mistakes, delusions, deception, and lies (Harrison, 2012, p. 19).

Science and belief are indispensable aspects of life, and it can be helpful to develop tools that assist us in deciding whether we can trust science and whether our beliefs are justified. Global warming is an issue. Is it a hoax, or is it real? And are humans causing it? What can we do to find out? How do we know what information to trust? CT allows us to confidently examine such questions that seriously impact our lives. The slew of decisions we make every day—whether about our health and foods, the entertainment options we choose, or even how we want to live our lives—are ultimately based on logic and evidence, and without CT it would be nearly impossible to properly evaluate all the factors that lead to those decisions.

All else being equal, "the quality of your life is determined by the quality of your decisions, and the quality of your decisions is determined by the quality of your thinking" (Schick & Vaughn, 2005, pp. 13, 331). Thus, CT matters because our lives are governed by our actions and choices, which in turn are governed by our beliefs and how we form them. If our beliefs are ungrounded, and our thinking is irrational, the quality of our lives is severely diminished.

Humans are instinctively obedient to authority (Cialdini, 2007, pp. 208–36; Hofling, Brotzman, Dalrymple, Graves, & Pierce, 1966; Milgram, 1963; Zimbardo, 2008), veridically accepting what we read online or see on TV and forming beliefs without examining the evidence. We do it all without attempting to find out if the person is an authority. Telltale indicators of authority, such as uniforms, titles, knowledge, and confidence, pacify System 2, putting System 1 in charge and not entirely without good reason. Deference to authority allows multilayered and complex societies to exist. Without authorities we wouldn't enjoy the comforts brought by business, trade, security, or social control, for example. By contrast, it's less natural to pause and doubt everything we see and hear: doing so requires *tons* of mental energy. Moreover, logic and CT are far less intuitive than we realize, and veridicality typically serves us well. Thinking critically is a learned skill, and just like most skills, the more you practice the less practice you need.[10]

Hello, Skeptics!

Earlier I used the word "metacognition," an underlying concept of CT: thinking about thinking. As conscious, sentient beings, we're uniquely positioned to critique our own thoughts; atoms contemplating atoms, so to speak. Metacognition enhances our lives by allowing us to change our beliefs based on evidence and in turn increases our awareness to *when* we're changing a belief and *why* we're changing it. Maintaining a high level of metacognition, however, isn't easy. The psychologist James E. Alcock explains,

> Just as speech develops automatically, and yet we have to study grammar and composition in order to become good speakers and writers, so too does thinking develop automatically, but we have to study logic and critical analysis if we are to become good thinkers. Yet,

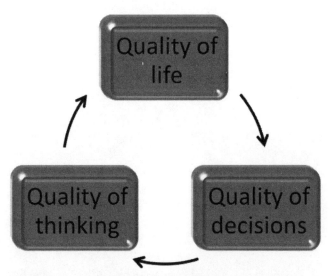

Figure 1.1.   Cycle of quality. *Schick, T., Jr., & Vaughn, L. (2005). How to think about weird things: Critical thinking for a new age, 4th ed. Boston: McGraw Hill.*

while the need for years of language instruction is well recognized by society, the need for thinking instruction—that is, in the ability to think critically—generally is not. (Alcock, 2018, p. 529)

Metacognition and skepticism are two sides of the same coin. The Greek and Latin roots of the word "skeptic" are themselves instructive. The Greek word *skeptikós* means "thoughtful" or "inquiring," while the Latin word *scepticus* can be translated as "reflective." Tracing its Greek roots further back, *sképtomai* means "to consider," while *skopéō* means "I view, I examine." Thus, skepticism is a thoughtful, critical approach to inquiry and examination.

Skepticism is a way of approaching the world, questioning everything like a curious child. It involves doubting our core beliefs, how we think, and even what we think we know. Skeptics and critical thinkers are pretty much the same thing, and I use the terms interchangeably. The point of skepticism is thinking before believing, asking questions, demanding evidence, employing logic and reason, and being honest about ignorance and comfortable with uncertainty. Being a skeptic is being sensible. While it may seem like nobody could object to such a positive worldview, many people do. Furthermore, although we tend to believe irrationality is common, we nevertheless assume that we arrive at our beliefs for rational reasons (Ehrilinger & Gilovich, 2005). CT teaches us that no one is safe from irrational thinking, even those who spend their entire lives dedicated to CT. Practice makes less practice.

Human nature is thinking we're more logical than we are, thinking we're too smart to be fooled, and thinking our memories accurately represent the past. Realistically, logic is extremely unnatural and takes a great deal of work. Our brains are massively deceptive and trick us all the time. Our memories are not the passive recorders we believe they are. Thus, skepticism focuses on discovering our shortcomings and finding a balance between knowing we'll be wrong and having the humility to challenge our convictions.

BELIEFS AND SKEPTICISM

In chapter 7, we'll examine belief thoroughly. For now, a note on how beliefs relate to skepticism in general will do. Beliefs develop from memory, personal experience, what we want to believe, gut feelings, and what we can just feel in our bones. Well, bones and guts don't think, and the organ that does is riddled with flaws and imperfections. Perhaps more than most, juries, defendants, and litigants of all stripes understand that something as simple as witnessing an event can be far from accurate. Witnesses frequently swear to something that never happened, but are they all lying? Can witnesses sincerely believe something that didn't occur? Could they be relying on memories they don't realize are inaccurate? If there's such a thing as an honest-but-mistaken witness, then there's a way to explain their mistaken testimony with compassion rather than hatred and reason rather than emotion.

The ancient Greek philosopher Epicurus taught that we can only experience reality through the lenses of our senses—our own "frame of reference"—which are both flawed and limited (DeBakcsy, 2019).[11] Because of how our brains process information, oftentimes sensory information cannot be trusted, so we must carefully decide how to form reliable beliefs. CT helps us navigate this process. Contrary to some views, science is not a belief; it's a method of testing an idea and learning whether the results agree with the idea or whether they provide some other explanation.

SCIENCE AND SKEPTICISM

Scientific facts remain open to revision because we don't always have all the information, and new ways of looking at the information we have could reveal previously hidden truths. Hence all scientific conclusions are considered *provisional*, in other words, temporary. Our ability to interact with the world, the state of our present knowledge, and the different ways we interpret information all place limits on our knowledge. That science only concerns itself with irrefutable facts is a common misconception. To the contrary, science is about what's probable, not what's certain. As the philosopher of science Thomas Kuhn (1922–1996) noted, "No theory can ever be exposed to all possible relevant tests, [so scientists] ask not whether a theory has been verified but rather about its probability in light of the evidence that actually exists" (Kuhn, 1962/2012, p. 144). Science cannot tell us that something is *absolutely* true; rather, science shows us how true something *probably* is.

Sometimes scientific truth is indistinguishable from absolute certainty, so we behave as if we're certain. For instance, that water is made up of one oxygen atom and two hydrogen atoms is a scientific fact that is practically guaranteed. Do we know that water isn't comprised of other atoms? To a high degree of certainty, yes, and this will probably always be the case. However, we may not know everything there is to know about water or atoms, so we can only say that there is a high probability that our **theory** of chemistry—which guides our knowledge about chemical compounds—is true.

Since the word "theory" is commonly misunderstood (Dawkins, 2009, pp. 3–18), let's take a detour to unpack the term. In science, a theory is a natural explanation for a body of facts (Amicus curiae, 1986) that are temporarily accepted as true until some other set of facts explains them better. A theory explains specific phenomena, such as how germs cause disease, how organisms evolve, or how gravity attracts. The philosopher of science Karl Popper (1902–1994) described scientific theories as "nets cast to catch what we call 'the world': to rationalize, to explain, and to master it." According to Popper, the aim of scientific discovery is "to make the mesh ever finer and finer" (Popper, 1959/2014, p. 59). Reaching the status of theory is the peak of scientific knowledge. Before an idea becomes a theory, it's called a *hypothesis*. Hypotheses become theories only after much supporting evidence has been collected and they survive rigorous attempts at falsification. When it comes to scientific facts, there's no such thing as "just" a theory unless, as the comedian Ian Harris quips, "you don't know the meaning of the word *just*" (Harris, 2016). Successful theories include gravity, plate tectonics, heliocentrism, evolution, climate change, the Big Bang, the germ theory of disease, and many others.

Applied Critical Thinking

Just as beliefs in sorcery and witches once replaced beliefs in succubae and spirits, today people believe in grand conspiracies and alien visitation (in extreme cases, combining the two). No good evidence was ever found for witches or succubae, and credible evidence for grand conspiracies or alien visitors continues to evade discovery. Moreover, belief in conspiracies, the paranormal, and other incredible claims correlates with low levels of analytical thinking (Lobato, Mendoza, Sims, & Chin, 2014), suggesting that CT can reverse these effects (Gervais, 2015; Pennycook, Fugelsang, & Koehler, 2015; Swami, Voracek, Stieger, Tran, & Furnham, 2014). Thus, skeptics sometimes find themselves on the front lines of highly contentious issues. Accordingly, skeptics have a responsibility not to demean or belittle others simply because they believe what others find ridiculous. As skeptic and professor of neurology Steven Novella observes, CT skills "are not weapons to attack other people and make yourself feel superior, they're the tools you need to minimize the bias, error, and nonsense clogging up your own brain" (Novella, 2018, p. 431). Believers in the incredible are not incompetent, unintelligent monsters that deserve ridicule; they're human beings that deserve to be shown a better way of thinking.

In addition to being cognitive misers, our prior beliefs and biases motivate us to believe certain things even before the evidence is in, so we routinely make up our minds before an honest inquiry can even be made. Thus, understanding our motivations is an important step toward CT. When motivated by a desire to believe certain conclusions, we are excellent at rationalizing false

> Absolute proof is for mathematical theorems and alcoholic beverages.
> —Michael E. Mann

beliefs.[12] Smart people are especially vulnerable to inventing reasons to justify irrational beliefs. In fact, smarter people will notice and retain evidence better than lesser minds, perhaps explaining why incredible claims attract people across the intellectual spectrum (Vyse, 2014, p. 59). Additionally, while higher intelligence doesn't prevent irrationality (Shermer, 2003b; Shermer, 2013, p. 6; Sternberg, 2004) and lower CT skills correlate with paranormal beliefs (Alcock & Otis, 1980), research shows that science education doesn't prevent paranormal beliefs, because learning facts doesn't teach one how to evaluate those facts (Trecek-King, 2022a; Walker, Hoekstra, & Vogl, 2001).

Open-Minded Reasoning

A common retort against skeptics who reject the "evidence" for, say, ESP or Bigfoot says that skeptics are just closed-minded. Skeptics already decided that neither ESP nor Bigfoot exists, and no evidence will convince them otherwise. However, being open-minded doesn't mean believing anything, and skeptics reject this sort of prejudice. Carl Sagan explained the skeptic's stance, stating, "I maintain this is not prejudice. It is postjudice. That is, not a judgment made before examining the evidence but a judgment made after examining the evidence" (Sagan, 2007, p. 135).

Open-mindedness isn't a willingness to believe anything or believing "anything is possible." Open-mindedness means (1) being willing to change your mind if the evidence demands and (2) applying the same standard of evidence and proof to all claims (Hare, 2009). Open-mindedness begins with the acknowledgment "I could be wrong." Are skeptics justified in rejecting the evidence for ESP or Bigfoot out of hand? No. We must examine the best evidence *first* before reaching conclusions. Being open-minded entails that your beliefs are *defeasible*, meaning they're open to revision or rejection based on new information or better arguments.

Thinking Straight requires thinking through the implications of a belief to find out if it's worth believing. Thus, reasoning means drawing conclusions or inferences from observations, facts, or assumptions (Wade & Tavris, 2017, p. 308). Beliefs that contradict facts are wrong—period. Analyzing beliefs takes brainwork, as it's much easier to compartmentalize our beliefs and separate them from one another to avoid the discomfort of refutation. This mental tension, known as *cognitive dissonance*, occurs when there's an inconsistency between one or more of our beliefs or when our behaviors conflict with our beliefs (Festinger & Carlsmith, 1959). We'll discuss cognitive dissonance further in chapter 7. For now, it suffices to say that one strategy of dealing with cognitive dissonance is to think about what else must be true if "X" is true.

Let's say you're one of millions who believe governments conspire to suppress and poison the masses with toxic chemicals dumped from airplanes (Radford, 2009a; Thomas, 2008). What else must be true for this conspiracy to "work" without any glaring contradictions (Loxton, 2017)? What do experts think (Cama, 2015; Shearer, West, Caldeira, & Davis, 2016; United States Environmental Protection Agency, 2000)? Who else must know about the conspiracy and who's covering it up? How do governments hide the massive spending required for the conspiracy? When conspirators quit, retire, or die, how do governments replace recruits that are willing to poison millions of people and are also trustworthy enough not to spill the beans? How do governments hide the conspiracy from civilian pilots and airport personnel, like air traffic controllers? How do the conspirators keep from being poisoned? How are accidental contaminations dealt with? How do governments work together to achieve this conspiracy, and is there proof that they do? Exactly how do governments hide the conspiracy? Where are the chemicals made? Who transports and loads them? Who builds the dumping machinery? How have average life expectancies around the globe risen from between twenty-five to forty years since the advent of the airplane (Pinker, 2018, p. 54)? What are the physical or biological effects of chemtrail poisoning? Is anyone studying those effects? If so, where is their research published? If not, how do you know any effects are caused by chemtrails (Hill, 2013)? Asking skeptical questions to challenge a previously held belief is exactly how CT works. Unfortunately, skepticism is the exception, not the rule, and most of us spend more time reinforcing our beliefs than questioning them, which is one of the ways we avoid cognitive dissonance. The reason is being right feels better than admitting fault, and so we invent creative and convincing ways to rationalize even the most outlandish of our beliefs.

What we know about the material world and our personal perspectives are ultimately limited, often in ways we're unaware of. As Donald Rumsfeld famously said, "We know there are some things we do not know. But there are [things] we do not know we do not know." Because of these limitations, another healthy exercise is checking with others and honestly seeking opposing evidence. Exposing our beliefs to others' opinions allows us to accept criticism and confront flaws in our thinking (Gilovich, 1991, pp. 53–56). Thus, healthy doubt is similar to the peer review process (chapter 3) researchers use to check their work and that of others.

The Benefits of CT

CT includes an array of benefits that make learning the subject much more than gaining a better understanding of ourselves and learning how to evaluate information (although each of these goals are crucial for anyone interested in reality!). With CT, we gain insight into ourselves and others, we deconstruct our biases, we learn to empathize and minimize judgment, we're better positioned to positively influence others, we motivate others to improve their lives, and we develop a set of skills that drastically improves the quality of our lives.

CT isn't always easy, but the benefits are nearly as endless as they are applicable. And regardless of the benefits, learning CT skills is a worthy endeavor in and of itself because learning how to think is interesting and exciting. The content I've covered ranges from mentally stimulating to downright breathtaking. Why do people go sky diving or climb dangerous rock cliffs? Because those adrenaline-inducing activities are fun and exciting in and of themselves. CT is like sky diving or rock climbing for the brain. Although it can be challenging, it's one of the most fun and exciting journeys you'll ever take your brain on.

Wrapping Up

Hopefully this chapter showed that critical thinking provides the bedrock for truth and knowledge. It guides our beliefs, informs our decisions, protects our well-being, and opens our minds to exciting new realities. Likewise, skepticism is an ethical worldview that embraces reality and values truth, while rejecting wishful and magical thinking as reliable sources of information. CT skills prepare us to face the world intelligently, confront an endless array of incredible claims, and look at ourselves as the source of both error and splendor. Without these skills, we're forever trapped deep within the confines of delusion and folly.

Although it's immensely powerful, the human brain is also massively flawed. As such, if we're going to maximize our life experience and base our decisions on the best information available, understanding these flaws and learning how they influence our actions is a must. CT offers many benefits, and the consequences of irrational thinking are deadly serious. Adopting a skeptical worldview, then, is a no-brainer. Accordingly, now could never be a better time to take your thinking seriously and treat yourself to the wonders of what clear, rational thinking has to offer. So let's get started!

CHAPTER 2

A Philosophical Detour

Science, CT, and philosophy are overlapping disciplines used to discover ultimate truths. Scientific discovery provides methods for dissecting the universe, nourishing our attempts to discover exactly how nature works. Similarly, philosophical inquiry probes the hidden depths of reality and helps us understand the ultimate meaning of things. Critical thinking is the process of evaluating the information we encounter along the way.

Our means of discovery are guided by certain philosophical concepts and principles that we couldn't do without. In this chapter, we'll discover the philosophical principles that allow scientists to navigate their hypotheses and theories. Additionally, we'll find out what knowledge is, what it means to know something, and what can go awry when knowledge becomes personalized.

Building Vocabulary

Naturalism:	Ontological:
The philosophical position that only natural (as opposed to supernatural or paranormal) laws and forces govern the universe.	The philosophical study of what exists; relating to reality or existence.

Epistemology:	Fact:
The philosophical study of knowledge; a specific theory of knowledge.	An actuality that has been proven to such a high degree that there is general consensus as to its reality.

Skeptical Roots

Before moving forward, a short discussion on another branch of skepticism is warranted. In chapter 1, we learned that scientific skepticism is a method for investigating truth claims, accompanied by a set of skills used for such investigations. Scientific skepticism subjects all truth claims to critical analysis in an effort to sort out what's true from what's not (Bunge, 2020). Philosophical skepticism, on the other hand, takes a completely different approach to truth and knowledge, holding that knowledge is unattainable; that is, that having knowledge is impossible. Philosophical skepticism rests on the assumption that true knowledge requires certainty and that since innumerable variables can undermine our certainty, knowledge is impossible (Vaughn, 2019, pp. 45–46). Take your own existence, for example. Recall that after deeply contemplating whether he could verify his existence, Descartes famously said, "I think, therefore I am." We talk, we feel, we see, and we interact with the world around us. For many, this is enough to confirm our existence, and so it was for Descartes. For philosophical skeptics, however, since each of these experiences can be questioned, even something as obvious as our own existence cannot be truly known. Responding to the evidence of our sensory experiences, the philosophical skeptic might argue that what we believe is our existence may be someone else's dream or a simulation that's so real it convinces us we exist.

As with many philosophies, philosophical skepticism suffers from self-contradiction. That is, if it's true that knowledge is unknowable, then it's false that knowledge is unknowable, and the idea that knowledge is unattainable is a claim to knowledge that, according to philosophical skeptics, cannot be known. A more reasonable position is that of the scientific skeptic. Where philosophical skeptics argue that knowledge requires proof beyond *all possible* doubt, scientific skeptics argue that knowledge requires proof beyond *all reasonable* doubt. Returning to the truth of our own existence, thinking we exist is more reasonable than thinking we don't exist because the fact of our existence rests upon far fewer assumptions and, as we'll discuss in chapter 4, claims resting on fewer assumptions are much stronger than those that rest on many assumptions.

As we'll see later in this chapter, if we believe a true proposition, and reliable mechanisms verify the truth of that proposition, then what we have is knowledge. Put simply, it's unreasonable to doubt what we have good reasons to believe, and we have good reasons to believe what can be reliably verified. Thus, scientific skepticism promotes a *reasonable* approach to truth and knowledge, rather than an *absolute* approach. If something is reasonable to believe, then we have good grounds for believing it. The quest of Thinking Straight, then, is discovering how we know what is reasonable to believe. Figuring it out requires some rules and principles that guide our thinking. One of those rules goes by the name of naturalism.

Science and Philosophy

Naturalism is the underlying philosophy by which scientists make assumptions and conduct research. Put simply, naturalism means that reality is the same for all observers and that the laws and forces governing reality are without supernatural influence. Naturalism is the platform upon which the methods and limits of science are defined. Philosophers recognize two forms of naturalism: *methodological naturalism* and *philosophical naturalism*.

Methodological naturalism holds that natural events have natural causes, making no claims about the existence or nonexistence of anything supernatural. Science follows methodological naturalism, assuming that nothing supernatural exists. Thus, the framework by which science operates rests on the assumption that natural effects have natural—that is, not supernatural—causes. Scientists adhere to methodological naturalism because, as dozens of scientists and scientific organizations explained to the U.S. Supreme Court, "science is devoted to formulating and testing naturalistic explanations for natural phenomena. It is a process for systematically collecting and recording data about the physical world, then categorizing and studying the collected data in an effort to infer the principles of nature that best explain the observed phenomena" (Amicus curiae, 1986). In other words, science operates by testing what exists in the physical world to try to discover how things work.

Philosophical naturalism is more forward in its claims about the supernatural. Philosophical naturalism maintains that nothing supernatural exists (Law, 2017). Where methodological naturalists may believe in the supernatural, philosophical naturalists reject the existence of anything beyond nature. As the philosopher Barbara Forrest explains, a handy way of distinguishing the two is "methodological naturalism is an *epistemology* as well as a procedural protocol, while philosophical naturalism is a *metaphysical* position" (Forrest, 2017, p. 38). In other words, methodological naturalism is a method for accumulating knowledge, whereas philosophical naturalism is an **ontological** belief about the universe.

While an argument is only as strong as the facts and assumptions it's based on (chapter 4), the assumptions which both schools of naturalism rest upon are well grounded. Methodological naturalists rely on the assumption that cause and effect are natural processes because without this assumption, epistemological progress would be severely hindered. Similarly, philosophical naturalists rely on the assumption that nature is all that exists because no convincing evidence for anything "beyond nature" has ever been presented.

Some scientists are indeed philosophical naturalists, but science itself operates under the framework of methodological naturalism. Science doesn't make believing that the material universe is all there is mandatory. However, science can only work within a materialistic framework since it can only explore what exists in nature. Anything existing outside of nature would be beyond science's ability to discover. Science is a search for knowledge and an understanding of the physical world and so cannot investigate claims that cannot be tested, regardless of how much we may want to understand the supernatural. As Carl Sagan elegantly explains, "Claims that cannot be tested, assertions immune to disproof are veridically worthless, whatever value they may have in inspiring us or in exciting our sense of wonder" (Sagan, 1996, p. 171). Supernaturalism cannot be tested, it's inconsistent, it fails to make testable predictions, and it cannot deliver on its promises. By contrast, naturalism consistently and accurately predicts the governing rules of reality. Epistemologically, naturalism is obviously superior to supernaturalism.

Science must also assume that the laws underlying nature are consistent, thus are ultimately knowable. Further, science assumes that an objective reality exists and that with talent and understanding, we can learn how it works. If this assumption is wrong and nature is unpredictable or doesn't consistently follow any rules, there'd be no way to interrogate nature or discover truths. As it turns out, relying on the assumption that reality abides by consistent rules has been hugely fruitful and contributes greatly to our understanding of the natural world.

Science and the Supernatural

As noted above, naturalism gives science its explanatory value but not to privilege the natural over the unnatural. Naturalism is simply an epistemological position reflecting the limits of scientific inquiry. **Epistemology** is a branch of philosophy that studies knowledge. The word "epistemology" also describes a specific theory of knowledge. Thus, when we speak about epistemology, we're simply talking about what we know and how we know it. Science is a demonstrably successful epistemology that provides the foundation for reliable knowledge. Accordingly, science, guided by methodological naturalism, drives our progression of knowledge ever forward.

What exists in nature is constrained by natural law and thus can be tested and confirmed (or refuted) by experimentation and observation. So for a claim to be considered scientific, at the bare minimum, it must be testable and exist in nature. By contrast, anything *super*natural is that which cannot be explained by—or violates the laws of—nature. If supernatural entities exist, they exist outside of nature and therefore aren't constrained by natural laws and could behave in unpredictable ways. None of this says anything about whether supernatural entities exist. It's just the nature of methodological naturalism that sets a fundamental limit to what science can probe. Nonetheless, what science *can* do is draw reasonable inferences by investigating claims made by proponents of the supernatural.

In chapter 3, we'll discuss independent and dependent variables. Briefly, in a scientific experiment, the independent variable is the variable that's being manipulated and controlled, whereas changes in the dependent variable *depend* on the independent variable. If we hypothesized that humans have souls and that those souls weigh something, assuming that souls leave the body at the time of death, if we wanted to weigh a soul, dying humans would be our independent variable. Our dependent variable would be the weight of the body at the time of death. If the same change in the independent variable (death) causes the same change in the dependent variable (variation in weight), then there's evidence for a cause–effect relationship between the variables. Here, if we consistently measured weight loss at the exact time of death, we could say dying causes X amount of weight loss.

So how does this pertain to the supernatural? If supernatural entities exist, existing *beyond nature* would mean they don't follow the laws of *nature*. Thus, no supernatural variables could even theoretically be isolated and controlled to test a supernatural entity's existence. Such an entity could manipulate variables in a way that either violated the laws of nature or existed outside the parameters of those laws, leaving scientists uncertain about the accuracy of the data. Accordingly, even if we consistently measured, say, 21 grams of weight loss at the exact time of a person's death, we still wouldn't have any evidence that such loss was caused by the departure of a soul. The weight loss could be the result of urination, defecation, or sweating that was caused by the rise in body temperature due to the lungs no longer cooling the body (Wiseman, 2011, pp. 63–64).[1] Therefore, we can infer that souls, if they exist, cannot be measured in this way (Grams, 2019).

Karma is another example. The karmic universe hypothesis says that karma trolls are responsible for doling out reward and punishment throughout the universe. If karma trolls exist beyond what we understand as nature, how might we test our hypothesis? No test could confirm whether karma trolls caused punishment or reward because, with no natural constraints, they could do anything they want, including interfering with the natural realm in ways that couldn't be scientifically detected. Karma trolls could exist, but because they're existence is untestable, their existence or nonexistence isn't scientific. Put differently, no observation or experiment would be incompatible with our hypothesis or an alternative hypothesis, such as that incorporeal gremlins dole out reward and punishment through undetectable dimension portals that disappear the moment one tries to detect them.

The philosopher Bertrand Russell once observed that, for all we know, we could have popped into existence five minutes ago with holes in our socks and hair that needed cutting. Russell was arguing that although we could be being deceived by an entity so powerful that it's convinced us into believing a false history, just because it's a logical possibility doesn't mean it warrants our belief. The idea goes back to Descartes's *Meditations on First Philosophy: Meditation I* (1641), in which he imagines that a "malignant demon, who is at once exceedingly potent and deceitful, has employed all his artifice to deceive me." Whether Descartes's "evil demon" is a god or a devil, an extraterrestrial or a computer simulation, it could've created us with memories we think are real, dreams we hope to accomplish, and messy hair to give an impression of a past. Descartes's claim could never be falsified, as any explanation offered could be refuted by saying that it too was caused by the "evil demon." Descartes's evil demon reminds us that when we hear supposedly scientific claims, we must decide whether they actually are scientific and remember that untestable, unfalsifiable claims are never scientific regardless of how they're presented.

So supernatural phenomena cannot be rendered scientific. However, another objection to supernaturalism is relevant to our discussion: for a supernatural entity to exist at all, it must *have* some properties, and those properties must *be* somewhere. The properties of a supernatural entity cannot exist in nature because, if they interacted with nature, those interactions would be detectable, thus natural, and not *super*natural. The notion that non-matter can interact with matter lacks a theoretical basis by which it *could* be possible. Accordingly, if supernatural agents somehow transcended realms, entering the natural universe, they'd be forced into abiding by the laws that govern our universe and thus would be constrained by those very laws. Such constraints would remove their supernatural status and whatever properties that came with it. Therefore, even if supernatural agents exist, without an ability to detect them, their properties would be off limits to science.

Belief and Knowledge

Do you want to know things? Are beliefs different from knowledge? Are some ways of knowing better than others? If so, what are they, and what makes them better? These are the kinds of questions that separate belief and knowledge, and they are the ones philosophers and scientists care about. Sometimes a disconnect exists between belief and knowledge, where believing *feels* just like knowing, and we mistake beliefs for knowledge. The firmer our belief, the more willing we are to defend it. But this situation is problematic. As the humorist Artemus

Ward once quipped, "It ain't so much the things we don't know that get us into trouble. It's the things we know that just ain't so." So, what does it mean to have knowledge? Can we know something that's false? If we believe something that happens to be true, does that count as knowledge?

Some suggest that knowledge progresses through four stages: *absolute knowledge*, *personal level knowledge*, *rules-based knowledge*, and *evaluative knowledge* (West, 2004). The more education one acquires, the more advanced one's knowledge and thinking skills become. Those in Stage 1 view knowledge as absolute, certain, black and white. Stem cell research is immoral; the right to bear arms is absolute; animals deserve the rights afforded to humans. Education reveals that experts often disagree, which leads to Stage 2. At Stage 2, people believe that since experts disagree, personal opinion is just as valid as an expert's. As we'll see, truth is not relative; that is, facts exist independent of human thoughts, so while everyone is entitled to their own opinion, nobody is entitled to their own facts. As education progresses, the relativism of Stage 2 gives way to Stage 3, where people abandon subjective opinion for systematic investigation. Here, people understand the value of rules and expert opinion. But not until Stage 4 are people ready to question rules and opinions that lead to accepted knowledge. Stage 4 involves, among other things, evaluating evidence both for and against a claim, criticizing sources, and questioning the conclusions and methods of those asserting claims. Stage 4, however, isn't the end of one's learning journey.

SCIENTIFIC KNOWLEDGE

The neurologist Robert Burton argues that the feeling of knowing, that is, the feeling that our judgment is correct, is something that happens *to* us, rather than something we arrive at intentionally (Burton, 2008). On this view, knowledge is as involuntary as digestion. Making his case, Burton presents the following paragraph with instruction to read it fully, at normal speed, and not to skip to the explanation:

> A newspaper is better than a magazine. A seashore is a better place than the street. At first it is better to run than to walk. You may have to try several times. It takes some skill, but it is easy to learn. Even young children can enjoy it. Once successful, complications are minimal. Birds seldom get too close. Rain, however, soaks in very fast. Too many people doing the same thing can also cause problems. One needs lots of room. If there are no complications, it can be very peaceful. A rock will serve as an anchor. If things break loose from it, however, you will not get a second chance. (p. 5)

What appears is a collection of disjointed sentences that makes no sense. Then, our whole conception of the paragraph changes with the word "kite." Primed with the word "kite," reread the paragraph and the disjointedness disappears, replaced by a "feeling of knowing." If I now suggest that the paragraph is just a random sample of sentences pulled from a newspaper or that it came about by instructing a computer generator to randomly string together fourteen sentences, your feeling of certainty would reject those ideas; it becomes physically difficult to imagine those as realistic alternatives. With the answer to the riddle, we can ask some questions about how we know that the word "kite" solves the riddle and shed some light on what knowledge truly is.

When you read the word "kite," did you consciously decide that "kite" correctly explained the paragraph, or did that decision seem to just come to you? When did you go from not knowing to knowing? Did you know "kite" was the correct explanation before, during, or after you reread the paragraph? Are you sure "kite" is the correct answer? If so, how do you know? Following Burton's reasoning, *knowing* that "kite" is the correct explanation happens *to* us. We no more have control over believing the answer is "kite" than we do over whether the sky is blue. On Burton's theory, knowledge is simply the brain's physical reaction to information. If you're like me, this conclusion seems a bit unsatisfactory. Certainly there's more to knowing than chemical reactions. So to better understand knowledge, we must explore its philosophical roots to uncover its meaning at a deeper level.

PHILOSOPHICAL KNOWLEDGE

We *know* 2 + 2 = 4, and we *know* a triangle has three sides; we even *know* that we *know* these things. But do we really *know* that electrons and black holes exist? How can we *know* the Earth orbits the sun if we've never seen it from afar? Typically, we make such distinctions without much effort (or even knowing we're making them)—I *know* the sun is bright; I *believe* it will rise tomorrow. Nevertheless, believing something that's false is possible, and so is believing something and not really *knowing* it. Knowledge differs from belief in that if you *know* a thing, then that thing *must* be true: since fire is hot, you cannot *know* it's cold. Perhaps paradoxically, one can believe something that's true without knowing it or having any evidence for it, such as believing your partner is in love with you simply because that's what you've been told. However, even if true, such beliefs shouldn't count as knowledge given that its truth is merely a coincidence. So what makes knowledge different from a correct belief? Until the second half of the twentieth century, the answer was justification.

In his dialogue *Theaetetus*, Plato argued that to have knowledge, three conditions must be met. First, to know something, you have to *believe* it, since you cannot know something you don't even believe: you cannot *know* a triangle has three sides if you *believe* it has four. Second, the belief must be true, since you cannot know something that's false: you cannot *know* that $2 + 2 = 5$. Lastly, your true belief must be justified. If you just *knew* the Patriots were going to win, and they subsequently won, we don't consider that true knowledge; your belief just happened to be true. This model of knowledge is known as Plato's justified true belief (JTB) model.

A correct belief one can justify with external, corroborating information is almost, but not quite, knowledge. It's the reason we can justify believing that electrons and black holes exist without having seen either—mountains of evidence support their existence. However, difficulties arise with Plato's JTB model, such that it leads to regress problems (chapter 5). That is, if justification is a requirement for knowledge, one would have to justify the justification and justify that justification, on and on, like the photographer who can never photograph the horizon. For example, to say that I *know* black holes exist, I'd have to justify that knowledge with Einstein's theory of relativity, then justify Einstein's theory of general relativity with the success of mathematics, and then justify the reliability of mathematics, ad infinitum.

Plato's model relied on reasoning alone. Then, introducing the power of observation into the equation and therefore solving Plato's regression problem, the American philosopher Edmund Gettier overturned two thousand years of epistemology in three pages (Gettier, 1963). Gettier's model, now referred to as *reliabilism*, is similar to Plato's JTB model and proceeds as follows:

1. The thing (P) must be true.
2. One must believe that P.
3. One's belief that P must be brought about by the fact that P via a reliable mechanism.

Since Plato's model mirrors Gettier's first two conditions, it's the third condition that separates the two. Where Plato's model leads to an infinite regress of justification upon justification, Gettier's reliabilist model halts the regress by requiring that one's true belief is supported by a reliable mechanism, defined as "a mechanism that tends to produce true beliefs" (Law, 2011, p. 143). Accordingly, to *know* the Earth is round, (1) the Earth must *be* round, (2) you must *believe* the Earth is round, and (3) there must be a *reliable mechanism* showing the Earth is round, such as satellite photos or empirical demonstrations.

When it comes to gauging knowledge or questioning incredible claims, "How do you know?" is oftentimes the best question. How do we know the things we believe are true? Sometimes we must be satisfied with just getting close. As Baggini explains, "The best we can hope for is to have good reasons for what we believe. Better reasons, at least, than those for believing the contrary" (Baggini, 2016, p. 278). As we've seen, absolute certainty is as impractical as it is unreasonable, so settling with proof beyond *a reasonable* doubt is more reasonable than demanding proof beyond *all possible* doubt.

The philosopher and neuroscientist Sam Harris argues that "when we distinguish between belief and knowledge in ordinary conversation, it is generally for the purpose of drawing attention to degrees of certainty" (Harris, 2011, p. 115). As Straight Thinkers, understanding the difference between supported and unsupported beliefs is more important than the item of knowledge itself, for skepticism allows us to evaluate which facts are supported and which ones are not.

> The magical thinking that becomes deeply ingrained whenever faith rules over facts warps all areas of life.—Victor J. Stenger

Fact vs. Faith

While the concept of **fact** originated as a legal term, the word "fact" is now widely used in everyday language (McKenzie-Mcharg, 2019). A scientific fact is a property of natural phenomena that always behaves as if it's true (Amicus curiae, 1986). Michael Shermer defines a scientific fact as "a description of a regularly repeating action that is open to rejection or confirmation" (Shermer, 1997, p. 33). The late Stephen Jay Gould said that "in science, 'fact' can only mean 'confirmed to such a degree that it would be perverse to withhold provisional assent'" (Gould, 1983, p. 255). Among other problems discussed below, faith often leads to circular reasoning where one believes something is factual by virtue of having faith in that thing. Unfortunately, the act of believing something isn't enough to justify believing it's true.

Facts help us accurately understand reality, and those interested in truth should remain willing to change their minds as our understanding of the facts improves. Overturning long-standing religious dogma, Pope Pius XII declared that there was no conflict between Darwinism and the teachings of the Catholic Church. Pope John Paul II agreed (Dawkins, 2006, p. 67) and further admitted that the Catholic Church was wrong to maintain that the earth was the center of the universe (Sagan, 1996, p. 146). Without hesitation, the Fourteenth Dalai Lama stated that if a fundamental tenet of Tibetan Buddhism were disproved by science, then Tibetan Buddhism would have to change (Sagan, 1996, p. 278).

A scientific fact begins its life as a hypothesis, which "is like a prosecutor's indictment; it's just the beginning of a long process" (Oreskes & Conway, 2010b,

p. 32). A jury doesn't decide guilt or innocence based on how nice the indictment looks or how much they like or dislike its claims. Juries' decisions are based on "the volume, strength, and coherence of the evidence" (p. 32) supporting the charge. Faith doesn't work that way. "Faith," as Mark Twain understood it, "is believing what you know ain't so." Like supernaturalism, faith *cannot* be scientific, as it treats facts as though they're irrelevant. Therefore, faith creates inaccurate epistemologies. Furthermore, faith is unscientific as it's usually only invoked when a conviction is unsupported by reason or evidence.

Faith, as an epistemology, is a claim to know something specific without evidence. When someone says they have faith, they're claiming to know—rather than trust or hope—that X is the case. We don't have "faith" that our surgeon

> I respect faith, but doubt is what gets you an education.—Wilson Mizner

can perform; that's trust, and it's earned by virtue of their expertise and training. Similarly, we have trust in science because its methods are proven, and its findings are earned. Therefore, many times when we say "faith," what we really mean is something like trust or hope, which aren't the types of faith that qualify as epistemologies. Faith in an afterlife is an example of faith as an epistemology. For example, when a jihad martyr says he has faith that Allah will award him a seat in paradise for murdering infidels, he isn't saying he *trusts* Allah to grant him a seat in paradise or that he *hopes* Allah will safeguard his seat in paradise; he's claiming to *know* his seat is secured. Faith, then, is the act of believing X regardless of whether X is true or false.

The American philosopher Peter Boghossian defines faith in two ways: (1) believing something without evidence or (2) pretending to know something you don't know (Boghossian, 2013, pp. 23–24).[2] If one has faith that their consciousness will survive the death of their body, they believe something without evidence, since, by definition, after-death experiences preclude the living from knowing about them. Actually, faith is more problematic than this description allows. Dr. Daniel Fincke says, "Faith is deliberately believing a proposition more strongly than the evidence warrants (either when you think that the proposition is not strongly supported by evidence or is even undermined by the best evidence)" (Fincke, 2011). When the faithful are challenged with arguments or evidence that contradicts something they have faith in, they double down and continue believing *in spite of* the disconfirming evidence. Of course, people are at liberty to believe anything they want. However, when beliefs go from personal beliefs to public proclamations, they're subject to the same standards of evidence and proof as anything else. Epistemologically, giving faith claims special pleading is a logical error. A claim to knowledge has factual support or not. Knowledge claims with factual support can be probed with GASPs to determine whether what we have is knowledge. Therefore, since faith is not a reliable mechanism, knowledge claims supported merely by faith cannot yield knowledge.

Additionally, faith cannot determine truth because disregarding reason and evidence is a virtue to the faithful. This view is highly problematic. The only reliable ways of acquiring knowledge are through evidence and reason, and since faith can lead to believing something that isn't true as well as unjustified beliefs, faith and knowledge are more like two opposing magnets than they are like two sides of the same coin. Indeed, if new facts or evidence challenge dogmatic beliefs, faith treats them as irrelevant or with outright hostility.

Claiming to know things you don't know is not virtuous, and nothing about having faith makes someone moral. Similarly, lacking faith doesn't make one immoral. For instance, people with tremendous amounts of faith who are positively immoral abound, and many who completely lack faith are positively moral. Divorcing faith from morality is important in the event someone tries to persuade you that having faith is a virtue. Faith may make someone feel good, but when people make claims about the world based on faith, they are claiming to know things they don't know or believing something without evidence; neither are moral positions.

Understanding which questions science can answer and which ones it cannot is helpful. Ultimately, science cannot answer many questions of judgment, values, ethics, morals, or aesthetics, though it may take an informative role (Harris, 2010; Harris, 2011; Shermer, 2004; Stenger, 2012). In the West, the unit of exchange is paper money. On the Micronesian island of Yap, islanders trade large, heavy, circular disks made from limestone. While science can inform which unit is, say, more efficient or economic, it cannot decide which one is more valuable. Faith's treatment of evidence and reason places it into a similar category, beyond the realm of science.

> We have never understood anything about the universe by assuming the supernatural, while assuming naturalism as a working hypothesis has moved our understanding ever forward.—Jerry Coyne

Biologist Richard Dawkins draws an analogy between science and faith using the question "Did an asteroid kill the dinosaurs?" (Dawkins, 2002). Dawkins's example contrasts a typical science paper written by a scientist with that of faith-based revelations. The science paper presents evidence, such as the iridium layer at the K-T boundary and a radiometric-dated crater in the Yucatán Peninsula. Conversely, the revelation relies on strong inner convictions, private revelation, childhood indoctrination, and official dogma. In science, the demands of evidence require more than inner convictions or private revelations.

Scientific knowledge doesn't come by intuition, by gut feelings, by hunches, or by asking psychics. It comes by testing hypotheses and evaluating evidence.

Faith is also impractical as an epistemology. Take a criminal trial, for example. Let's say that in broad daylight, on camera, and without wearing a mask, a man is caught in the act of robbing a bank and is shot during a shoot-out with police. Are juries more likely to believe the video footage, police testimony, and hospital records all attesting to the man's guilt, or the heartfelt testimony of the man's character witnesses who have faith that he didn't do it? More than likely, the jury will accept facts over faith, and the world is better for it. Imagine a world where mechanics had faith rather than mechanical training or doctors had faith rather than medical degrees or pilots had faith rather than a pilot's license.

Just as we want juries basing their conclusions on facts and evidence, we want experts to have more than faith in their abilities. How would knowledge progress if our understanding of nature was based on indoctrination rather than empirical evidence? How could science evolve if knowledge was accepted on the basis of private beliefs? If inner convictions were accepted as knowledge, then anything could be true.

Science is imprecise, but don't worry. We use language to express our thoughts, yet no matter how thoroughly we explain something or how many words we use, no one will ever know exactly what our inner experience is, though it's possible to get close. As with language, so it is with science. We use science to explain what happens in the world. Regardless of how much evidence we accumulate or how well our scientific models explain reality, because we experience reality indirectly through our flawed senses, we can never fully understand it on its own terms. Accordingly, when dealing with varying epistemologies, CT calls on us to recognize a claim based on facts, evidence, and reason and gives us the tools we need to recognize those that are not.

Relativism

Despite the rigors of scientific research, some maintain that what qualifies as a "fact" is not determined objectively, that what's true for one may be false for another. This philosophy goes by the name relativism, which holds that facts or truths are contextually relative. As we'll see, relativism confuses the differences between fact and fiction, between belief and knowledge, and between truth and opinion. In a nutshell, relativism implies that something can be true (or false) simply by believing (or disbelieving) it. Does believing you can fly make it so? No, just as someone believing she can breathe underwater will surely drown if she tries. Moreover, if truth is determined by beliefs, then there's no such thing as right or wrong, just or unjust, good or bad, and Thinking Straight is a pointless endeavor. Why examine what's true if truth depends on who's asking? People's beliefs cannot change facts and truth. Thus, this section is dedicated to explaining relativism and how it fails to explain reality.

THE APPEAL OF RELATIVISM

Some things actually are relative. For instance, some people prefer van Gogh over Rembrandt. It can be "true" for one observer that van Gogh's paintings are the better and "true" for another that Rembrandt's are. However, this is not Truth-with-a-capital-"T" because the properties of a good painting are rooted not in the painting itself but in the subjective opinions of those viewing it.

Furthermore, just because people disagree on a fact doesn't mean there's more than one fact. In other words, a disagreement about the truth isn't the same thing as there being multiple truths. For example, if I believe O. J. Simpson killed Nicole Brown, it cannot be true for me that he killed her and true for you that he didn't. Regardless of how tirelessly we dispute the evidence, O. J. either killed Nicole or he didn't, and our opinions cannot change Nicole's fate. Similarly, it cannot be true for one culture that demonic possession causes seizures and true for another that a seizure is a neurological episode.

What about the "discovery" of America, for example? Some people say that Christopher Columbus discovered America. Others say the ancestors of the Amerinds did. Is it true for Europeans that Columbus discovered America and true for Amerinds that he didn't? Sure, as long as we don't clearly define what a "discovery" is. Making a discovery is the act of uncovering something for the first time. If we can agree that Columbus wasn't the first human to uncover the fact that the land we now call the Americas exists, then clearly Columbus didn't discover America (Baggini, 2003, pp. 24–26).

The notion that all truths are relative is attractive because people prefer when their "facts" fit their worldview. For example, if Tamera believes in the "fact" of astral projection (a type of out-of-body experience) because she experienced it for herself, the fact that her experience can be explained scientifically will conflict with her worldview. Using our Street World metaphor, explaining astral projection as an experience of brain function rather than out-of-body travel would be like demolishing a familiar boulevard, only to replace it with a road less (or never) traveled. Thus, relativism appeals to our need to be right and our need for consistency by insulating our beliefs from disproof or rational examination. However, "If everything we think to be true is true, or there is no difference between the two, then that means we can never be in error" (Baggini, 2003, p. 27).

Relativism also appeals to political correctness in that it supposedly fosters tolerance by considering all views equal. If all views are equally valid, the argument goes, we're more likely to respect the views of others. This view, the *argumentum ad passions* (appeal to emotion) fallacy, is logically flawed, since it attempts to use emotions as the basis for its conclusion. Imposing one's views as *the* truth (such as the notion that Columbus discovered America) may be politically incorrect, but that doesn't mean one truth doesn't exist. However, a deeper problem lies with this false claim of tolerance. For example, relativism holds that a near-death anorexic's opinion that she is healthy is true for her or that it's OK to let one's children die slowly and painfully from preventable diseases if denying the efficacy of modern medicine is true for others. The problem is, if one accepts the concept of health as that which maximizes well-being and accepts that maximizing well-being is better than minimizing it, then it cannot be true that life jeopardizing malnourishment or suffering from preventable disease is healthy or good.

RELATIVISM AS A PHILOSOPHY AND AN EPISTEMOLOGY

Relativism also holds that truths and facts are governed not by their relationship to reality but rather by the context in which they're perceived. In its crudest form, relativism entails that there are no absolute or universal truths and that whether something is true depends on one's perspective. Relativists see the world subjectively, believing that what's true for one can be false for another. For the relativist, if one believes something is "true for them," that's the end of the story because there's no such thing as an objectively true fact. At one point in history, people believed Tourette's syndrome was caused by witches or demonic possession (Hines, 2003, pp. 83–84). If truth were relative, then those people suffering from Tourette's must have *really* been witches or were *really* possessed by demons.

Realists oppose relativism and believe that reality exists independent of one's perspective. According to realists, facts are facts and human thoughts cannot change them; even if every person in the world thought otherwise, a triangle has three sides, fur keeps rabbits warm, and there is a fundamental difference between lightness and darkness. As philosopher Peter Boghossian explains, "If everyone—including you—were to disappear, the universe would continue to be what it is" (Boghossian, 2013, p. 181). Realists believe, for example, that the sun is hot regardless of what people think about suns or heat. The sun being hot is an objective statement about the nature of suns and the meaning of heat, qualities that exist independent of human thought. So if relativism claims that reality depends on our thoughts about it, can it be a reliable epistemology?

Science assumes that nature can be understood. As Albert Einstein put it, "The most incomprehensible thing about the world is that it is at all comprehensible." Science answers questions about the natural world by asking those questions systematically. When scientific results are repeatedly verified, we say that, to a high degree of certainty, we've discovered something true about nature. If truth were relative, if truth depended on human thoughts and opinion, it would be impossible to discover scientific facts, since what is true in one lab could be false in another, and we'd be without a good explanation for why that is. Since we know that scientific facts adequately explain what happens in the world and that we can use those facts to make accurate predictions about the future, the notion that there are no absolute truths fails as a reliable epistemology.

Despite its popularity, relativism is all but dead in philosophy. Relativists have about as much credibility in philosophy as flat-earthers have in geology. In fact, in his dialogue *Theaetetus*, Plato destroyed the concept of relativism by pointing out that if relativism is true, it is only relatively so. Thus, relativistic statements such as "All truths are relative" are internally inconsistent and violate the law of noncontradiction. Explaining the *law of noncontradiction*, Boghossian says that "a thing cannot be both X and not X" (Boghossian, 2013, p. 110), meaning that something cannot exist and not exist or have a property it doesn't have or be both true and false. For instance, the statements "All bachelors are unmarried" and "Red is a color" are *necessary truths*, truths for which their legitimacy is implied by their meaning. A relativist cannot claim that "red is not a color" or that "all bachelors are married," as these statements are *necessary falsehoods* (Schick & Vaughn, 2005, p. 17). Moreover, if one could make something true just by believing it, then someone else could just as easily make it false by disbelieving it. So the phrase "Never say never" cannot be both true (one should *never* say never) and untrue (it's *occasionally* OK to say never). This statement is a *logical paradox*, a statement that, despite its appearance, is self-contradictory or logically flawed. If we turn the statement "All truths are relative" in on itself, we see that it cannot be both true and untrue. Either all truths are relative or all truths are not relative. Realists don't suffer from these logical problems.

> If yesterday we appraised a statement as true which today we appraise as false, then we implicitly assert today that we were mistaken yesterday; that the statement was false even yesterday—timelessly false—but that we erroneously "took it of true."
> —Karl Popper

Let's take the statement "Everything I say is a lie."[3] If this statement is true, then it's also a lie, since if everything I say is a lie, my statement that

"Everything I say is a lie" would be both true and false. Again, a thing cannot be both X and not X. Applying Plato's reasoning, the notion that all truths are relative cannot be sometimes true and sometimes false. When it comes to matters of fact, a thing is either true or false, and it's an error of logic (i.e., special pleading) to suggest that all truths are relative with the sole exception being the statement that "all truths are relative."

None of this is to say that reality can only be represented one way; that's what philosophers call *absolutism*. In their book *How to Think about Weird Things*, Theodore Schick and Lewis Vaughn use the different ways of mapping an area as a metaphor to demonstrate how reality can be accurately represented in multiple ways. "Consider . . . road maps, topographical maps, and relief maps. These maps use different symbols to represent different aspects of the terrain, and the symbols that appear on one map may not appear on another. Nevertheless, it makes no sense to say that one of these maps is the correct map. Each can provide an accurate representation of the territory" (Schick & Vaughn, 2005, p. 89). Absolutism, therefore, doesn't logically follow from realism. Just as observations of the cosmos in infrared or ultraviolet light are equally valid representations of the cosmos, a reality that exists independently of our thoughts can be accurately observed in different ways.

A last note about relativism might help to dispel a common myth that's often lodged against science. In some circles, it's frequently said that scientific facts change from one moment to the next. One day scientists say meteors are impossible, and the next day meteors are a proven fact. One day scientists claim continents can't move, and the next day moving continents are called plate tectonics. The changing nature of scientific facts may make them seem relative after all. If the best supported facts change from day to day, can't we conclude that facts are relative? The answer is no, because facts don't change. Meteors always existed, and the continents—that is, the Earth's crust—have been moving for billions of years. What changed is our interpretation of the facts. In other words, if we once thought something impossible that is now a known reality, our minds have changed, but the facts didn't.

Relativism fails as a philosophy because its self-defeating posture represents a logical impossibility, which cannot exist (chapter 11). Relativism's self-contradictory nature fails the basic philosophical test of internal consistency and relies on special pleading as support. Similarly, relativism fails as an epistemology because an epistemology is a claim to knowledge, and relativism claims that knowledge is relative. If knowledge (facts, truths, etc.) is in fact relative, then the saying "knowledge is relative" is logically inconsistent, like saying a circle is not a circle.

Wrapping Up

In this chapter, we learned the differences between philosophical and scientific skepticism. Whereas philosophical skeptics believe that knowledge cannot be obtained, scientific skeptics believe that science and CT are the tools we use to discover the secrets of knowledge. We also learned that since science operates under the assumption that what exists in nature is all there is, this assumption places some limits on what science can investigate. Even so, these limits do not mean that science is unreliable or that facts cannot be discovered, only that the facts science discovers are provisional.

We also discovered what knowledge is, what it means to know something, and that the most effective path to knowledge is through reliabilism. If a belief is to be knowledge, three conditions must be met: (1) it must be true, (2) you must believe it, and (3) it must be supported by a reliable mechanism. Thus, a true belief that is reliably supported is more than a belief; it is knowledge. And finally, truth, facts, and knowledge do not lie in the eye of the beholder, nor do they depend on human opinions. They are objective states that are knowable so long as we know how to know anything at all. As science fiction author Philip K. Dick rightly said, "Reality is that which, when you stop believing in it, it doesn't go away."

CHAPTER 3

Piercing the Scientific Veil

Science can seem like a mysterious realm that outsiders dare not enter. This belief is partially true; sometimes science is extremely complicated, which is why scientists do it! Nevertheless, the rules upon which science is based are fairly straightforward and easily understandable. This chapter is about learning some of those rules, along with how they apply to our everyday lives. If we don't understand or respect the scientific process, we lack a reason to trust its findings. Since the findings of science are our most reliable guides to reality, it's extremely important to understand how science is done.

Mistrust in science and belief in the incredible endures partly because few people understand science, although many of us trust science until it contradicts our beliefs. Thus, teaching how science works builds trust in the scientific process and so is just as important as teaching what science knows. Unwarranted mistrust and misunderstanding of science are around every corner, and we cannot even begin to change that without understanding how science works.

Building Vocabulary

Random assignment:	Independent variable:
A technique used for randomly assigning research participants to different groups, for example, by using a chance procedure or a random number generator.	The variable a researcher purposely changes or manipulates (controls) in scientific experiments but that is not changed by other variables.

Confounding variable:	Dependent variable:
A variable that is unknowingly, unavoidably, or inadvertently introduced into a research project that could compromise the outcome of the research.	The variable a researcher measures but does not manipulate in a scientific experiment; the variable one expects to be affected by changes to the independent variable.

Why Science Matters

Why does it matter what goes on in the obscure laboratories of countless scientists we've never heard of and will never meet? What's so important about science? What does learning about scientific methods do for me? Why care? Does learning about science change anything? Is learning about science useful information?

From financial waste and animal extinction to our very lives and well-being, the harms caused by scientific illiteracy abound. Whether it's through the technology we use or the medicines we take (or don't), science impacts our lives every day. Even matters far removed from us, like public funding of scientific research, affect us from what medicines and technologies become available to whom we vote for or support for public office. While these are all important reasons to take science seriously, there are numerous ways that science reaches us as individuals.

For instance, many who oppose stem cell research—which has the potential to improve and save the lives of hundreds of millions of people—are the same ones who bring snowballs onto the Senate floor to contest global warming (Mann & Toles, 2018, p. 95).[1] Many who seem bent on destroying our planet by publicly denying the fact of global warming are the same ones who think magical spirits control the hundreds of billions of galaxies in the universe. The science demonstrating the unreliability of eyewitness testimony is (see chapter 6) is the exact same science that shows the diminishing returns of mass incarceration (Mauer, 2005) and the benefits of educating prisoners (Duguid, Hawkey, & Pawson, 1996; Ross, Fabiano, & Ewles, 1988). The problem here is that those making the decisions on important issues oftentimes don't understand or trust science (Stanovich, 2013, pp. 111–12), and the people electing them are oftentimes equally as untrustworthy and uninformed.

People are remarkably capable of disregarding facts, ignoring logic, and clinging to completely irrational and evidence-free beliefs. Science is the only known process that can unveil these biases and show how to overcome them. Science took the penal system from public quarterings and chain gangs (Foucault, 1995) to electronic tablets and advanced drug treatment. With science, we've saved billions of lives and nearly eradicated world hunger, poverty, and many diseases (Pinker, 2018, pp. 62–78, 87–90, 115–18). Without it, we're stuck in the dark ages of bloodletting and authoritarian rule. With science, we understand how bacteria evolved into hundreds of millions of different organisms (Dawkins, 2005). Without it, we wouldn't understand how a zipper works. So, if we care about advances in medicine, the future of incarceration, protecting the planet, and making informed decisions based on the best available evidence, our best move is to bet on science. If we fail to understand science for ourselves, if we cannot explain how scientists determine facts, if we are as uninformed

> **Disconfirmation:**
>
> - Truth seeking
> - Falsification
> - Replication
> - Career protecting
>
> Scientists proceed with a hypothesis by subjecting it to rigorous tests attempting to falsify it. This method has proven to be extremely effective at weeding out the bad ideas and discovering novel truths about nature.

about science as the people who display their ignorance on the Senate floor and those who vote them into office, we're consigning ourselves to the rule of those very senators and their scientifically illiterate constituents.

In a culture where scientific progress isn't valued, people visit shamans rather than hospitals, quacks instead of doctors, and spiritualists instead of psychiatrists and take homeopathic sugar pills instead of evidence-based medicine. Imagine a world where the leaders considered themselves experts on all matters and didn't turn to scientists to ask questions such as "What are the consequences of a nuclear war?" or "What will happen if we continue burning fossil fuels?" During the ninth century, when science was thriving throughout the Arabic world, Muslims and Arabs named Ptolemy's work—which demonstrated that the Earth is round—*Almagest*, which means "The Greatest." In the twentieth century, the religious leader of Saudi Arabia proclaimed that the Earth is flat and is orbited by the Sun (Sagan, 1996, p. 325). A society that devalues science can easily crumble into the depths of the dark ages.

Science for the Masses

You don't need to know how to read to appreciate a good story, just as you don't need to understand refraction and the electromagnetic spectrum to appreciate the beauty of a rainbow (Dawkins, 2000; Dawkins, 2012). Similarly, we needn't understand everything about science to trust its findings. In fact, most people trust science on most matters. Hardly anybody disagrees that the Earth is round or that it orbits the sun, even though neither *seem* true to the casual observer. What about black holes, electrons, moving continents, or the distant past? No person has ever witnessed these objects or events, yet no one seriously doubts their reality. At one point electricity and magnetism were invisible and mysterious. Then James Clerk Maxwell explained them mathematically, and Michael Faraday showed that they were two closely related phenomena (Wolfson, 2000). All this leads to the question, "Why do people accept science on matters that seem to contradict our senses, yet reject the science on other matters, such as human-caused climate change or evolution?" After all, the science that proves the existence of electrons is the exact same science that proves human activity is warming the planet.

One reason is scientific illiteracy. Science is hard to do, and it can be even harder to explain. Among the thousands of scientists, the talented few who can both do and explain science become household names, such as Carl Sagan, Bill Nye, or Neil deGrasse Tyson. But despite the efforts of the great science communicators, scientific understanding is sorely lacking, and our lack of scientific literacy is an unfortunate cause of rejecting science.

Political ideology and religious belief also drive the rejection of science. Climate change denial is closely aligned with right-wing politics (Mann & Toles, 2018; Oreskes & Conway, 2010a), while evolution denialism is closely aligned with new-Earth creationism (Dawkins, 2009; Shermer, 2007). Understanding that scientific findings are politically, ideologically, and religiously neutral is one step toward scientific literacy. Just as there's no such thing as Chinese math, Hindu logic, or Brazilian chemistry, the theory of evolution by natural selection is neither true nor false based on who is looking at the evidence. The theory of evolution by natural selection is true because the evidence shows it's true. The same reasoning applies to other issues such as climate change and the safety of vaccines (chapter 13). If scientific literacy were more common, it's likely that such controversies wouldn't exist, which is why it's so important to understand how science is actually done.

Doing Science

One of our main objectives is learning how to understand reality by basing our beliefs on evidence and reason. To accomplish this goal, we need to gain an accurate understanding of evidence collection, the strengths and weaknesses of evidence, and how to best interpret the available evidence. Since any hypothesis is only as strong as its supporting evidence, our challenge is to familiarize ourselves with different kinds of evidence and discover how to determine whether the available evidence is actually reliable.

Just as there's no one method that defines science (Feyerabend, 1975/2010), there's no one type of evidence that scientists and researchers use. (I use the words "scientist" and "researcher" interchangeably.) Nonetheless, some basic types of evidence are commonly found in particular fields. For example, some studies require scientists to generate experimental evidence, while others demand observational evidence. The type of evidence required ultimately depends on what sort of hypothesis is being tested and what methods are used to test it.

Reliable evidence cannot be gathered in an unorganized, haphazard manner, since any interference with or influence on that evidence could potentially compromise the conclusions being drawn. Therefore, evidence collection that isn't careful and systematic is wholly unreliable. Sometimes scientists begin by seeking evidence that supports a hypothesis but not always. For three main reasons, science is oftentimes a pursuit to falsify one's theory rather than an attempt to confirm it. First, good scientists are interested in what *is* true, not what they *believe* is true. If a scientist pursues a hypothesis and the evidence is lacking, something valuable has still been gained. For example, for more than forty years, SETI (the search for extraterrestrial intelligence) researchers have scoured the skies looking for signs of intelligent life without success. Among other valuable insights, their failures suggest how far away intelligent life must be. The unsuccessful experiment that attempted to discover the ether—the medium in which light supposedly traveled through space—gave credence to Einstein's theory of general relativity.

Second, if a hypothesis can be proven false in the lab, it would behoove those proposing it to exercise some caution and humility by conducting some preliminary experiments that might prove their ideas wrong. Generally, new scientific ideas are introduced in scientific journals where they can be evaluated and critiqued by experts (see below). Sometimes, however, scientists get ahead of themselves and rush to the press with what they believe is incontrovertible evidence for their work. This method of "science by press release" was never more embarrassingly demonstrated than when, in 1989, electrochemists named Martin Fleischmann and Stanley Pons reported that they had created an apparatus that produced the long sought-after "cold fusion" (Ball, 2019; Hines, 2003, pp. 28–32). During a widely covered news conference, Fleischmann and Pons told the world that they had discovered cold fusion, a hypothetical type of nuclear fusion that's achieved at or near room temperature. If true, Fleischmann and Pons could have won the Nobel Prize for their discovery, as their work would've solved many of our energy and pollution problems. Unfortunately, their announcement was premature. Other scientists trying to replicate their results were unable to do so, and to this day cold fusion remains undiscovered.

Lastly, scientists make a name for themselves by novel discoveries and would rather not become known (or fail to become known) by their mistakes. Imagine how famous the woman who proved Einstein or Darwin wrong would be. Good scientists attempt to falsify (disprove) their hypotheses or risk their funding or, worse, their reputations. This point was excellently articulated by psychologist Keith E. Stanovich. According to Stanovich,

> Areas of investigation arise and are expanded or terminated according to a natural selection process that operates on ideas and methods. Those that lead to fruitful theories and empirical discoveries are taken

up by a large number of scientists. Those that lead to theoretical dead ends or that do not yield replicable or interesting observations are dropped. This natural selection of ideas and methods is what leads science closer to the truth. (Stanovich, 2013, p. 187)

At one point ESP was a live topic in several reputable scientific journals. However, ESP failed to continue capturing the attention of serious researchers specifically because the research led to dead ends. On the other hand, hypnosis, once thought purely pseudoscientific, has slowly found its way into mainstream science because serious researchers have not ruled it out as a genuine phenomenon (Hines, 2003; Lynn, Gautam, Ellenberg, & Lilienfeld, 2018; Wegner, 2002).

Falsification

Which brings us to falsification. Falsification—exploring how a hypothesis could be wrong—is an extremely important scientific concept (Popper, 1959/2014). First, falsified hypotheses are wrong and should be treated accordingly. Second, if a hypothesis can be falsified, someone will eventually figure out how, usually to the embarrassment of the person proposing it. Third, falsification helps examine how well a hypothesis explains the evidence. Finally, falsification minimizes confirmation bias—the tendency to pay attention to and accept information that aligns with our beliefs and ignore and reject information that doesn't.

Falsification is also important to science because proving or disproving unfalsifiable claims is impossible. To test whether a claim is falsifiable, simply ask the "F-Question" (Smith, 2018, p. 97): "What would it take to falsify the claim?" We can ask the F-Question by comparing natural selection with flood geology. Flood geology is the incredible claim that real geological evidence exists for a near-recent (by geological standards) global flood, a claim that contradicts evidence from every field of geology and biology (Rice, 2020). Asking the F-question would show that flood geology cannot be falsified by the fossil record (Shermer & Davies, 2007).[2] If fossils didn't exist, flood geologists would say, "Not enough time has passed for fossils to form." If fossils were randomly scattered throughout the geologic columns, flood geologists would say, "See, the flood killed all creatures at the same time." If fossils were found in order of complexity, with simple organisms found in the lower layers and more complex organisms found above, flood geologists would say, "This is due to the rate at which bodies decompose and sink," or "This is due to the fact that different ecological zones drowned at different times," or "This is because complex animals were smarter and thus escaped to higher ground." The arrangement of the geological column wouldn't matter, as any arrangement would "fit" the flood geology hypothesis (Law, 2011). Therefore, merely "fitting" the data isn't enough, as numerous theories that "fit" the data have turned out wrong (Laudan, 1981).

Natural selection, on the other hand, is entirely falsifiable by fossil records; one fossil of a complex organism where only simple organisms existed would destroy evolutionary theory. As Darwin himself explained in *On the Origin of Species* (1859), "If it could be demonstrated that any complex organ existed, which could not possibly have been formed by numerous, successive, slight modifications, my theory would absolutely break down."

Contrary to the evidence-free conspiracy theory that scientists suppress information and monopolize knowledge, scientists constantly subject established theories to disproof, which is one way scientists earn their stripes: by falsifying theories, especially widely accepted theories. They understand that "the cure for a fallacious argument is a better argument, not the suppression of ideas," as Carl Sagan believed. When a scientist disproves an accepted theory, she is rewarded with fame, tenure, Nobel Prizes, money, peer respect, and so on. As Sagan noted, when a young scientist disproves the accepted ideas of an older, well-respected scientist, the former inherits the latter's reputation and the respect she is paid (Sagan, 1996, pp. 255–56).

A hypothesis that cannot withstand attempts at falsification could never gain traction as a respectable theory. If a chimpanzee gave birth to a human, Darwin's theory of evolution by natural selection would be instantly destroyed. Hypotheses that withstand attempts at falsification are taken seriously, for if no one can falsify it, it might be true. As Michael Shermer is fond of saying, "If you don't seek contradictory data against your theory or beliefs, someone else will, usually with great glee and in public form, for maximal humiliation" (Shermer, 2008a, p. 92; Shermer, 2011a, p. 279).

Peer Review

Hypotheses are also subject to *peer review*—the process of opening claims to public review—prior to publication. Peer review serves the purpose of "making mistakes for all to see," according to Daniel Dennett, "in the hopes of getting others to help with the corrections" (Dennett, 1995, p. 380). Qualified experts in relevant fields, known as *referees*, are selected to carefully review articles before they're published. A referee's job is to identify flaws in the paper under review. Referees demonstrate their own talents by applying GASPs while identifying errors in logic and methodology. Whether the paper was written by an extremely intelligent researcher who's highly reputable, well-funded, or well-connected is of no concern, for science progresses by discovering what's true, and truth doesn't care about IQ scores or social status. For example, a nine-year-old girl named Emily Rosa tested the claims of what's known as Therapeutic Touch (TT), whose practitioners claim that they can feel a "human energy field" (chapter 12) without physically touching their patient. The results of Emily's study were published in the prestigious *Journal of the American Medical Association* (Rosa, 1998; Rosa, Rosa, Sarner, & Barrett, 1998), making her the youngest person ever published in a peer-reviewed journal.

Peer review helps ensure that (1) quality research is being published, (2) the findings are supported by the evidence presented, (3) the researchers properly understood and cited prior researchers' work, (4) all ethical standards and requirements are met, and (5) the research is original and meaningful (Benos, Kirk, & Hall, 2003). Good researchers are extremely meticulous and careful when gathering evidence and designing research methods, and they *cautiously* submit their findings when it's appropriate to do so. When researchers fail to exercise caution, they do so at the expense of their career or the respect of their colleagues.

Once a paper is written, the researcher(s) submits it to a journal for publication. An editor of the journal then reviews the paper to see if it can be accepted or if it needs certain revisions. If acceptable, the editor sends the paper to referees *anonymously*, blinding the referees to the identity of the author(s). The referees evaluate the paper looking for certain indicators, such as those listed in the box below (Ernst, 2018, p. 86), and return it to the editor with comments and suggestions, noting any weaknesses or **confounding variables** that may have contributed to the author's findings. The editor, usually someone familiar with the field of the paper, will then do one of the following: (1) recommend the paper be accepted; (2) accept the paper on the condition that the author makes some improvements; (3) reject the paper but suggest changes that might qualify it for resubmission; or (4) reject the paper outright (Oreskes & Conway, 2010a, pp. 154–59). Most peer review journals require that any disconfirming evidence be presented in the paper. Skepticism is rewarded, while false claims are punished.

In a criminal trial, the jury discusses all of the evidence together and must come to a consensus; otherwise there's a *hung jury*. With peer review, the process is completely different (the justice system could take several lessons from the peer review process). To prevent biases and social influence, the referees don't communicate with each other about the paper, have no knowledge of who else is reviewing it, and aren't allowed to see other referees' evaluations until the initial peer review process is completed.

The peer review process is careful, methodical, well-intended, and even peer reviewed! Nevertheless, like everything else in science, peer review cannot eradicate every mistake or bias. The articles editors choose to accept for publication, and the critiques and recommendations referees make, are still pursued by fallible humans. A paper that aligns with a referee's beliefs is less likely to receive

Good research includes the following:

- Author is from a respected institution
- Article is published in a respected journal
- A precise research question
- Full description of the methods used so replication is possible
- Randomization of subjects
- Use of placebo
- Double blinding
- Clear definition of a primary outcome measure
- Representative sample size
- Adequate statistical analyses
- Proper presentation of data
- Self-critical analysis
- Cautious conclusions based on the data
- Full disclosure of ethics approval, informed consent, funding sources, and conflicts of interest
- Reference list that includes contradictory findings

negative critique and more likely to receive positive recommendation. The opposite is true if a paper opposes a referee's beliefs. Something as unavoidable as the findings of a paper can influence how referees evaluate it (Mahoney, 1977),[3] meaning the same paper can receive contradictory reviews depending on the reviewer. What eventually overcomes this *publication bias* (see below) is the ability of other researchers to replicate the data.

I should also note that just because something has been published in a peer-reviewed journal doesn't mean the science is settled. Even today, junk science infiltrates respectable journals (Vyse, 2019a). In 2015, researchers found that fewer than 40 percent of findings from one hundred psychology papers published by reputable scientists could be replicated (Bohannon, 2015). Moreover, in a widely cited 2005 paper (Ioannidis, 2005), Dr. John Ioannidis[4] showed that not only is most published research false, but for most study designs, they are *more* likely to be false than true! According to Ioannidis, a substantial number of published findings are preliminary, contain small sample sizes, are contradicted by other research, go unchallenged, or are confounded by financial complications or other biases. His conclusion: identify solutions that will improve the research process (Begley & Ioannidis, 2015).

So should we trust a process that's so riddled with errors? The answer is a resounding yes. No one study ever defines a science, and no scientific fact rests on one study alone. Such a fact wouldn't be scientific, since science demands that findings are reproducible and falsifiable. By the same token, no theory or established scientific fact is falsified by a single study. As Popper observed, "We shall take [a theory] as falsified only if we discover a *reproducible effect* which refutes the theory" (Popper, 1959/2014, p. 86). Science is about what's most likely true, not discovering the absolute truth. "Science is not a simple line leading to *the* truth" according to the philosopher of science Ian Hacking. "It is more progress *away from* less adequate conceptions of, and interactions with, the world" (Hacking, 2012). Peer review accomplishes the goals of science by leading us away from inaccurate ideas about how the world works, even if every published paper isn't *the* truth.

Methods of Research Design

Now that we know what happens to a research paper after it's written, let's see how they originate. Many of us understand that there are different ways of looking at the same information. Leading questions, duress, anxiety, self-interest, or any number of biases can cause us to interpret the same information differently. This tendency has even been given a name, the *Rashomon effect*, after the 1950 Japanese film *Rashomon*, in which four different people tell contradictory stories about a rape and murder they all witnessed. As we'll see throughout this book, everyone has their own perspective, their own beliefs, their own memories, and their own ways of constructing reality, virtually guaranteeing that what one person thinks, remembers, or believes cannot be the whole story. Given that scientists are people too, the same holds true for science itself.

The human element in science can make discovering facts difficult. Sometimes this task is even more challenging in the presence of contradictory evidence. Therefore, it's extremely important to gather and examine all of the available evidence before coming to any particular conclusion. Our brains want to draw conclusions quickly, and the result is often drawing incorrect or inaccurate conclusions. So properly evaluating scientific facts requires understanding how to weigh and compare the different kinds of evidence that scientific research generates.

Although research studies are the bedrocks of science, individual studies, especially preliminary (or pilot) studies—those studies with small *sample sizes*, little to no blinding or placebo control—are almost never conclusive, and so after a study is complete, the next step in the process is to assess that study, try to replicate it, and compare it to similar studies. This process helps establish whether the findings are the result of some anomaly or bias and builds confidence in the truth-seeking process.

Several types of studies exist, including case studies, observational studies, and longitudinal studies. Overall, studies can be generalized as *experimental* or *correlational*. In experimental studies, scientists try to isolate and manipulate an **independent variable** in hopes of measuring how their intervention affects a **de-**

pendent variable. The reason for isolating one variable and observing its effects on another is to determine some degree of statistical certainty that changes to the independent variable did or did not cause specific changes to the dependent variable (Scollon, 2018).

An easy way to visualize the interplay between independent and dependent variables is by mixing paint. Let's say we want to find out what mixture of colors causes purple paint. The different colors are our independent variables, and purple paint is the dependent variable; that is, the resulting color *depends* on what colors are mixed. After a long process of mixing different colors (i.e., eliminating variables), we discover that mixing red with blue makes purple. This process is a clear-cut example demonstrating that one effect (purple paint) has a specific cause (mixing red and blue paint). Scientific experiments aren't always as clear-cut, as human error and the number of confounds that can enter the fold are virtually limitless.

Now for a real example. In an experiment that attempted to measure happiness, Professor Elizabeth Dunn gave participants $20, instructing them they must spend it by the day's end. Dunn split the participants into two groups, telling the first group to spend the money on themselves and the second group to spend the money on others. Dunn gave them self-report questionnaires at the end of the day to measure their levels of happiness, finding that those who gifted the money away reported being happier than those who spent the money on themselves (Dunn, Aknin, & Norton, 2008). Dunn controlled how the money was spent (the independent variable) to measure happiness (the dependent variable), thus manipulating one variable (money spending) to see how it affected another (happiness). One of the keys to understanding how Dunn attempted to isolate one variable is **random assignment**. When participants are randomly assigned, on average, they're similar in all respects because with larger sample sizes, chance factors are more likely to equal out (Stanovich, 2013, p. 89). Thus, having what's known as a *representative sample* is hugely important. A representative sample is a group that is as similar as possible to the population being studied by (1) having all the same relevant characteristics and (2) having those characteristics in equal proportion to the target group (Vaughn, 2019, p. 266). In other words, representativeness describes how similar a sample of a population is to that population. If we wanted to study how a certain pain medication affected teenagers with cancer, we wouldn't select senior citizens with dementia for our study. Random assignment increases representativeness and helps ensure that any differences between and within a group are random at the outset, so those differences tend to average out. If participants are randomly assigned, at the end of an experiment, researchers trust that any differences between groups are a likely result of the experiment rather than differences between participants (Scollon, 2018, p. 18).

Here it's important to understand the difference between random assignment and random sampling. Random assignment is a process by which research participants have an equal chance of being assigned to a control group or an experimental group. *Random sampling* is the process of selecting a sample of people from a population so that each member of the population has an equal chance of being selected. Random sampling can be used for correlational studies but isn't necessarily a requirement for true experiments. On the other hand, random assignment is imperative to any experimental study (Stanovich, 2013, pp. 107–10). Experimental studies are considered the gold standard for measuring cause–effect relationships because they allow researchers to more accurately isolate and control variables, which in turn allows them to confidently infer that specific outcomes are the result of their intervention rather than confounding variables. Experimental studies also allow researchers to perform statistical analyses by comparing other groups within their study. For all their benefits, however, experimental studies are not flawless.

Problems with Research Methodology

For instance, experimental studies can be impractical. Detonating a nuclear device to study the effects of radioactive fallout or severing someone's *corpus callosum*—the band of neurons that connects the left and right brain—to study brain activity is neither practical nor ethical. Accordingly, experimental studies are constrained by certain limitations. Additionally, experimental studies don't always represent real-world experiences; that is, they may lack *external validity* (think about how the coronavirus behaves in a petri dish versus how it behaves in a human body). The results of studies with external validity can be generalized to situations other than those involved in the study (Mehl, 2018). For example, if we wanted to measure effective decision making by inviting prisoners to attend a group that teaches CT skills, even if the participants began making better decisions after taking the group, the efficacy of their decision making might not accurately represent how *ex*-convicts make decisions in the real world because conditions in prison are drastically different than those in the free world. Perhaps prisoners can apply CT skills while inside prison, but those skills are forgotten as soon as they're released. Or maybe they pretend to use them in prison to impress the parole board, abandoning those skills as soon as they get out.

By contrast, in correlational studies, researchers don't intervene to manipulate variables, or, when intervention is necessary, it's kept to a minimum. Researchers using correlational studies passively observe and measure. In Dunn's experiment on happiness, she also asked participants how much money they *typically* spend on others or give to charity, later asking them how happy they were.

Dunn's intent was to passively observe (observing how participants spent their money) for the purpose of measuring a particular phenomenon (how spending money on others affects happiness).

Correlational studies help researchers gather huge amounts of information at relatively low cost because they're usually looking at existing data. Correlational studies make it possible to compare different groups, and since intervention isn't required, there's virtually no risk that the study doesn't re-create real-world situations.

However, correlational studies aren't foolproof either, as they too have multiple weaknesses. With correlational studies, one can study no more and no fewer than two variables at a time (Scollon, 2018, p. 20). Also, since the researchers don't interfere to control variables, confounding variables (see below) can find their way into the data, meaning correlational studies can typically detect associations between variables but cannot determine cause–effect relationships.

Cohorts and Confounds

In addition to being experimental or correlational, studies can also be *prospective* or *retrospective*, and knowing the difference can be useful. Prospective research involves selecting *cohorts*—others who share defining characteristics—to observe and examine. Retrospective research analyzes events and outcomes in the past, like looking at people who served years in solitary confinement to see if they developed permanent psychological problems. Since one cannot control the past, retrospective studies are more susceptible to confounds than prospective studies, but they're cheaper, they require less time to conduct, and they're better for analyzing multiple outcomes.

Prospective studies are *longitudinal*, meaning they track the same cohorts over time to repeatedly observe the same variables. If we wanted to study the long-term psychological effects of solitary confinement (Arrigo & Bullock, 2007; Gawande, 2009; Tietz, 2012), we'd conduct a longitudinal study. Longitudinal research is great because it isn't susceptible to as many confounds as retrospective studies are, and it generates huge amounts of information. On the other hand, longitudinal studies are expensive, complicated, and suffer from high *dropout rates*.[5]

CONFOUNDS

The term "confounding variables" is one that nonscientists rarely hear, so to familiarize ourselves with it, we'll go over a few examples of various *confounds*. Our first confound is *participant demand*,[6] a bias in which participants consciously or unconsciously alter their behavior because they know they're part of a study. For example, knowing they're part of an experiment, participants may try being a "good participant" and behave according to how they believe the experimenter wants them to (Orne, 1962). Also, providing false information to purposely sabotage the experiment isn't unheard of (Nichols & Maner, 2008).

Even hardcore skeptics can be led to have otherworldly experiences on account of participant demand. Our friend Michael Shermer, whom we've already met several times, had just such an experience (Shermer, 1999; Wiseman, 2011, pp. 218–20). Shermer was invited to try on the "God Helmet," a device that projects magnetic fields around the skull, which causes about 80 percent of people who try it to report having strange, out-of-body or spiritual experiences, some even sensing a ghost-like presence. The God Helmet received much fanfare, which is why Shermer fully expected to have a strange experience by putting it on.

Later, when better controlled experiments were conducted, researchers discovered that psychological factors, not magnetism, cause such strange experiences (French, Haque, Bunton-Stasyshyn, & Davis, 2009; Granqvist et al., 2005). Furthermore, the amount of magnetism participants were subjected to was five thousand times weaker than a refrigerator magnet, not nearly enough to penetrate the skull and influence neuronal activity (Aaen-Stockdale, 2012).

THE (IN)FAMOUS PLACEBO

Perhaps the most popular confound is the *placebo*, from a Latin word meaning "I shall please." Many of us experienced our first placebo when Mom relieved the pain of a knee scrape with a Band-Aid and a kiss. Of course, we didn't know we were given a placebo, which is precisely why it worked! Science understands a placebo as an inactive (or inert) substance, treatment, procedure, or activity that can produce physiological or psychological benefits to a patient who has an expectation that the placebo is effective (Smith, 2018, p. 168). Accordingly, the so-called placebo effect occurs when someone (a) is given an inactive treatment and (b) has an expectation of relief and (c) experiences symptomatic relief.

> The scientific method consists of the use of procedures designed to show not that our predictions and hypotheses are right, but that they might be wrong.—Carol Tavris

Research shows that although placebos sometimes help with pain management, they cannot produce clinical improvements (Hróbjartsson & Gøtzsche,

2001; Kienle & Kiene, 1997; Madsen, Gøtzsche, & Hróbjartsson, 2009). Even so, sugar pills (big colorful pills work better than small white pills), sham surgeries (Carroll, 2014; Cobb, Thomas, Dillard et al., 1959; Moseley et al., 2002; Sihvonen et al., 2013), fake acupuncture (Interlandi, 2016), and even sham injections (Ernst, 2018, p. 25) have all been successfully used as placebos. Suggesting the psychological significance of placebos, placebos work best in hospital settings with enthusiastic "doctors" (Harrison, 2015, pp. 199–200) wearing lab coats and administering the placebo to unsuspecting recipients (Cha, Hecht, Nelson, & Hopkins, 2004; McKinstry & Wang, 1991; Rehman, Nietert, Cope, & Kilpatrick, 2005). People have even reported becoming addicted to placebo pills, developing a tolerance, and requiring more and more of the placebo to maintain their health (Ernst & Abbot, 1999). Moreover, expensive placebos have a greater effect than inexpensive placebos (Waber, Shiv, Carmon, & Ariely, 2008). Interestingly, placebos don't work in reverse: for example, you cannot "expect" heroin not to work and remain sober—you need naloxone for that—nor can you "believe" your beer is nonalcoholic and pass a breathalyzer.

The term "placebo effect" is really a misnomer since the placebo *itself* doesn't actually provide relief; the recipient's mental and emotional response does. Moreover, what researchers call the placebo effect is really a mixture of multiple variables, including the natural progression of the illness and regression to the mean (Ernst, 2018, p. 23). Thus, a more appropriate title for the placebo *effect* is the placebo *response* (Alcock, 2018, p. 335). Importantly, the placebo response is not a "mind-over-matter" cure-all as many people believe. As the skeptic and clinical neurologist Steven Novella explains, "There's no compelling evidence that the mind can create healing simply through will or belief. However, mood and belief can have a significant effect on the subjective perception of pain" (Novella, 2018, p. 228). The belief that placebos cure ailments is false. When researchers say a placebo "worked," all they mean is that the administration of a placebo has been shown to cause a temporary relief of symptoms.

People respond to placebos in numerous, often illusory, ways. Studies show a link between receiving a placebo and the release of pain-relieving endorphins (Clement-Jones, McLoughlin, Tomlin et al., 1980; Levine, Gordon, & Fields, 1978; Offit, 2014, pp. 230–31), so, at least in cases of the subjective experience of pain, placebos, combined with expectation, can produce a positive physiological response. Other variables, such as reduced anxiety, blood pressure, strain on the heart, or levels of stress hormones, can cause positive physiological response to a placebo (Novella, 2010b). With or without a sugar pill, reducing anxiety, blood pressure, and stress is likely to make almost anyone feel better!

These facts raise the question, "Is using a placebo to treat ailments worthwhile?" This is a bit like asking, "If mediums can't really communicate with the departed, is it OK for them to take peoples' money pretending they're talking to the departed, given that it might make the customer feel better?" Setting aside ethical questions concerning giving people ineffective treatments, we should obviously avoid using placebos as treatments. If acupuncture relieves anxiety, isn't it useful? If chiropractic relieves back pain, how is that bad? Well, acupuncture is not without serious risks (Ernst, Lee, & Choi, 2011; Gnatta, Kurebayashi, & Paes da Silva, 2013; Xu et al., 2013), and while chiropractic may relieve pain, it can also cause permanent injury (Ernst, 2007; Ernst, 2010a; Ernst, 2010b; Hall, 2019d) and even death (Lee, 2016; Chiropractic, 2013). Judging something's worth by how it makes people feel is a step in the wrong direction. Malnutrition eases the anorexic, and delusional boasting warms the narcissist, but neither comfort maximizes human well-being. Isaac Asimov said it best: "If solace and comfort are how we judge the worth of something, then consider that tobacco brings solace and comfort to smokers; alcohol brings it to drinkers; drugs of all kinds bring it to addicts; the fall of cards and the run of horses bring it to gamblers; cruelty and violence bring it to sociopaths. Judge by solace and comfort and there is no behavior we ought to interfere with" (Asimov, 1989a). Humans progress—as a society and as individuals—by basing decisions on what works, not on what doesn't, so risking health and injury for temporary relief doesn't exactly qualify as progress.

Feeling better is not the same as *being* better. Let's say you have a bacterial infection that requires antibiotics. If you unknowingly take a placebo to treat the infection, chances are nothing will happen, since the infection requires antibiotics. Now, if you unknowingly take the proper antibiotics, chances are your infection will go away. Even if you knowingly took a placebo, you may temporarily feel better, but you wouldn't likely rid yourself of the infection because placebos cannot cure real ailments.

IDEOMOTOR EFFECTS

Like participant demand, *experimenter expectancy* is when researchers' biases or foreknowledge causes them to misinterpret findings or subconsciously influence their participants in some way. The most famous example of experimenter expectancy is the story of "Clever Hans," a horse that was said to successfully do arithmetic and intelligently answer other questions posed by humans (Alcock, 2018, pp. 269–71). Hans's owner, a math teacher named Wilhelm von Osten, would ask basic math questions and Hans would tap the correct answer. Hans's fame came at a time when people were just beginning to accept Darwin's theory of evolution. During that period, there was a lot of interest in animal intelligence because, people thought, if humans are closely related to animals, perhaps animals have undiscovered intelligence, and how cool that would be! When von

Osten—himself a strong proponent of evolution, who also believed animals shared humans' intellectual capacities (Wiseman, 2011, p. 244)—claimed his horse could perform mathematics, public interest was greatly aroused, and the whole spectacle received extensive media exposure.

One of the reasons Hans's performance was so convincing was because it was honest; that is, von Osten wasn't lying: he genuinely believed his horse was strangely intelligent. After the philosopher and psychologist Carl Stumpf performed a few controlled experiments, greater minds prevailed. Stumpf convincingly showed that Hans, while especially observant, was no mathematician. Stumpf, along with his assistant Oskar Pfungst, himself a biologist and a psychologist, showed that under conditions where Hans couldn't see von Osten (or any questioner) or von Osten didn't know the right answer, Hans couldn't answer questions correctly. At this point it was the behavior of von Osten that became the center of Pfungst's focus. Armed with the hypothesis that von Osten's expectations could influence how Hans answered, Pfungst showed that Hans was tapping in response to the posture, facial expressions, and unconscious muscular movements of von Osten, such that von Osten relaxing would cue Hans to stop tapping (Hines, 2003, pp. 120–21).

IDEOMOTOR MADNESS

The involuntary and unconscious movement of muscles is a now well-understood phenomenon known as the *ideomotor effect*. The psychologist Ray Hyman summarized this effect:

> Under a variety of circumstances, our muscles will behave unconsciously in accordance with an implanted expectation. What makes this simple fact so important is that we are not aware that we ourselves are the source of the resulting action. This lack of any sense of volition is common in many everyday actions. (Hyman, 2001)

The ideomotor effect explains phenomena from table-turning and Ouija boards (Loxton, 2018a) to dowsing (chapter 12) and facilitated communication (FC) (see below). Such phenomena were once quite mysterious and thought inexplicable. Or they were attributed to "demons, angels, and even entities from the future or from outer space, depending on their personal contact with cultural theories about such effects" (Wegner, 2002, p. 113). The ideomotor effect was firmly established in 1852 when the physicist Michael Faraday (1791–1867) investigated table-turning (Wiseman, 2011, p. 150). Faraday's research was replicated in the 1890s when the psychologist Joseph Jastrow (1863–1944) showed the same principles shed light on the enigmatic Ouija board (Wiseman, 2011, p. 157).

Psychologist Gary Wells likened the ideomotor effect to police uncon-sciously sending witnesses nonverbal messages during investigations. Wells believes that the ideomotor movements by police detectives can subtly influence people to choose a particular suspect in a lineup (Wiseman, 2011, p. 247). If true, Wells's research calls into question the very validity of witness identifica-tion using lineups (Cutler & Penrod, 1988). Furthermore, Wells argues that "mistaken identifications by eyewitnesses are responsible for more actual cases of wrongful conviction by juries than all other causes combined" (Wells & Olson, 2002). Imagine a police detective showing a potential witness pictures of crime suspects. If the detective knows who the prime suspect is or, worse, would prefer arresting someone in particular, the detective might put that sus-pect's picture in the center of the table; make subtle, even unconscious gestures toward it; or ask the witness more questions about that picture than all the others. These "hints" can influence the witness in subtle ways that neither the witness nor the detective is aware of (Cialdini, 2007). In one study, participants watched a video of a supposed terrorist planting a bomb. Although the video clearly showed the suspected bomber's face, when the participants were later shown photos and asked to identify the bomber, all the participants identified who they believed the bomber was, even though the bomber wasn't in any of the photos (Alcock, 2018, pp. 93–94).

Facilitated communication is the latest pseudoscience involving ideomotor action that, if true, "would require that we overturn basic knowledge in fields as diverse as neurology, genetics, and cognitive psychology" (Stanovich, 2013, p. 126). FC is a pseudoscience that makes the incredible claim that autistic people are not *cognitively* incapable of communicating with the outside world but instead are *physically* incapable. Proponents of FC claim that through a third party, people with autism can communicate their thoughts to the world. Since its inception, research has shown that FC is the latest version of the Clever Hans story yet with more serious consequences (Hines, 2003, pp. 403–7; Stanovich, 2013, pp. 97–100). The basic process is similar to the Ouija board. First, a question is asked of a person with autism. Then a "facilitator" guides the hand of the person over a keyboard, periodically typing keys. Through this process the question is eventually answered. When the idea was first developed, autistic individuals began communicating everything from basic sentences to songs and poetry. Children who had been unresponsive for years and were never taught how to read were doing book reports in grammatically intact English. And it appeared as though the facilitators genuinely believed that their patients were the ones doing the typing.

Comparable to the tests Pfungst performed with Clever Hans, the research on FC found that if the patient had information that the facilitator didn't, the patient couldn't communicate that information; that is, the patient's "answers" resulted from ideomotor cues given by the facilitator (Wegner, Fuller, & Sparrow, 2008). For example, if an individual with autism was asked a question the facilitator couldn't hear or was shown a picture while the facilitator was shown a different picture, the individual couldn't answer the question and instead would describe the picture the facilitator saw (Jacobson, Mulick, & Schwartz, 1995). Thus, some have said FC is not facilit*ed* communication but rather facilit*or* communication (Schlosser, Balandin, Hemsley, Iacono, Probst, & von Tetzchner, 2014).

FC presents several concerns, not least of which is the fact that pseudosciences such as FC undermine public confidence in science. As I mention in chapter 1, one of the consequences of FC was several prison sentences based on allegations from patients of physical and sexual abuse (Hagen, 2018). In 1994, there were sixty known cases of abuse reported by FC, a number which has surely increased over time. Through FC, people with autism have also "consented" to sexual relationships with their facilitators (Todd, 2012). Even more concerning, parents and caregivers of people with autism were "creating in their own mind a human being who did not exist, and [were] ignoring the very real human being who did exist and very likely had very different needs and wants" (Hines, 2003, p. 406).

Accuracy and Validity

One way researchers minimize confounds is by *blinding* or *double-blinding* researchers and participants. A blind study keeps participants ignorant about certain aspects of the experiment, such as "blinding" them to whether they received a drug or a placebo. Double-blind studies keep researchers *as well as participants* blind to information about the study. Nearly all modern researchers use blinding or double-blinding to increase the quality of their research and to reduce the chances of inadvertently introducing confounds into their study. Blinding prevents subconscious or unintentional biases, such as those mentioned above. Researchers have found that blinding helps significantly reduce erroneous results. For example, fingerprint analysts typically know what a "match" entails as far as who the suspect is or what he or she is alleged to have done. If an analyst knows the prosecutor who sent in the print, her findings are more likely to support the prosecution's case than the defendant's. If the analyst were blinded to the details of the crime, the suspect, or the source of the print, those potential biases would be eliminated.

Another way scientific research makes progress is by researchers trying to find the right balance between a study's internal and external validity (Mehl, 2018, p. 31). *Internal validity* refers to how strong or reliable causal inferences drawn from a study are. In other words, if a study shows a probability greater than chance that the independent variable caused a change in the dependent variable, it's said to have internal validity. Usually, the variables in laboratory experiments are controlled carefully enough that researchers are confident in the cause–effect relationships between variables. Like our earlier example of measuring effective decision making by inviting prisoners to attend a group on CT, if participants' ability to make effective decisions is reportedly higher after taking the group, there is a strong correlational relationship between taking the group (cause) and the participants' decision making (effect).

> We live in a society absolutely dependent on science and technology and yet have cleverly arranged things so that almost no one understands science and technology. That's a clear prescription for disaster.—Carl Sagan

This example, however, doesn't necessarily reflect effective decision making outside the setting of the group. In addition to the group taking place in a controlled environment, there could also be something special about this particular group that causes participants to make better decisions: maybe the instructor is attractive or especially pleasant, or perhaps the group setting made the participants more accepting of the information. Such potential confounds could account for a participant's decision making, so while the study is internally valid, it still lacks external validity. Our CT group example is not externally valid because it doesn't necessarily represent decision making in real-world settings. Creating experiments where variables are closely controlled and apply generally to real-world experiences is the difficult balancing act of scientific research.

Reading the Data

Studies are extremely helpful in compartmentalizing information and providing a platform from which inferences can be drawn. Conducting studies is one of the most effective and reliable tools for obtaining knowledge. This process spans from the time we're babies learning that letting go of a toy causes it to travel down and not up, to Edwin Hubble discovering that our universe is expanding by reading the *Doppler shift* of distant galaxies. If humans quit conducting studies, the progress of our knowledge would come to a dramatic halt. But caution is still needed. Studies are human endeavors, their results are interpreted by humans, and it's important to remember that no human endeavor is flawless or perfect. But just because science isn't perfect, that doesn't justify abandoning science altogether. Doing so is a form of fallacious reasoning called the *nirvana fallacy*, which holds that if something isn't perfect, it's essentially worthless. The justice system and children are two examples of things that aren't perfect, but we don't do away with them on account of their flaws; we try to improve them. Science is the attempt to continually improve our understanding of how the world works, and so far it's done an excellent job.

Once studies are conducted, the data from those studies are analyzed and compared. However, simply having a study that reaches a conclusion proves nothing. Studies must meet certain criteria. Let's say we wanted to know how many people use marijuana recreationally. To find out, we stand outside of a marijuana dispensary and ask ten people, "Do you use marijuana recreationally?" If nine out of ten people answered yes, concluding that 90 percent of people use marijuana recreationally would be a mistake (specifically, the hasty generalization fallacy discussed in chapter 5), since ten respondents located near a dispensary is hardly representative of most people. Studies with small sample sizes are more likely to contain errors or unusual qualities than larger ones, since they provide fewer opportunities for random effects to average out.

> Everyone is entitled to their own opinion but not their own facts.
> —Daniel Patrick Moynihan

Obviously our survey is flawed. Yet even in well-controlled scientific studies, when scientists identify something statistically significant in their data, they still hesitate to draw definitive conclusions because they understand that any undetected biases or flaws in the research can influence the results. Departures from chance are statistically expected, so researchers hesitate to draw definitive conclusions based on anomalous results. Continuing with our dispensary example, examining how the data were collected, where the data were collected, and how many people were questioned reveals obvious flaws. Any statistically significant correlation—whatever it was—would likely be inaccurate because our survey lacks external validity, the sample size isn't representative, our sampling wasn't random, and conclusions should rarely be drawn from statistical data alone.

Far from Perfect

To dispel another misconception, science isn't a perfect process. This misconception is dangerous, as it can lead to the false conclusion that if science isn't perfect, we shouldn't use it at all. Such misconceptions also lead to other erroneous beliefs, such as if agreement among experts on a given issue isn't unanimous, we shouldn't accept the findings of their research as true, or if science isn't perfect, it is therefore false. Such illogic is an error in deductive reasoning (chapter 4) as it contains false and hidden premises and thus is invalid. For one, as is evident from all the problems that plague the scientific process, the premise that science is a perfect process is wrongheaded to the extreme. Assuming there must be 100 percent agreement among scientists for us to accept something as factual is poor reasoning. Lastly, the argument makes the false assumption that scientists' research is the only way to measure accuracy. As easy as it is to pick this argument apart, it remains popular nonetheless. Rather than accept such fallacious arguments veridically, skeptics question such arguments, understanding how to put them in their proper place.

Science progresses by interrogating the natural world: the scientist as detective, nature as the suspect. Both are trying to give probable answers to questions of fact. What the scientist interrogates, what questions he asks, what methods he uses, and what tools are at his disposal will all depend on the current state of science and technology. The scientist proceeds by following GASPs, since GASPs tend to produce accurate results. In law, those GASPs are codified into "rules of evidence" or "rules of criminal or civil procedure." Scientific rules aren't codified, and scientists aren't all required to follow the same ones. Nevertheless, researchers use a number of GASPs to interrogate the natural world and investigate claims, such as empirical testing, expertise, replication, falsification, citations to credible sources, peer review, operational definitions, open access publishing, the strictures of logic, the minimization of confounding variables, consistent evidentiary standards, statistical analyses, and so on.

> It is not what the man of science believes that distinguishes him but how and why he believes it.
> —Bertrand Russell

Replication is such an important part of the scientific process because conducting studies that are completely free of confounds where all possible variables are controlled is difficult (if not impossible). If other researchers can perform the same experiments under the same or similar conditions and reach the same conclusions, they can be much more confident in their findings about whatever phenomenon is being studied. Conversely, if others attempting to replicate a study come up with completely different findings, we know something is wrong with the process or that no correlation exists between the variables being examined.

Medical studies are especially plagued with subtle confounds, which is one reason why medical research can go on for years or even decades. How were the participants chosen? Are they representative of a large enough population? Are they similar in every way possible? Did those who received positive results continue with the study while others did not? What was the dropout rate of the study? These question bear on whether any *sampling biases* were inadvertently introduced into the study.

Researchers work hard trying to account for anything that could affect a study's results or how researchers interpret them. What researchers want to know is essentially whether all the data are being taken into account because creating false results by not accounting for all the data or, worse, unknowingly introducing confounds is all too easy. Careful attention to detail and the ingenuity of researchers envisioning novel studies have carried science as far as it has and are why science has become such a successful path to knowledge.

Healthy Skepticism

In the public eye, science progresses by great leaps, with discoveries constantly being overturned in dramatic fashion. As is evidenced by Einstein's theory of relativity and Darwin's theory of evolution, some science makes monumental leaps forward. But most science progresses by the slow processes of scientific methodology, namely by scientific studies. Scientific studies are the main vehicles that drive the progression of scientific knowledge. Societies don't advance much without them, yet the studies themselves aren't quite capable of telling the entire story. As we've seen, studies can be flawed or misinterpreted, and many reasons exist for why no single study definitively settles fundamental questions about nature.

A Note on Replication

I noted earlier that replication is an important aspect of science, and it's important to stress that point. Peer review is used to vet research for biases, methodological problems, or other confounds, but it is replication that renders facts scientific because sound science can still yield false results. Findings that are not or cannot be replicated by other researchers and are continually touted as factual are pseudoscientific because science works under the assumption that the laws of nature are constant. This means that if all variables are held constant, different researchers using the same (or better) methods should reach the same conclusions. Otherwise, we assume something was different about the research, rather than assuming that something was different about the laws of nature. As such, findings are considered scientific only if other researchers can duplicate them (Stanovich, 2013, p. 10).

One Study Doesn't Cut It

Replication is extremely important because statistically, false-but-seemingly-significant findings will occur by chance alone. That is, even studies that don't suffer from methodological flaws will sometimes still yield false results (Hines, 2003, pp. 217–18). This prediction has nothing to do with how well-controlled the study is or how honest the researchers are; it's just the nature of chance. One way of understanding probability is by using *p-values*—or "probability" values—to measure whether a given correlation is merely a statistical fluke. Some researchers—such as psychologists or biomedical researchers—don't consider a correlation statistically significant unless the probability of its being due to chance is 1 in 20 (or $p < .05$) or better, meaning that "even if all clinical trials were conducted perfectly, 5 percent of studies would appear to be 'positive' through random chance alone" (Gorski, 2018). For example, a p-value of .01 means that the odds are only 1 in 100 that a correlation happened by chance, so such findings would invite further research. However, the more studies conducted, the better the chance of finding false results. If we conducted 100 well-controlled studies looking for, say, beneficial effects of chiropractic, we should expect to find positive results in roughly 5 studies. The same would be true of any pseudoscience.[7] Now, imagine that pseudoscience proponents touted only those well-controlled studies that found positive results and ignored the rest. They could present a good case that the science supports the practice of chiropractic, and the reader, ignorant to the other 95 percent of research findings, could be forgiven for believing such incredible claims.

Studies claiming to prove something are accepted as provisionally true only if other researchers can repeat those studies and reach the same conclusions using the same or similar methods. Scientific facts are those that are subject to confirmation or refutation by the processes of replication, falsification, and experimentation. Accordingly, when studies are published, other scientists rigorously attempt to replicate and/or falsify their findings. If the original findings withstand these attempts, they're accepted as *provisionally* true, since there's always a chance that those attempting to falsify the facts are missing something. If repeating an experiment fails to yield the same results, scientists doubt that the original study's findings were accurate or explained any new phenomenon.

For example, in 1998, Dr. Andrew Wakefield published a paper (Wakefield et al., 1998; Editors of the Lancet Retraction, 2010) in the prestigious medical journal *The Lancet*. Although the paper itself claimed that "we did not prove an association between measles, mumps, and rubella vaccine and [autism]," Wakefield began touting his research in press conferences and warning against using the vaccine (Hall, 2009b). Once other scientists began conducting independent studies trying to replicate or falsify Wakefield's findings, the "link" between the MMR vaccine and autism vanished (Offit, 2015, pp. 238–39).[8]

Study after study found no evidence that vaccines cause autism (Foster & Ortiz, 2017) or that the MMR vaccine influences the development of autism (Honda, Shimizu, & Rutter, 2005; Jain et al., 2015; Taylor, Swerdfeger, &

Eslick, 2014). In fact, the investigative journalist Brian Deer revealed that Wakefield fraudulently changed and misrepresented information in his study, performed procedures on participants without approval from his *institutional review board* (i.e., an ethics committee that oversees research projects), lied about conflicts of interest, and accepted over $400,000 from lawyers who were suing manufacturers of the MMR vaccine (Deer, 2004; Deer, 2006; Deer, 2011; Randi, 2017b). Even one of the referees of Wakefield's fraudulent paper received over $40,000! Thus, single studies shouldn't be counted as evidence that something is factual (see especially Ioannidis, 2005). Actually, the findings of single studies shouldn't be overly interpreted or thought particularly important outside of their relevant fields (Wilkinson, 1999) because scientific knowledge is achieved by the ongoing collection of information across multiple disciplines (Wilson, 1998).

Often the first question a good skeptic asks about a study is "Has it been independently replicated by other scientists?" If the answer is no, we can confidently say that, whatever the issue, whether it establishes a fact or not isn't settled. On the other hand, if it has been replicated, what were the results? Are the findings consistent, or are there important differences? Were the different researchers conducting the studies using the same methods, controlling the same variables, or looking to explain the same phenomenon? Do the methods used comport with GASPs? Is there a consensus among experts? Are those conducting the research competent experts in the relevant fields? Or do the researchers have a conflict of interest where those financing the study have a stake in a particular outcome?

Publication Bias

Another precaution we can take is to watch out for what's called a *publication bias*. As we'll see in chapter 5, since news is about what happens rather than what doesn't, people frequently commit the *availability error*, assuming something is more common than it really is. News is also about what's interesting, not what isn't, which is one reason reports of mysterious crop circles enjoyed plenty of media exposure, while the confessions that they were hoaxed didn't (Ridley, 2002). The publication bias is similar in that researchers, like the media, prefer publishing studies they think people will find interesting, will boost their careers or reputations, or will get their work—and by extension themselves—into the public eye (Rosenthal, 1979). This bias openly invites the availability error (see chapter 5) and the assumption that the research on any given topic is one-directional, when actual results may vary widely. One study cannot establish a fact. However, repeated findings by independent researchers using the same or better methods establish grounds for holding those findings as provisionally true. The problem is, since journals selectively publish positive research, it's almost impossible to tell how much research has been conducted in a particular field.

So while journalists emphasize negative stories because they're more unexpected and vivid and thus are more memorable (Bohle, 1986; Combs & Slovic, 1979; Galtung & Ruge, 1965; Miller & Albert, 2015), scientists prefer publishing positive results because negative results aren't scientifically interesting. Consequently, media coverage of scientific research is largely disproportional. For instance, studies showing that "saccharin causes cancer" received significantly more coverage than the subsequent studies that found that humans cannot get cancer from consuming saccharin (Dybing, 2002; Whysner & Williams, 1996). Studies purporting to show that listening to Mozart makes you smarter (Rauscher, Robinson, & Jens, 1998; Rauscher, Shaw, & Ky, 1993; Rauscher, Shaw, & Ky, 1995) received far more publicity (Bangerter & Heath, 2004; Knox, 1993) than subsequent studies showing it didn't (Chabris, 1999; Steele, Bass, & Crook, 1999). Research showing silicone breast implants can cause serious health problems made news, while findings that there was no replicable evidence establishing this went unnoticed. In a city of five million people, a story titled "Man Brutally Assaults His Wife" makes the news because the story "4.99999 Million People Are Getting Along" isn't interesting.

The publication bias isn't surprising. If journals only published research that's been replicated, they'd have little to publish, which would also diminish the usefulness of peer review. Moreover, if a journal refused to publish research that hadn't been replicated, other journalists would beat them to it; yet another reason they spice up their stories. Moreover, to the victor go the spoils. The person who first makes a discovery gets the acclaim, not the person following in their footsteps. Charles Darwin is a household name, while hardly anybody has heard of Alfred Russel Wallace, the man who independently co-discovered evolution. Accordingly, we must remain cautious when hearing about scientific research, remembering that there's good research, bad research, preliminary research, overly exaggerated research, and everything in between.

Researching Research

One way scientists examine all of the literature on a topic is by conducting what's called a *meta-analysis*. A meta-analysis brings together multiple studies on a particular topic to examine statistical information on the data. An easy way to remember what a meta-analysis does is to imagine the carnival game where people guess the number of gumballs in a jar. The guesses that are too high will cancel out the ones that are too low, and what's left over is probably close to the number of gumballs in the jar. So meta-analyses are basically used to discover statistical trends that are spread out across many studies.

However, meta-analyses have their own sets of potential risk factors. If the studies the meta-analysis collected were somehow flawed, the meta-analysis would carry those flaws into its statistical data, thus rendering the meta-analysis unreliable. So conducting a meta-analysis on several junk studies is like the kid who keeps digging through a pile of horse manure on Christmas because "there's got to be a pony in there somewhere." Thus, researchers also perform *systematic reviews* where they examine all the evidence and quality of each study on a particular research question to determine how strong the evidence for something is. In some instances, such as medical studies, systematic reviews are considered the gold standard of available evidence (Ernst, 2018, pp. 11, 75). Even so, systematic reviews can't escape the presence of human bias and error. For example, the inclusion criteria used to determine which studies are reviewed, and what methods are used to gather those studies, invite bias and error and therefore can affect the outcome of the systematic review.

Wrapping Up

This chapter isn't a complete description of how science is done or an exhaustive list of research methods. Several scientific techniques exist, including some that haven't been discovered. The methods described herein are complementary types of evidence, each of which has its own strengths and weaknesses and provides different methods of discovery and observation. And, while no process is perfect, due to the amount of effort, creativity, ingenuity, and genius that goes into scientific research, combined with science's self-correcting nature, science is a highly reliable method for discovering truths about nature.

If you don't remember all the terms introduced in this chapter, remember this: science is careful, self-critical, and guided by empirical evidence, though it is far from flawless. If we can agree that science is the most reliable process for discovering truths about reality, and we understand the pains it takes to ensure a rigorous, high-quality examination before reaching provisional conclusions, then we can agree that committing to a scientific worldview is our best chance of understanding the truth about reality.

CHAPTER 4

The Language of Logic

With an understanding of how scientific knowledge is gathered, our next step is to learn the rules for gathering and examining information. The discipline of logic is exactly what we need. Logic and argument are like the R&D (research and development) of truth and knowledge, and they're one of the many GASPs I introduced in chapter 3. When someone makes a claim that violates the laws of physics, we don't relax the laws of physics to accommodate the claim. We employ logic to either validate or invalidate it. Likewise, when the rules of logic are violated, we don't relax the rules of logic because "anything is possible." We look to GASPs to decide what's most reasonable to believe.

Logic and argument are the tools we use to uncover truths and expose falsities, to probe and investigate, and to evaluate and discover. Both are rigid disciplines, and they're much more technical than we normally suppose. In this chapter, we tackle both, learning about the importance of—and how to use—logic and argument. I'll also discuss some tools of the logician, how to spot common logical errors, the value of argument and reason, what a proper argument is, some of the rules and limits of logic, and how these limits relate to Thinking Straight. And we're going to do it all over the course of the next two chapters, so let's get started.

Building Vocabulary

Syllogism:	Argument:
A form of deductive reasoning in which the conclusion is arrived at through two or more propositions that are taken as true.	An organized series of statements where the final statement is a conclusion that follows logically from the preceding statements.

Rationalize:	Reason:
To think about or describe something in an effort to justify that thing.	The power of the mind to think and understand in a logical way.

Logic and Its Importance

Figure 4.1. Ancient Greek philosopher Aristotle (384–322 BCE). *Wikimedia.org.*

The ancient Greek philosopher Aristotle (384–322 BCE) is credited with the advent of logic and analytically evaluating **argument**. Since Aristotle, logic has taken many forms. For our purposes, we can think of logic as the science of argument. Logic and argument are thus inseparably wed, bound together like two stars in a binary system. In a valid argument, between the premises and the conclusion, there's a specific connection called the "logical support" for the argument. Understanding what having logical support or being logical or doing something logically means is one of the aims of Thinking Straight.

A dictionary definition of logic will yield something like:

- a method of human thought that involves thinking in a linear, step-by-step manner about how a problem can be solved
- a proper or reasonable way of thinking about or understanding something
- the science that studies the formal process used in thinking and reasoning
- making sense of or examining the way facts or events relate to one another

Whatever definition you prefer, logic boils down to a set of tools that helps us gain knowledge and separate fact from fiction. In his 1877 essay *The Ethics of Belief*, after discussing the various dangers of gullibility and false belief, the distinguished mathematician William Clifford concluded that "it is wrong always, everywhere, and for anyone, to believe anything upon insufficient evidence" (Law, 2011, p. 141). Or, using Edmund Gettier's reliabilist model (see chapter 2), we can say that it's wrong to believe something that isn't supported by a reliable mechanism. We evaluate claims using the rules of logic because they work. Additionally, "the rules of logical reasoning and evidence do not change simply because that which is subject to scrutiny is asserted to be beyond their reach" (Forrest, 2017, p. 38). If Joe makes the incredible claim that he has an angel on one shoulder and a devil on the other, we don't ignore the rules of logic simply because Joe claims that those rules don't apply to his companions. When we evaluate a claim, our goals should be to avoid error and discover the truth. We do so by employing logic and reason, promoting the use and understanding of evidence, along with the capacity to critically analyze the beliefs we have. Thus, logic is the glue that binds evidence and knowledge together.

The process of reasoning is paramount to argument and has practical applications as well. Take the following example (adapted from Hurley, 2012). You are at a job interview and are asked what your strengths are. You say you're energetic, enthusiastic, and willing to work long hours. You're also creative and innovative and have great leadership skills. You are then asked what your weaknesses are, and after a moment of pause, you reply that your reasoning skills are not very good.

The interviewer says this is a big problem. She states that reasoning skills are imperative to good judgment and without that your creativity will lead to projects that make no sense. Your leadership skills will lead others in circles, your enthusiasm will undermine everything the company has accomplished, and working long hours will compound these errors. This example shows how reasoning extends to all areas of our lives and thus is a worthy concept to thoroughly understand.

Tools of Logic:

- Evidence
- Reason
  - Inductive
  - Deductive
  - Abductive
- Inference
- Argument
- Judgment

The logician uses many tools, some being more useful than others.

The Methods and Limits of Logic

EVIDENCE

Sometimes the relationship between a piece of evidence and what it is evidence for is logical. For example, if you see snow on the ground, the snow itself logically entails that your belief that it snowed is true. On the other hand, evidence doesn't always logically support that for which it is evidence. Buying bullets doesn't logically prove I own a gun, only that I have some reason for purchasing bullets. Normally, evidence provides grounds for supposing a belief is true. Notice that you may have good evidence for believing something (why would someone buy bullets if they don't own a gun?) yet still be mistaken (the bullets may be a gift). Since evidence alone doesn't provide a logical guarantee that a belief is true, it's important to understand that there are limits to evidence and the inferences we draw therefrom. We should never assume we know a thing just because there is something we regard as evidence for that thing. The *strength* of the evidence is what's important.

In fact, just as the courts have "rules of evidence" that are intended to secure fairness, minimize expense and delay, and promote the growth and development of the law,[1] skeptics also have rules of evidence that guide our beliefs and decision making. For example, evidence should be available to everyone so it can be critiqued. Evidence gained by reliable means (such as scientific methodology) carries more weight than evidence that is not (such as by revelation). Evidence can be good or bad, and our duty as evidence consumers is understanding how to tell the difference.

INFERENCE

When presented a piece of evidence, our brains automatically make inferences about that evidence to arrive at some belief about it or its implications. When we wake up and see snow on the ground (a piece of evidence), we draw the inference that it snowed (the conclusion) because, in our experience, snow on the ground usually means it snowed (the logical support). Despite this logical support, we cannot know for certain that it snowed; there could be a jolly old billionaire who loves the winter, devised a gigantic snowmaking machine, and put it to use at some time during the night. To *know* that it snowed, we need additional, supporting evidence, such as news reports or video footage of it snowing the night before.

As our snow example (and buying bullets example) shows, since an inference is merely a tool we can use, we must understand that there are no guarantees and that we can be right or wrong about the inferences we draw. We use logic because we care about the truth, and because we prefer to separate what is true from what is false. So when we endeavor to get closer to reality, using all the tools at our disposal provides our best chance of getting closer to the truth.

REASON

All too often, opinions are mistaken for reason, and people never say, "I'm quite unreasonable, but hear me out." We all think the decisions we make are sound and based on evaluating evidence. The other option is to admit we're biased and uninformed, and most of us aren't willing to make this concession. If opinions are a dime a dozen, reasoning skills are in a seller's market, and it's likely to stay that way. As German writer and statesman Goethe (1749–1832) put it, "Let us not dream that reason can ever be popular. Passions, emotions, may be made popular, but reason remains ever the property of the few." Logic relies heavily on the power of reason to discover ultimate truths, so it is important to have a thorough understanding of what reason is as well as what it is not. Reason isn't the end-all arbiter of truth as many people believe (Baggini, 2017). Reason is one tool among many that we use to consciously make sense of things. We employ reason to establish and verify facts, change or justify practices and beliefs, or develop new ways of looking at information. Reason is the means by which we think about cause and effect, truth and falsehood, right and wrong, and so on.

One of reason's biggest strengths is being truth sensitive, which just means it favors true conclusions over false ones. While many methods of persuasion exist, and indeed some are effective means of garnering support for a cause, competing methods are not concerned with what's true. The English philosopher Stephen Law explains that "you can use emotional manipulation, peer pressure, censorship and so on to induce beliefs that happen to be true. But they can just as effectively be used to induce the belief that Big Brother loves you, that there are fairies at the bottom of the garden, and that the earth's core is made of yogurt" (Law, 2011, p. 204). A concern for what is true requires using reason toward that end.

When engaging in a reasoned discussion, one should follow the discussion wherever it justifiably leads, even if this means changing your mind on a firmly held opinion or belief. We often have trouble doing so because we are compulsively biased and predisposed to follow arguments and conclusions we prefer and dismiss those we reject. This error, known as a *belief bias*, is a product of System 1 thinking. Belief biases cause us to heavily misconstrue arguments and warp reason to suit our own ends. As the psychologist Daniel Kahneman explains, "When people believe a conclusion is true, they are also very likely to believe arguments that appear to support it, even when these arguments are unsound" (Kahneman, 2013, p. 45). Employing reason honestly requires System 2, but System 2 is a bit of a crapshoot. Kahneman again: "System 1 is gullible and biased to believe, System 2 is in charge of doubting and unbelieving, but System 2 is sometimes busy, and often lazy" (Kahneman, 2013, p. 81).

> He that will not reason is a bigot; he that cannot reason is a fool; and he that dares not reason is a slave.—William Drummond

Kahneman recounts evidence that we're more likely to be influenced by "empty persuasive messages, such as commercials," when we're tired and drained. Furthermore, research shows that when we're relaxed, feeling good, and our guards are down, we're much more likely to believe incredible claims than when we're suspicious or stressed (Gilbert, 1991; King, Burton, & Hicks, 2007). We may feel like we can reason effectively even when we cannot. Research also shows that emotional depletion can lead us to express prejudices we normally keep hidden from the world (Chugh, 2004; Crandall & Eshelman, 2003; Tavris

& Aronson, 2019, pp. 82–83). Being tired, angry, frustrated, stressed, or intoxi-cated can limit the ability of System 2 to keep the prejudices we keep from the world in check.

Law argues that we can apply reason by thinking of our brains as a basket in which both sensible and insensible beliefs fall into. How we determine which beliefs are sifted through is a matter of judgment (see below), and Law suggests that we "allow through only those beliefs that have a high probability of being true" (Law, 2007, p. 194). By Thinking Straight about our varying beliefs, we are essentially using a mental strainer, shaking the basket and filtering through likely beliefs and discarding those that make no logical sense.

JUDGMENT

Reason, like evidence and inference, has its limits as well. We think and act rationally when our beliefs are grounded on the best information available, yet we rarely have all the available information. Moreover, sometimes having all the available information isn't enough because all the available information may be false, or we may not be qualified to assess the information (and we can be completely ignorant to this inability). Thus, we must use our judgment if we are to ever believe anything at all. The trick is in understanding *how* to use judgment effectively.

We know reason isn't the whole story because if people always followed reason to its logical conclusions, as most of us *believe* we do, beliefs would follow the best line of reasoning with people constantly changing their minds when new and better information is available. However, this is typically not what happens. For example, when a cult member hears a well-reasoned and strongly compel-ling argument *against* the teachings of their cult, she doesn't say, "Well, I guess I should leave this cult." By the same token, when a skeptical outsider hears a well-reasoned and strongly compelling argument *for* the cult's teachings, he doesn't abandon his skepticism and proclaim, "Well, I guess I should join this cult now." If the argument is compelling, each party usually thinks something along the lines of, "That's an interesting argument. There must be something wrong with it." Accordingly, there must be something more to reason than cold logic and rigid contemplation.

> You must accept the truth from whatever source it comes.
> —Moses Maimonides

Following the footsteps of Nobel Prize–winning psychologist Herbert Simon, who taught that "judgment deviates from rationality" (Bazerman, 2018, p. 245), this disconnect led British philosopher Julian Baggini to proclaim that "judgment is . . . philosophy's dirty secret" (Baggini, 2017, p. 61). Nevertheless, philosophers prefer thinking their discipline is guided purely by reason, yet Baggini, building from Simon, shows that judgment is an inseparable aspect of the reasoning process. While logic relies heavily on reason, it doesn't completely eliminate the human tendency to pass personal judgment. And while judgment may be an essential component to the reasoning process, its presence provides those who endeavor to Think Straight reason enough to approach any given claim skeptically, regardless of how logical it sounds. So we don't have a license to replace reason with opinion; we have a responsibility to use judgment reasonably. When used properly, judgment is still based on the best facts and evidence available, and in this sense, "there is nothing dirty about it" at all (Baggini, 2017, p. 73).

ARGUMENT

As we saw in chapter 2, philosophers often debate what it really means to know something, though most of us are oblivious to such debates. Argument is the strongest tool the logician has, so we might say that logic and argument are not only wed but that they're mutually dependent on one another. As we'll see, like anything else, argument is a human endeavor that we should approach carefully and cautiously.

Oftentimes arguments are settled based on people's attitudes toward the facts rather than on the facts themselves. The Greek philosopher Epictetus (55–135) taught that "people are disturbed not by events alone but by the view they take on them." As it is with events, so it is with arguments. The attitudes we adopt can be influenced merely by how an argument is framed. Authors often attempt to bolster their arguments with tactics like using punctuation, italics, scare quotes, or intonations such as "merely," "nothing more than," "only," or "just" (Baggini, 2017). For example, the statement "The notion that *mere* molecules spontaneously turned into organic molecules is *nothing more than* one theory among many" turns into "The notion that molecules spontaneously turned into organic molecules is one theory among many" without the rhetoric. Even if you disagree with the latter, you see that it doesn't sound as prejudiced as the former.

Logical Fallacies

The type of logic we've been discussing mainly concerns the classification and structure of argument. While using logic isn't a foolproof path to truth, understanding logic isn't simply knowing how to properly argue. When our arguments are invalid or our reasoning is faulty, we can both perpetuate as well as become prey to logical errors. Fallacious reasoning doesn't jump out and announce itself as fallacious. Actually, logical fallacies are psychologically persuasive and can seem perfectly logical. In fact, the word "fallacy" is derived from the Latin word *fallere*, which means "to deceive."

The great body of logical fallacies is the culmination of years of great thinkers dissecting the many ways that reasoning can go awry. Fallacious arguments are characterized by (1) unacceptable premises, (2) irrelevant premises, or (3) insufficient premises (Schlecht, 1991). Accordingly, unacceptable premises are as doubtful as the claim they supposedly support. Irrelevant premises don't relate logically to the truth of the conclusion. Finally, insufficient premises fail to prove the conclusion beyond a reasonable doubt. With this understanding, we're ready to discuss logical fallacies.

THE FALLACY ZOO

Put simply, a logical fallacy—what I'm calling pseudologic—is a defect in an argument not limited to inadequate premises. Pseudologic is used to rationalize conclusions, as opposed to employing reason to reach them. While no umbrella theory describes every logical fallacy, they can be generalized by fallacies with irrelevant premises and fallacies with unacceptable premises. Thinkers have catalogued dozens, if not hundreds, of distinct fallacies, and I could hardly hope to cover them all. Accordingly, my goal here is to familiarize readers with some of the most common, raise awareness of their existence, and explain how to spot one when we encounter it. This next section deals with irrelevant premises; that is, premises that provide no logical connection to their conclusion.

THE NON SEQUITUR

Fallacious reasoning begins with the non sequitur, a Latin term that literally means "it does not follow." A non sequitur occurs when a link is omitted from what is otherwise a logical chain of reasoning. As a result, the conclusion does not logically follow from the argument that is said to support it. Consider this example:

> If Frank was a prisoner, then he remained in prison until he left (true premise). Frank remained in prison until he left (true premise).
> Therefore, Frank was a prisoner (does not follow from the premises).

The conclusion may or may not be true, but it doesn't follow from the premises. It doesn't follow that Frank was a prisoner simply from his having stayed in prison until he left. If Frank was a guard or a caseworker, nothing is logically wrong; in other words, all of the premises would still be true, even if Frank returns to prison again and again.

ARGUMENT FROM AUTHORITY

The *argument from authority* is committed when one claims that an authority's opinion constitutes proof that something is true or false. If Billy, the yard's shot-caller, says Steve's a snitch, others who believe Steve's a snitch *because Billy said so* are committing the argument from authority fallacy. Steve either *is* or *is not* a snitch regardless of what Billy says, and Billy can't make Steve a snitch just by saying so.

Relying on authorities for specialized knowledge and then saying an authority's opinion doesn't count may at first seem paradoxical. So how do we escape this bind? We're justified in trusting an authority's opinion only if it comes from someone who is qualified to speak on the issue at hand, and that person has evidence to support his claim. A person is qualified if he can make *reliable* judgments about whatever it is he's speaking on. Climate scientists currently working in relevant fields are qualified to speak about climate change; politicians, conspiracy terrorists, or talking heads hired by oil companies are not.

Reasons for skepticism about an authority's opinion are everywhere. For instance, an honest, credible authority is still human and could be wrong. Likewise, authorities often speak on matters in areas beyond their expertise. Also, advertisers aware of this thinking error use celebrities to promote products because we judge people's credibility based on how much we admire or respect them. Lastly, most of us are bad at figuring out who counts as an authority. Fake college degrees and police uniforms can fool us into thinking someone has authority where they don't.

ARGUMENTUM AD HOMINEM

Our next fallacy is closely related to the argument from authority. Literally translated "to the person," this fallacy occurs when an argument is rebutted by attacking the person making it instead of attacking the argument itself. For example, many people dismiss advice offered by prisoners, mistakenly thinking prisoners can't give good advice, reasoning that if they could, they wouldn't have ended up in prison. Prisoners may very well be unable to give good advice, but their past isn't an indicator one way or another; advice must stand alone and must be judged on whether it's good advice, not who or where it came from. Bertrand Russell, Nelson Mandela, and Thomas Paine were all prisoners at one point, and they all gave great advice. Moreover, a prisoner may be unjustly imprisoned, and to shun his or her advice would be compounding an injustice. People can still be mistaken, but saying someone has made a misstatement of fact is not an ad hominem attack because it addresses the argument itself rather than the person making it.

POST HOC ERGO PROPTER HOC

Superstitious sports players and their like-minded fans often fall victim to our next fallacy. Literally translated "after this, therefore because of this," this fallacy states that because some event (e.g., winning the Super Bowl) followed another event (e.g., wearing a particular T-shirt), the preceding event caused the later event. Research shows that the desire for control is a major contributor to belief in superstition (Vyse, 2014, pp. 16, 153–61, 242–43), and, given the strong desires involved in sports players and their fans, it makes sense that they should attempt to control the outcome of a game with silly rituals. When an event follows another event, it's tempting to think the former caused the latter, but doing so is an error in reasoning.

In an argument, beware of the following:

- False premises
- Hidden premises
- Personal biases
- Assumptions
- Rationalizations
- Logical errors

If we are not even aware of the premises of an argument, we can't examine its truth or falsity.

Let's say a prisoner named Jason is seen talking to the cops in a back office, and an hour later, several of Jason's friends' cells are searched and the cops find a batch of hooch. It's all too easy to say, "After we saw Jason talking to the cops in the back office, we got searched. Jason was the only other person to know about the hooch. He must have told the cops." Maybe Jason did tell the cops about the booze, but it doesn't follow that he told them simply from the fact that the hooch was found after he was seen talking to the cops. This remains true *even if Jason was the only other person who knew about the booze.* The shakedown could've been random, a higher-up could have ordered it thinking that Jason's friends looked suspicious, or maybe the guards just don't like Jason's friends and decided to harass them with a shakedown. We can imagine several possibilities that don't include Jason telling on his friends, and it's an error in reasoning to conclude that he did so based strictly on the order of events.

ARGUMENT FROM IGNORANCE

One of the most widespread fallacies committed in the promotion of paranormal and supernatural beliefs, the *argument from ignorance* occurs when someone begins an argument with an absence of evidence and then makes a specific claim about the thing that lacks evidence. "What's that fast-moving light in the night sky?" quickly turns into "It's an alien spacecraft that the government is trying to keep a secret." Additionally, arguing from ignorance occurs when someone uses one's inability to disprove a claim as proof of the claim's accuracy, or vice versa. Psychologist Ray Hyman calls this "loopholism," which he defines as dismissing a skeptical investigation because it cannot account for every detail of a supposed paranormal event (Hyman, 2001). Thus, loopholes are exploited to protect one's cherished beliefs.

For example, after unsuccessfully identifying the source of a sound in a "haunted" house, a paranormal believer might conclude that, since the source is unexplained, the sound came from a ghost. However, going from an absence of knowledge (the source is unknown) to a specific claim about the very thing you lack knowledge about (the sound must be caused by a ghost) is flawed thinking. The source of the sound is unknown. Invoking ghosts to solve the problem merely complicates it and makes solving it impossible.

Believers in the paranormal aren't the only ones that argue from ignorance. Conspiracy theorists (chapter 13) thrive on the argument from ignorance, searching for flaws, then claiming those flaws are positive evidence for something specific. For instance, since the government doesn't regularly brief the public on what happens at Area 51 (an absence of knowledge), their silence is viewed as positive evidence that the government is covering up alien activity (a specific claim). Area 51 is a top-secret U.S. Air Force base. The government chose the site specifically to develop and flight-test combat aircraft in seclusion (Harrison, 2012, pp. 382–88; Harrison, 2013, pp. 159–61; Prothero & Callahan, 2017, pp. 39–60). Governments have obvious and legitimate reasons to keep what happens there confidential: secret military operations are useless if your enemies know about them.

Skeptics haven't missed an opportunity to weigh in. "Believers in Area 51 mythology get tripped up," says Guy P. Harrison, "when they confuse plain *secret aviation activity* for *secret aviation activity involving aliens*." Such believers "also mistake *secret military aircraft* with *extraterrestrial spaceships*." Harrison concludes that "*seeing* unknown aircraft near a secret base that flies secret aircraft

"I THINK YOU SHOULD BE MORE EXPLICIT HERE IN STEP TWO."

Figure 4.2. This Sydney Harris cartoon captures the argument from ignorance beautifully. It satirizes the popular intelligent design movement's insertion of miracles as genuine science, which defies the purpose of science itself. *Copyright © 2021 by Sidney Harris, http://sciencecartoons plus.com/.*

is not strange" (Harrison, 2013, pp. 159, 161). Perhaps Michael Shermer put it best when he said, "Terrestrial secrets do not equate to extraterrestrial cover-ups" (Shermer, 2011a, p. 337).

Argument

Argument wears many faces. Some of them are obvious, like someone's angry face. Some of them are discreet, like the face of jealousy. We argue ourselves, we listen to arguments from others, and we're exposed to arguments without even realizing it. Thinking Straight requires understanding what an argument is, how to spot one, how to mount a proper argument, and how to tell a good argument from a bad. In this next section, I briefly cover argument in everyday life before moving on to argument in its more mechanical sense—argument in logic.

> Unfortunately, rational arguments never convince fanatics.—Edzard Ernst

Argument is important for several reasons. First, an argument's flaws aren't always easy to identify. Second, the purpose of an argument is to influence someone's thinking; that is, to persuade people that they should believe what you're saying or buy your product or whatever. Therefore, learning how to make a good argument and learning how to identify a bad one, can help us immensely. Following the rules of logic and argument offers a new approach for accomplishing these goals. Moreover, understanding these rules protects us against coercion, fraud, faulty logic, or negative influences and provides constructive alternatives to aggressive disagreements that too often are settled by force or intimidation.

EVERYDAY ARGUMENT

When we hear the word "argument," many of us think of screaming and violence or at least heated disagreements. The last one standing is thought to have "won" the argument. In the intellectual world, winning an argument isn't that simple. Who wins an intellectual argument is judged by one's peers and is determined by who presents the best factual case in support of their claim. Formal arguments are not what one typically encounters every day. However, having an accurate understanding of what a proper argument looks like is a good way to avoid quarrels and resolve disputes with brains rather than brawn.

But arguments don't just prove points. We argue for many reasons, and although argument has a reputation as a negative endeavor, we make arguments for almost everything. Arguments can be subtle, such as an advertisement, or they can be obvious, like a yelling match. A typical definition of argument is "any text—written, spoken, aural, or visual—that expresses a point of view" (Lunsford, RuszKiewicz, & Walters, 2013, p. 5). With this definition, arguments easily vary in structure and form. For example, people use *arguments of fact* to prove or disprove something, whereas *arguments of definition* seek to determine the nature of things. When people take a moral or religious stance on the issue of abortion, for example, they are often using arguments of definition to determine what exactly a human being is.

> He who strikes the first blow admits he's lost the argument.
> —Chinese proverb

Depending on the goal, people use different arguing strategies—and not always to establish a truth. Thinking Straight helps us determine a person's goal and assess how much stock to put into their argument. People have personal motivations and employ clever arguments for reasons that vary from trying to sell something (such as a conversation with dead relatives) to getting people to align with their cause (like political promises).

PERSUASION

A common practice, especially in politics, is to use emotive words to stimulate an emotional reaction, like liberals referring to fracking as "raping" the planet, conservatives calling criminals "vermin," or military leaders calling comrades "brothers." Using emotionally charged language is a form of persuasion, a particular type of argument with a rich history. Persuasion is an ancient form of art called *rhetoric* that traces back to the Greek philosopher Aristotle. Nowadays, persuasion theorists recognize two main routes to persuasion: the *central* route, which invokes System 2, and the *peripheral* route, which invokes System 1 (Levine, 2018, pp. 517–18). The central route counts on a motivated audience to deliberately weigh pros and cons of logical arguments and is intended to create lasting agreement. For example, this book is an exercise in central persuasion as it seeks to persuade its audience that Thinking Straight is an effective path to discovering truths and making better decisions. The peripheral route appeals to System 1 and seeks to bypass the audience's deliberative processes with distractions such as attractive salespeople, cute animals, or celebrity endorsements. People often make decisions based on how attractive the source of information is, which is why the peripheral route is a favorite of cult leaders and politicians. The phrase "sex sells" wonderfully captures the peripheral route to persuasion.

Professor of psychology Robert V. Levine presents what he calls the "Triad of Trust," which describe the characteristics of trustworthiness. That is, when

> When assessing information, do the following:
>
> - Gather evidence
> - Consider the source
> - Consider the motives of the source
> - Question anecdotes
> - Weigh the evidence
> - Look for logical inconsistencies
> - Examine alternative explanations
> - Understand what's being stated
>
> Understanding how to assess information is a critical tool for critical thinking.

we think someone exhibits authority, honesty, likability, or any combination of those characteristics, we are more easily fooled into buying things we don't need, agreeing to their requests, or falling for their nonsense ideas: in other words, not *thinking critically about the facts* (Levine, 2018, pp. 519–21).

As shown in table 4.1 (adapted from Lunsford, RuszKiewicz, & Walters, 2013, p. 14), Aristotle classified the occasion for argument by time—present, past, or future. According to this model, different arguments are used depending on what purpose they serve, although some overlap between them still exists. For example, a district attorney uses forensic arguments (concerning what happened in the past) to convince the jury that her epideictic argument (her argument about what she believes the defendant did) supports her deliberative argument (namely, what punishment is appropriate). In other words, a district attorney argues that the evidence she presents shows the defendant is guilty, which gives cause to justify the punishment sought or required by law.

Table 4.1. Occasions for Argument

	Past	Future	Present
What is it called?	Forensic	Deliberative	Epideictic
What are its concerns?	What happened in the past?	What should be done in the future?	Who or what deserves praise or blame?
What does it look like?	Court decisions, legal briefs, legislative hearings, investigative reports, academic studies	White papers, proposals, bills, regulations, mandates	Eulogies, graduation speeches, inaugural addresses, roasts

THE APPEAL OF ARGUMENT

According to Aristotle, when used to persuade, argument takes on one of three forms. He called these forms *pathos*, *ethos*, and *logos*. Pathos, the Greek word for "suffering," is an emotional appeal that can be extremely powerful because many of our decisions are based solely on emotions. If you're the defendant in our district attorney example and you're arguing that your punishment should be mitigated, you may appeal to the emotions of the judge or jury, focusing on the person you are, the circumstances that justify a lesser sentence, and so on. Ethos, Greek for "character, custom, and habit," is an appeal to ethics. If one's goal is to instill trust, their argument would center on the ethos of their character. For example, when people run for public office, they often appeal to their trustworthiness or prior acts of kindness to establish their credibility and invite voters to empathize with them. Lastly, logos, Greek for "discourse, calculation, and reason," is an appeal to logic. Depending on your audience, persuading others to accept a conclusion works if your argument shows that doing so is reasonable.

Whether arguing for facts or persuasion, it's also important to understand the dynamics of timing. You wouldn't use humor as a persuasion tactic at a parole revocation hearing, for example. Politicians will attempt to gain public support for going to war by appealing to emotions (pathos) following tragic events, for example, 9/11 or Pearl Harbor. In court, criminal defendants don't appeal to logic (logos) when it's their character that needs defending. The Greek work *kairos* means seizing an opportune moment, and sometimes that's what argument is: capitalizing on the moment your argument will have the most effect. It should also be remembered that people have their own agendas and exploit opportune moments to turn others to their cause.

ARGUMENT IN LOGIC

In logic and philosophy, as well as in science and mathematics, the meaning of argument is quite technical. We know argument is important, so just how important is it? What types of logical arguments are there? And is there a way to argue *properly*?

The two most popular and useful logical arguments are *inductive* and *deductive*. Before delving in, however, it's important to understand that an argument attempts to prove something, while statements, opinions, disagreements, explanations, and assertions do not (Vaughn, 2019, pp. 13–18). For example, an explanation is intended to show *why* something is the case, whereas an argument is intended to show *that* it is the case.[2]

STRUCTURING MEANINGFUL ARGUMENTS

Whether to prove a point, to get something we want, to defend a position, to spur a reaction, or to play the devil's advocate, everyone has experienced arguments. Unformalized arguments usually entail collecting whatever information we have and haphazardly laying it on the table, then waiting to destroy the rebuttal. However, this is a disagreement, not an argument. A formal argument's purpose is specific. An argument presents a *claim*, marshals *support* for that claim, and offers what is *warranted* because of the support for the claim (Rottenberg, 1995, pp. 9–13). For example, the death penalty is inhumane and doesn't deter crime (the claim), as is shown by its tortuous history and the fact that crime rates in states where the death penalty is still in effect remain high (the support). Thus, the death penalty should be abolished (the warrant). [3]

Today we take formal argument for granted, although it didn't always exist, much like toilet paper, electricity, and fast food. Argument was most likely formalized in the direct democracy of Athens, Greece, where skills in argument and debate were both highly prized as well as necessary to win the favor of crowds and gain political success (Law, 2007, pp. 24–25). As old as the rules of logic and argument are, they're still human-operated, thus they can easily be twisted to suit someone's goals. Accordingly, "logic and arguments should be used as a tool, not a weapon" (Novella, 2018, p. 60). The goal of a formal argument is to get to the truth, not to win, so to Think Straight, we must be careful to argue according to rules rather than desires.

ARGUING FOR A CAUSE

An argument is a set of statements wherein one or more statements seek to provide support for another. An argument can have one or more *premises*—which seek to support the *conclusion*—but only one conclusion. A *premise* lets us take facts or assumptions as true to build our argument. Arguments based on false premises are either *invalid* or *unsound*, depending on what kind of argument it is. Accordingly, to better understand the claim being made, we examine all the premises and identify any false or hidden premises. In doing so, we avoid needless disagreements and are better prepared to proceed intelligibly.

> It is better to debate a question without settling it than to settle a question without debating it.
> —Joseph Joubert

Assumptions can be used as premises for the sake of an argument, though arguments that begin with assumptions are weaker than ones that begin with established facts. When it's unknown if an assumption is true or when it might be incomplete or might not fully capture the situation, the argument is weakened since an argument is only as strong as the premises it's based on. Furthermore, hidden premises can complicate an argument. For example, if I argued that the moon illuminates the night, typically I wouldn't have to explain what "illuminates" means or bicker about whether the moon is made of cheese. However, if I argued that SUVs contribute more carbon dioxide to our atmosphere than other vehicles with the same engine type (the premise), therefore SUVs should be banned (the conclusion), you'd notice that something was missing. The conclusion follows from the premise only by means of the hidden premise that anything that excessively contributes carbon dioxide to our atmosphere should be banned. Now our argument is complete, and the premises can be evaluated on their own terms.

For an argument to have any purchase, it must minimally satisfy two criteria: (1) it must be structured properly, and (2) it must work. Accordingly, a proper argument must follow the rules of logic, and its premises must produce a valid conclusion. Naturally, this leads us to ask "What is a proper argument?" "How is a proper argument structured?" or "What's a valid conclusion?" Here's where the structure of argument gets technical.

A *valid* argument is one in which the logic works; that is, the conclusion logically derives from the premise(s). A *sound* argument is an argument in which all the premises are true and the logic is valid. So, by definition, the conclusion of a sound argument must be true. As such, an unsound or invalid argument could be either true or false, and merely showing an argument is unsound or invalid is not enough to establish a false conclusion (albeit this showing helps demonstrate that the argument's conclusion isn't supported by its premises). As Baggini explains, "If a conclusion is not supported by an argument, that means only that support for a conclusion is lacking, not necessarily that the conclusion is wrong" (Baggini, 2003, p. 73). Since unsound or invalid arguments can have true conclusions, showing an argument is invalid isn't the end of the story. If we say, for example, that amusement parks are expensive, therefore Earth is a planet, the conclusion is true while the logic is clearly unsound, since there's no logical connection between the premise and conclusion.

A common fallacy committed in the attempt to present a logical argument is what's known as *affirming the consequent* (Smith, 2018, pp. 66–67). Let's use the following argument to explain this fallacy.

> Premise 1: If my lawyer and the district attorney are in cahoots, then
>   I'm doomed for a conviction.
> Premise 2: I was convicted.
> Conclusion: Therefore, my lawyer was in cahoots with the district
>   attorney.

Here we have an initial premise with two parts: an "if" and a "then." The "if" part is called the *antecedent*, which simply means "comes before." The "then" part is called the *consequent*, meaning it's the consequence of the "if" provided the "if" is true. In this example, the conclusion is just a restatement of the consequent. Because the consequent is built into the initial premise, the conclusion is foregone, thus the argument proves nothing. If we make the argument slightly more absurd, the problem reveals itself more easily.

> Premise 1: If purple cows can fly, then humans are robots.
> Premise 2: Purple cows can fly.
> Conclusion: Therefore, humans are robots.

In the aforementioned argument, we've proven nothing by affirming the consequent. Since one of the goals of arguing is to establish what is factual, when two people disagree about a matter of fact, they cannot both be right: either one side or both must somehow be wrong. Arguing properly requires examining the arguments to identify any assumptions in the premises, any false premises, or any logical errors. Although working together to identify assumptions or errors of logic is challenging, it helps resolve differences and arrive at more accurate conclusions.

Saying a claim is true is to say it's factually correct. We don't say arguments are factually correct because arguments are neither true nor false; their conclusions are. Due to our incomplete level of knowledge and the limited ways we can interact with the world, Straight Thinking avoids saying something is true with 100 percent certainty. Such uncertainty is healthy, for if we thought we had all the answers, we'd stop looking for the truth. As the philosopher Alfred North Whitehead put it, "Not ignorance, but ignorance of ignorance, is the death of knowledge" (Burton, 2008, p. 216). Certainty is the enemy of discovery, so instead of saying something is certainly true, we say it's been established to a point where we can treat it as a fact. As Michael Shermer puts it, "There are no final answers in science, only varying degrees of probability" (Shermer, 1997, p. 124).

Logical arguments are used to describe what is most likely true by providing reasons for accepting a conclusion. The Austrian-British philosopher of logic Ludwig Wittgenstein (1889–1951) taught that the process of arguing and analyzing is much more important than the conclusions reached (Dennett, 1995, p. 109). Fast-forwarding to present day, Dr. Steven Novella advises us to emotionally detach from factual claims by valuing the process of being skeptical over the facts themselves (Novella, 2018, p. 56). Given that emotional distress heightens our propensity toward magical thinking and biased reasoning (Keinan, 1994; Keinan, 2002), it seems sound advice to separate emotions from fact-finding.

Properly arguing isn't *intrinsically* more valuable than a true conclusion. Instead, true conclusions are best achieved through reliable processes. Therefore, by valuing the process over winning an argument, we can rest assured that if (when) we're wrong, we're wrong for good reasons. This outlook includes a willingness to change our minds about a conclusion or change a conclusion when new information or a better argument is presented. As Aristotle wisely advised, we must honor truth above our friends, to which we might add our biased selves.

The terms "**rationalize**" and "reason" were defined at the beginning of this chapter, and I want to risk belaboring them to clear up any potential confusion. Rationalizing begins with a conclusion (my client is innocent) and then marshals anything that supports it (that DNA belongs to his identical twin brother). Asking whether we're mounting a proper argument or rationalizing through one takes practice and discipline, and taking this step is crucial. Rationalizing is also the process we engage in after making some decision that's discordant with our beliefs. When smart people make stupid decisions, moral people commit immoral acts, or sane people do insane things, they rationalize their actions by offering "good" reasons for why they engaged in "bad" behavior. Humans are far better at rationalizing than being rational, despite what we tell ourselves. Moreover, our biases and other flaws in thinking make it hard to see when we're reasoning and when we're rationalizing.

> Those who are most convinced that they are absolutely right are often those who are most terribly wrong.—Julian Baggini

Reasoning, on the other hand, begins with a premise (education reduces recidivism) and is concerned with whether the conclusion (more education is needed in the criminal justice system) follows logically therefrom. Though near opposites, rationalization and reasoning often seem identical, which is the whole point of rationalizing. We rationalize to make our decisions and conclusions seem reasonable, whereas we reason to guide our beliefs toward the best possible understanding of reality.

Although doing so can seriously cloud our judgment, part of human nature is becoming emotionally attached to our preferred conclusions, which we end up defending at all costs. Our desire to avoid cognitive dissonance (chapter 7), as well as our inclination to avoid admitting fault, reinforces this emotional drive. One reason we shouldn't get emotionally committed to a particular conclusion is to avoid the dissonance we experience when new information disconfirms our preferred conclusions. When we're not emotionally tied to a conclusion, it becomes easier to change our mind to fit the evidence (reasoning) than it is to change how we view the evidence to fit what is in our minds (rationalizing).

INDUCTIVE REASONING

In his groundbreaking work *Novum Organum*, Francis Bacon (1561–1626), the seventeenth-century English polymath, proposed a new approach to the scientific endeavor: the method of induction. Bacon advised scientists to begin with observations,[4] using them as a basis for producing general theories. But science isn't the only place where induction is useful. Most people use induction every day. For example, using statistical data on recidivism to predict how likely one is to return to prison is inductive, as is assuming that seeing footprints in the snow means something walked through the snow. Following his near-contemporary Nicholas Copernicus (1473–1543), Bacon called on scientists, rather than authoritarians, to determine the structure of the natural world, a structure he described with a legal metaphor—among other things, Bacon was a lawyer—as the "Law" of nature. What makes an argument inductive is the way it determines what is *most likely* true, as opposed to what must be true. Assuming that since your alarm clock sounded every day in the past it will likely sound tomorrow is an example of induction.

Induction has been compared to faith (chapter 2), though induction has a far better success rate. If I suppose that when I walk on the sidewalk the concrete won't collapse underneath me, I'm not claiming to *know* the concrete won't collapse. I am making an assumption based on numerous bits of evidence, such as past experiences. I *hope* the ground will not collapse, I *think* the ground will not collapse, but if it did collapse, nothing about how I view the world would be upset. My reasoning here is inductive because while it is *likely* that the ground will not collapse underneath me, what has happened in the past in no way *guarantees* what will happen in the future, à la sinkholes. The Scottish philosopher David Hume (1711–1776) went so far as to argue that what has happened in the past provides us with no clue as to what will happen in the future![5] Just because we've managed to avoid getting sick in the past doesn't mean we'll never get sick in the future.

Faith, on the other hand, requires no evidence at all and thus cannot point reliably to an objective fact. Faith thrives where there's an absence of evidence: as far as faith is concerned, the less evidence, the better. If Marcus has faith that he can speak to dead relatives, for example, he isn't claiming that he *hopes* or *thinks* he has these powers. He's claiming that he can, in fact, communicate with the dead. And while my understanding of the world wouldn't be ruined if the ground collapsed beneath me, if Marcus underwent controlled experimental conditions that unequivocally demonstrated that he couldn't communicate with the dead, a huge part of his worldview would be shaken to its core.

An inductive argument asserts that the truth of a conclusion is supported to some degree of *probability* by its premises. With a representative sample, we can make certain inductive predictions (see chapter 3). For instance, given that about 68 percent of U.S. prisoners released in 2005 were arrested for a new crime within three years and 77 percent were arrested within five years[6] (premise = true), we are justified in supposing that, all things being equal, it's likely that at least 65 percent of prisoners will reoffend within five years (conclusion = likely true). Arguments that involve predictions are always inductive, as the future is uncertain.

Suppose a jury hears the following evidence in a murder case:

> Three eyewitnesses saw the defendant fleeing from the victim's apartment shortly after hearing gunshots [Premise 1]. The victim's blood was found on the defendant's clothes [Premise 2], and a gun matching the weapon used to kill the victim was found in the defendant's car [Premise 3]. Under questioning, the defendant confessed to the crime [Premise 4].

Furthermore, suppose that the defense presented no evidence that contradicted or explained these premises. If the premises are true, is this argument inductive? You might be tempted to believe it's deductive because if all the premises are true, the chances are remote that the defendant is innocent. However, the argument is inductive because the premises cannot provide a logical *guarantee* that the defendant is guilty. The eyewitnesses might have misidentified the defendant, the lab could've made a mistake about the blood, the gun could've been planted, and the confession could've been coerced. Unlikely, but while court is about reasonable doubt, induction is about possible doubt.

Inductive arguments assume that the unobserved past, present, and future resemble the observed past and present. Since induction deals in likelihoods rather than in certainties, an inductive argument is either *strong* or *weak*, with a strong inductive argument being one in which, assuming the premises are true, the conclusion is likely true. A weak inductive argument is one in which the conclusion doesn't follow as a likely consequence of the premises. When an inductive argument is logically strong and has true premises, the argument is *cogent*, meaning it offers logical support for accepting the conclusion. If an inductive argument is weak or any of its premises are false, the conclusion is likely false, rendering the argument *uncogent*.

DEDUCTIVE REASONING

In contrast to induction where the truth is a *likely result* of the premises, the truth is a *logical consequence* of the premises of deductive arguments. If the premises of a deductive argument are true, the conclusion must be true.

Deductive arguments are *valid* or *invalid*, rather than strong or weak. An argument's validity is determined by the relationship between its premises and conclusion. Whether the premises are true or false has nothing to do with validity, so at this stage, whether the premises support the conclusion is all that matters. In a valid argument, the premises support the conclusion, and in an invalid argument, they don't. So if one assumes the truth of a premise in a deductive argument, notwithstanding whether the premise actually *is* true, the conclusion *must* be true. These rules can be confusing, so let's look at some examples.

If premise 1 says all people are mortal and premise 2 says all mortals die, then the conclusion that all people die is true. This is also an example of a **syllogism**, the oldest form of which can be traced back to Aristotle (Baggini, 2017, p. 62). Deduction allows one to advance an argument that is logically valid but still has a false conclusion. Take the following example:

> Premise 1: Hillary Clinton is a space alien.
> Premise 2: All space aliens are locked inside secret government prisons.
> Conclusion: Therefore, Hillary Clinton is locked inside a secret government prison.

The argument is ridiculous, though it's valid because it follows the rules of deduction: the premises guarantee the conclusion. However, the logic is not *sound* because the premises are false. An argument is sound when it is valid *and* all the premises are true. Accordingly, a deductive argument is *unsound* if it's invalid, if one or more of the premises are false, or both.

To determine validity, the structure of the argument is essential, rather than the truth values. Consider the following argument: since bats can fly (premise = true) and all flying creatures are birds (premise = false), bats are birds (conclusion = false). If we assume the premises are true, the conclusion follows necessarily, thus the argument is valid. If a deductive argument is valid and its premises are all true, then the argument is sound. Otherwise, it is unsound, as this example shows.

Sound deductive arguments produce conclusions that are objective statements of fact rather than value judgments pertaining to what we believe. For example, there's no way to logically prove M. C. Escher's art is better than Salvador Dalí's, nor could any evidence prove this claim. Which artist is better is a matter of judgment, not a matter of objective fact, and when an argument includes value judgments in the premises, identifying them might not help resolve the argument, but at least you'll know the disagreement pertains to values rather than to facts.

As a final point of concern, it's often misstated, even within science and skepticism, that inductive arguments go from a particular to a general proposition, whereas deductive arguments go from a general to a particular proposition. Despite there being several otherwise credible sources supporting this view

(Novella, 2012, p. 56; Rottenberg, 1995, pp. 252, 260; Shermer, 2011a, p. 291; Smith, 2018, pp. 65, 67; Vyse, 2014, pp. 143–44),[7] it's now considered outdated (IEP staff, n.d.; Lander, n.d.). *Particular statements* are those that make claims about one or more members of a particular class. *General statements* are those that make a claim about all members of a class (Hurley, 2012, pp. 38–39). Some inductive and deductive arguments do indeed follow this general-to-particular/particular-to-general order. However, both deductive and inductive arguments can proceed from the particular to the general, particular to particular, and general to particular. Here's an example of a deductive argument that proceeds from the particular to the general:

> 2019 was the first year of the coronavirus pandemic.
> 2020 was the second year of the coronavirus pandemic.
> 2021 was the third year of the coronavirus pandemic.
> Therefore, the first three years of the coronavirus pandemic are 2019, 2020, and 2021.

Here's one that proceeds from the particular to the particular:

> Joshua is a mailman.
> Joshua has a gun.
> Therefore, Joshua's gun is the gun of a mailman.

And here's an inductive argument that proceeds from the general to the particular:

> All asteroids ever observed consist of rock and water.
> Therefore, the next asteroid we discover will consist of rock and water.

How an argument proceeds shouldn't be used as a criterion for determining whether it's inductive or deductive. Telling the difference is a matter of focusing on how strongly the premises relate to the conclusion (Hurley, 2012, pp. 33–39). If the premises *guarantee* the conclusion, the argument is deductive. If the premises show the conclusion is *likely*, the argument is inductive.

Wrapping Up

Figure 4.3 (adapted from Hurley, 2012, p. 53) sums up what we've learned about the structure of logic and argument. Argument is a powerful tool that can enhance our experience and understanding of the world. Good arguments—i.e., sound, cogent, valid arguments—provide reasons to accept a conclusion, whereas bad arguments—unsound, uncogent, or invalid arguments—provide reasons to reject their conclusions.

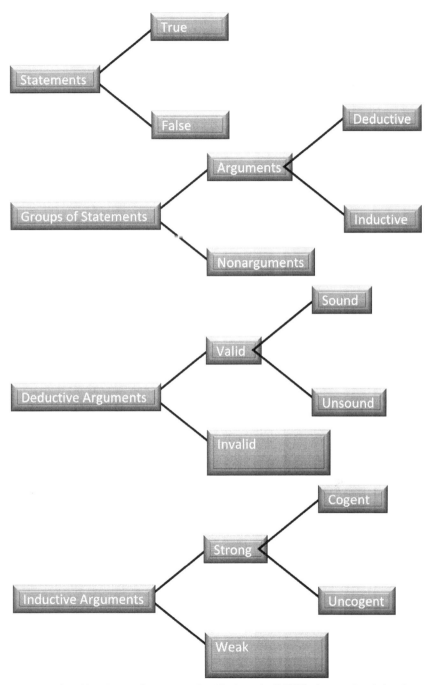

Figure 4.3.   Structure of arguments. *Hurley, J. (2012). A concise introduction to logic, 11th ed. Boston: Clark Baxter, p. 53.*

Logic is used to develop methods and techniques that distinguish good arguments from bad and help us separate fact from fiction. If the premises of an argument are weak or false, we have fewer reasons for believing the conclusion they claim to support. Moreover, while limits to logical reasoning endure, logic remains one of the most reliable mechanisms for investigating reality. While far from perfect, these methods are important tools for the Straight Thinker's tool kit, so long as we understand their strengths, their limits, and how to utilize them properly.

In their seminal book *How to Think about Weird Things*, Theodore Schick Jr. and Lewis Vaughn write, "Since most people have never learned the difference between a good argument and a fallacious one, they are often persuaded to believe things for no good reason" (Schick & Vaughn, 2005, p. 165). Most people also can't tell the difference between a good scientific study and a bad one or an accurate press release from an inaccurate one and end up forming beliefs based on the charisma of the person doing the talking. Understanding logic and argument and thinking critically about the arguments we confront provide a good defense against these common errors.

CHAPTER 5

Building Your Skeptical Tool Kit

Chapter 4 provided an overview of some of the tools needed for Thinking Straight. We saw that logic is imperfect and that oftentimes reasoning goes wrong. This chapter delves a little deeper by covering more logical fallacies, exploring cognitive biases, and learning more tools of the CT trade. Pseudologic and mental biases are the cognitive vaults for which CT is the combination. Our goal is to discover the combination by defending ourselves against logical and cognitive errors, while gaining knowledge that helps us become better thinkers in our daily lives.

Building Vocabulary

Heuristic:	Bias:
A cognitive shortcut (similar to an algorithm) that quickly, although not always optimally, categorizes information.	A usually unwitting, unfair, and disproportionate inclination toward one thing over another.

Cognitive priming:	Confirmation bias:
An unconscious process by which exposure to one stimulus influences a response to a subsequent stimulus.	A cognitive bias toward confirmation of existing beliefs and suppression of disconfirming evidence.

The Importance of Pseudologic

One of the reasons to learn about pseudologic is to spot it when it's used against us, or to support some dubious claim. Additionally, understanding pseudologic helps us in numerous aspects of our lives, be it personal, professional, financial, or spiritual. However, spotting our own fallacious reasoning is harder than spotting others'. Accordingly, understanding pseudologic raises our awareness of how fallacious thinking influences our approach to information, regardless of its source. But remember: logical fallacies can also be overused, such as when someone dismisses legitimate scientific claims by invoking the *argument from authority* fallacy. Facts must come from somewhere, and arguing that something isn't factual *because it comes from an authority* is mistaken. This type of argument is known as the *fallacy fallacy* and, like many fallacies, is an easy trap to fall into. We'll begin with several fallacies of unacceptable premises before moving to fallacies with irrelevant premises.

THE SLIPPERY SLOPE

The *slippery slope* fallacy happens when one asserts that taking some small first step leads to some significant negative consequence. The "legalize marijuana" debate offers a good example. Legalizing marijuana, the argument goes, is a small step toward social chaos. Next, the government will legalize crack and heroin, and society will crumble under the effects. However, nothing about legalizing marijuana implies that other drugs will become legal or that legalizing other drugs will ruin society. Similarly, allowing same-sex marriage doesn't mean marrying a toaster will soon become legal, and placing restrictions on guns doesn't mean the second amendment will soon be abolished.

Sometimes when you give someone an inch, they take a mile. Nonetheless, overestimating a risk without evidence that some negative consequences follow from that risk, or even that a negative consequence is likely, is faulty thinking.

THE FALSE DILEMMA

Our next fallacy is the false dilemma, aka the "either/or" fallacy. The either/or fallacy is perhaps best captured by the slogan "If you're not first, you're last," popularized by John C. Reilly's character Cal Naughton Jr. in the movie *Talladega Nights*. The false dilemma occurs when something is falsely claimed to have only two options when in fact there is a wide range of other options. Obviously there are more options than first or last, just as there are more choices than being with us or against us, in or out, and so on. One could be indifferent, on the fence, or sometimes "with" us and sometimes "against" us. Hardly anything boils down to a simple either/or choice, and it can be helpful to recognize when someone is luring us into a false choice.

Pseudomedicine abounds with examples of this fallacy. Suppose you suffer from arthritis, and after trying on a copper bracelet, you felt much better. You might be tempted to conclude that either the copper bracelet worked or it was a divine miracle. Of course, these two options aren't the only explanations for overcoming an ailment, and science doesn't support the claim that copper bracelets have any effect on arthritis (Hines, 2003, p. 336). Moreover, even if copper could be absorbed into your skin (it cannot), it would poison you, not heal you. The feeling of improvement is much better explained by regression to the mean, spontaneous remission, misdiagnosis, the waxing and waning of arthritis pain, or the body's natural healing abilities.

MOVING THE GOALPOST

Moving the goalpost occurs when one changes the criteria for accepting a claim whenever evidence disproves the claim. Paranormal believers and conspiracy theorists often fall victim to such pseudologic. For example, when conspiracy theorists doubted President Obama's birthplace, after Obama posted his birth certificate online, they simply discounted its authenticity (Berinsky, 2012; Uscinski, 2019c, p. 20). A historical example of moving the goalpost is the widespread belief that extraterrestrials crashed in Roswell, New Mexico, in 1947. A close examination of the facts demonstrates the full scope of this fallacy.

The Roswell "incident," by far the most popular of all incredible claims of extraterrestrial contact, probably isn't going away anytime soon (Kloor, 2019). The story originated when, on June 14, 1947, Mack Brazel, a rancher from outside of Roswell, found about five pounds of debris on his property. Brazel described the debris as "bright wreckage made up of rubber strips, tinfoil, a rather tough paper, and sticks" (quoted in Hines, 2003, p. 290). Brazel ignored the debris. However, on July 5, 1947, after visiting Roswell, Brazel learned that just days before (June 24), a civilian pilot named Kenneth Arnold reportedly witnessed "flying saucers" (chapter 8). **Primed** with this new information, Brazel collected the debris and turned them over to local authorities. When the debris made its way to the Roswell Army Air Field, a then intelligence office major, Jesse Marcel, inspected it, turning it over to higher authorities. A press release by Walter Haut, an official who hadn't inspected the debris himself, said they recovered a "flying disk," which the media later sensationalized, wrongly connecting this debris with Arnold's flying saucer story. Once people learned the details about what Brazel found, the story was forgotten.

However, the Roswell story resurfaced thirty-one years later, when Marcel, now a TV repairman, began this mythical legend. Throughout the course of numerous tellings, Marcel's story grew and grew as he recounted "ever-growing distortions, made-up events, deliberate misinterpretations, and previously unknown 'witnesses' with ever more fantastic stories to tell" (Hines, 2003, p. 291). Unfortunately, the debris came from a secret government program called Project Mogul, which aimed to detect sonic booms from nuclear blasts through sensing devices placed in what researchers falsely called "weather balloons" (Harrison, 2012, p. 123). Project Mogul was declassified in 1972, and in 1977 the release of the movie *Close Encounters of the Third Kind* reignited the public interest in UFOs. At that point, Ufologists put continued pressure on Congress to investigate the Roswell incident.

After two exhaustive investigations, investigators concluded that the debris was from a felled Project Mogul balloon. Furthermore, faded memories reported more than thirty years after the event likely contributed to the story's popularity. A UFO sighting during the Cold War, alien or otherwise, would be a top priority to U.S. defense initiatives, yet no military personnel reported any sightings, no military warnings were issued, there were no reported alerts, and there was never an increase in operational activities. In other words, nothing you'd expect to happen if a UFO entered U.S. airspace during the Cold War happened. Nonetheless, Ufologists simply move the goalpost, dismissing the reports and evidence as another layer to a government cover-up (Harrison, 2013, pp. 119–24; Kloor, 2019; Prothero & Callahan, 2017, pp. 61–79; Sagan, 1996, pp. 85–86).

What about the reports of bodies falling from the sky and alien autopsies? Actually, those reports didn't surface until the 1980s (Harrison, 2012, pp. 125–26). Nonetheless, they may still have some merit. Research during the 1950s led the U.S. Air Force to throw 250-pound test dummies dressed in silver pressure suits out of planes, and the dummies would later be carried away on stretchers (Harrison, 2012, pp. 125–26; Prothero & Callahan, 2017, pp. 39–60). Onlookers, witnessing a body dressed in a silver suit falling from the sky and military personnel recovering it, could be forgiven for suspecting foul play, especially given that those activities went unreported by the press. Moreover, reports of fallen bodies could have come from witnesses who inadvertently saw documented military accidents where soldiers were actually killed and injured. Given what we know about memory and how close in history the Roswell incidents and Air Force research were, it shouldn't surprise anyone if someone's memory placed the two events together.

SPECIAL PLEADING

Special pleading is invoking explanations after knowing what's needed to explain something. Combining our reluctance to admit being wrong with the tendency to tie ourselves to certain conclusions, we often engage in special pleading to make something an exception to a rule.

Imagine a friend says to you, "Stealing is wrong, and thieves should be imprisoned." A few weeks later, you catch your friend stealing office supplies from work. "I thought you said stealing is wrong," you say. "Why is it OK for you to steal but not others?" Your friend insists, "Nobody will miss these supplies, and I haven't victimized anybody." By minimizing the value of office supplies, she has engaged in special pleading by making an exception to the rule "stealing is wrong."

HASTY GENERALIZATION

The *hasty generalization* fallacy occurs when someone rushes to a conclusion without first considering all the evidence. Moreover, this fallacy also promotes hidden biases, such as prejudice or bigotry. Take the following argument: "Roy is wearing blue shoelaces on white shoes. People dress like that to represent a gang. Roy must be a gang member." This argument would be a hasty generalization since obviously everyone who wears blue shoelaces on white shoes isn't a gang member. Furthermore, drawing conclusions about an entire population based on one sample commits this logical error.

> There are many things we cannot explain, but not understanding them does not make them supernatural.
> —Bruce Hood

COUNTLESS COUNTERFEIT

The *countless counterfeit* fallacy is committed when someone argues that there must be some truth to "X" because it's unlikely that all the evidence supporting "X" is false (Johnson, 2018). The mistake here is misunderstanding how probability relates to evidence. A thousand pieces of bad evidence is still just a pile of bad evidence. Vaccine contrarians, arguing that there must be something to all of the adverse events reported to the Vaccine Adverse Event Reporting System (VAERS), completely miss the point that the reports themselves must be worthwhile (Skeptical Raptor, 2022). The existence of thousands of magic tricks is no more evidence for real magic than millions of stories of alien abductions, ghost sightings, or demon possessions are evidence that "they can't all be wrong." Actually, they can. As David K. Johnson—the person who coined the phrase "countless counterfeit"—said, "A bunch of stories about people seeing UFOs is not a reason to think that at least one is real: it's confirmation of what we already knew: people are easily fooled" (Johnson, 2020).

"TU QUOQUE"

Literally meaning "you too," the *tu quoque* fallacy, or the *appeal to hypocrisy*, is an attempt to counter an opponent's argument by stating that the opponent is guilty of the very thing they are arguing against. The saying "Do as I say, not as I do" reflects this common fallacy. However, the actions of someone making a claim are irrelevant to the claim itself.

Let's say you tell your children, "Don't smoke tobacco because it's bad for your health, it'll make you stink, and it's expensive." If you're a smoker, your child might retort, "But dad, *you're* a smoker!" So long as you're a smoker, your child's objection is perfectly legitimate. Unfortunately, the objection is irrelevant since your behavior has nothing to do with the facts; that is, smoking is bad for your health, it stinks, and it's expensive. Accordingly, appealing to someone's hypocrisy doesn't counter their argument.

THE STRAW MAN

The *straw man* fallacy is an extremely easy logical trap to fall into, and it's difficult to realize when we're committing it ourselves. It occurs when someone alters an argument and responds to the altered version instead of the original argument. People sometimes commit this fallacy because they don't understand the original argument or because the altered version is easier to knock down (hence the "straw man").

Let's say I argue that overpopulation is a huge contributor to global warming. In response, my detractor counters that efforts to control human populations are expensive and immoral and cannot be done without violating human rights. Even if the counterargument is true, it's a straw man, because it addresses an altered version of my original argument.

Cognitive Biases

Biases are inescapable. They dominate our cognitive lives and influence most, if not all, of our decisions and behavior. Unfortunately, biases don't disappear just because we discovered them (West, Meserve, & Stanovich, 2012). Many of our tendencies, such as why we avoid changing our minds or admitting fault, why we think we're better than others, why we become victims to misinformation, or why we defend our beliefs, opinions, and ideologies, result from biases. Biases are the reason people tell pollsters they believe in giving to the needy but disagree with welfare. Giving to the needy and supporting welfare is exactly the same thing, but to some, the former sounds more respectable. In a study on group influence (Cohen, 2003), people were significantly more willing to give to the needy if they thought their political party endorsed the program. But the brain is tricky. The participants said their political preference didn't influence their decisions, although they generally agreed that their political opponents would've been influenced. Biases affect the way we argue, behave, and think, which can be especially dangerous with biases we aren't even aware we have (Pronin, Lin, & Ross, 2002).

Identifying and recognizing our biases, especially the ones that trouble us, helps immeasurably. For example, our legal system must be especially sensitive to biases, such as by courts prohibiting leading questions. Thus, trials are structured so that two sides control what evidence is presented to the jury. Since the lawyers who control what the jurors see and hear have a vested interest in a trial's outcome, court rules focus on limiting the number of biases that enter the process.

Take jury selection. Jury selection is the process of asking potential jury members questions that might expose biases, prejudices, or experiences that could compromise their judgment. Do you dislike the police? Do you know the victim? Have you ever been a victim of a similar crime? Are you a specialist in a relevant area? Biases tend to make us stick to our guns or draw preemptive conclusions, often without any reasoning or considering evidence, arguments, or other useful information.

We dress up for court, weddings, parties, and job interviews because we don't want people's biases negatively affecting us based on our attire. Since biases lead us to invalid conclusions (e.g., "Look at those face tattoos; he's probably guilty"), when it comes to judging someone's guilt or innocence, minimizing biases is in everyone's best interest. Research agrees. In 2013 (Korva, Porter, O'Connor, Shaw, & ten Brinke, 2013; Shaw, 2017, pp. 34–35), researchers had photos of people examined, asking participants to rate how trustworthy the person in the photo appeared. Later, those photos were randomly matched with a crime and a short description of the crime and given to mock jurors to decide guilt or innocence. As expected, those whose photos were judged less trustworthy were found "guilty" significantly more often.

> An adopted hypothesis gives us lynx-eyes for everything that confirms it and makes us blind to everything that contradicts it.—Arthur Schopenhauer

STORYTELLING BIAS

As intensely social creatures with the added ability of verbal language, humans have a strong bias for stories, especially emotional ones. Listening to, telling, and creating stories is one of our favorite pastimes. People began trying to explain the world with stories. Nowadays, the film industries of the United States (Hollywood) and India (Bollywood) together gross over $12.5 billion a year. We spend huge amounts of time and money watching purely fictitious stories and being captivated by celebrities to the point where some know more about their favorite actor than they do about close family members. Marketers structure commercials and other advertisements as short stories to capitalize on the storytelling bias.

Sometimes the storytelling bias isn't all that useful. Take the story of a man who was unfortunate enough to drive down a road when a sinkhole opened beneath him, killing him instantly (Dobelli, 2014, p. 38).[1] Naturally, the news describes who the man was, where he was going, what family he left behind. The news, however, doesn't tell us anything useful, such as what caused the sinkhole or what the city is doing to prevent deaths from sinkholes. Conversely, some people use this bias for good. Charitable organizations send pictures with a story of, say, an impoverished child who survives by donations. The organization knows people will care more if they see the child and hear their story because emotional stories prompt us to react emotionally, thus impulsively. And while supporting poverty-stricken children is as worthy a cause as any, the point is playing on this bias works, even on the most seasoned skeptics (Shermer, 2008a, pp. 137–38).

THE CONFIRMATION BIAS

The significance of our next bias cannot be overemphasized nor can the dominating role it plays in our lives. Perhaps the most captivating bias of them all, the **confirmation bias** is our compulsion to unwittingly notice and remember information that confirms what we already believe or want to believe (Nickerson, 1998) and to systematically ignore, reject, forget, suppress, or minimize information that contradicts what we believe or want to believe (Edwards & Smith, 1996). If cognitive dissonance (chapter 7) makes conflicting beliefs uncomfortable, the confirmation bias makes agreeable beliefs feel amazing.

Like blinders that keep horses looking ahead, the confirmation bias is like tunnel vision for what we already believe and prefer. It's driven by emotions rather than logic and evidence. Additionally, the more deeply entrenched a belief and the more emotionally invested we are, the more our biases protect that belief (Westen, Blagov, Harenski, Kilts, & Hamann, 2006). You can think of the confirmation bias in terms of Street World. The more invested a city is in maintaining its infrastructure and the more commuters enjoy it, the more the city will take steps to protect it. Buying insurance, allocating funds, and establishing committees are all ways of protecting their investment. Similarly, to keep our beliefs firm and unchanging, we make friends with like-minded people, we listen to podcasts we already agree with, or we watch news that aligns with our political preferences. Thus, the confirmation bias imprisons us in entire belief systems without our awareness. Confirmation bias "is one of the primary reasons why somewhat-reasonable people are able to hold totally unreasonable positions with great confidence" (Harrison, 2015, p. 179), illuminating how some really smart people can believe some really stupid things.

Suggesting that the confirmation bias is responsible for a number of bogus prison sentences isn't unreasonable. During a murder investigation, detectives found a cigarette butt at a crime scene but discounted the butt as relevant evidence because their main suspect, later exonerated by DNA evidence, didn't smoke (Harland-Logan & Morin, n.d.). In a 1994 study, participants were exposed to an audio recording of an actual murder trial. The researchers found that rather than listening to the evidence and formulating an opinion as to guilt or innocence, the participants did the opposite; they formed an opinion, and then cherry-picked the evidence that supported their preconceived notions (Kuhn, Weinstock, & Flaton, 1994; Shermer, 2008a, p. 91; Shermer, 2011a, p. 260; and see Gilovich, 1991, pp. 33–37).

We remember information that substantiates our beliefs and ignore or suppress information that doesn't, especially information that contradicts a belief. Francis Bacon argued in *Novum Organum* (1645) that once people have adopted an opinion, they "draw all things else to support and agree with it," driving "all superstitions, whether in astrology, dreams, omens, divine judgments, or the like." Bacon's beliefs were confirmed in a study that measured confirmation bias with respect to capital punishment.

Researchers placed several participants holding differing views on capital punishment at a table while another researcher instructed them that they'd each be given two randomly selected research papers on whether capital punishment deters crime (Lord, Ross, & Lepper, 1979). Researchers told the

> Whenever we try to propose a solution to a problem, we ought to try as hard as we can to overthrow our solution rather than defend it.—Karl Popper

participants to evaluate each paper and report whether they supported the case for capital punishment. Before beginning, however, the participants were given one of two statements: one showing that research supported the deterrent theory of capital punishment and the other showing that research did not support it. After reading the statement and the first research paper, a researcher asked the participants how strongly they leaned for or against the deterrent theory and how the research influenced their belief. Then, participants were shown commentary from referees that peer-reviewed those papers, along with responses to the peer reviews, before being asked to write whether the study was pro- or anti-deterrent and to revisit their beliefs. The same process was followed for the reading of the second paper.

This study revealed that confirmation bias drove the participants to interpret the studies in a way that aligned with their prior beliefs, regardless of whether the study supported them. Participants criticized the study that didn't support their beliefs as flawed or methodologically problematic, while judging the study supporting their prior belief as scientifically sound. By the end of the experiment, the participants held their beliefs more firmly than they did at the beginning, having found empirical evidence to support them.

Another example of the confirmation bias is the so-called *lunar effect*, the notion that a full moon influences human behavior. People everywhere believe in the lunar effect, often claiming that during a full moon, more babies are born; more crimes are committed; and more people are admitted into psych wards, commit suicide, and engage in aggressive and sexually promiscuous behaviors (Hines, 2003, p. 223). Unsurprisingly, the lunar effect traces its roots to writings from the pre-scientific Assyrians and Babylonians, long before electricity illuminated the night. In fact, the word "lunatic" comes from the Latin word *lunaticus*, which literally means "moonstruck."

One study showed that nurses who believed in the lunar effect took more notes of patients' "unusual" behavior during a full moon than skeptical nurses. Moreover, believers, compared to skeptics, were aware of when a full moon would occur and therefore expected and looked for instances that confirmed their belief (Hines, 2003, p. 225). Additionally, studies conducted by biologists, physicists, astronomers, and psychologists found no correlation between a full moon and any human behavior (Foster & Roenneberg, 2008; Rotton & Kelly, 1985; Vyse, 2014, p. 46). And, like astrology (see chapter 9), no theoretical mechanism by which the moon *could* influence human behavior exists.

Despite the overwhelming evidence against it, this myth survives in the twenty-first century, largely because of the confirmation bias. When there's a full moon and someone identifies some strange behavior, that's confirmation of the lunar effect. When no strange behavior is identified during a full moon, it's forgotten because it's a non-event. Similarly, when the moon isn't full and people behave oddly, believers forget because it doesn't confirm their belief.

Internet personalization—algorithms that megacorporations use to expose us to specific content—feeds the confirmation bias (recall the filter bubble discussed in chapter 1) and saps creativity. And since System 1 does it all for us, we're none the wiser. Moreover, confirmation bias combines with other cognitive errors, creating potentially dangerous situations that can dramatically impact our lives. For example, cognitive dissonance combines with the confirmation bias when an attorney questioning a witness is friendly, provided the answers confirm the attorney's beliefs. However, if the answers deviate from what the attorney believes, the treatment of the witness becomes hostile.

Confirmation bias also affects us in arguments, causing us to act less virtuously than we otherwise would, behaving as badly as those we're arguing against. Seeking truth requires honesty, humility, objectivity, and asking questions that are likely to reflect as much. This approach to what's known as *virtue epistemology* is advocated by the skeptic and philosopher Massimo Pigliucci, who provides the following checklist that a virtuous skeptic should keep in mind. When challenging someone's arguments, avoiding the confirmation bias requires asking questions like:

- Did I carefully consider the argument, or did I dismiss it out of hand?
- Could my interpretation of what was said have been more generous?
- Did I entertain the possibility that I might be wrong?
- Are either of us experts on what we're arguing about?
- Did I independently research or consult any experts on the subject, or did I just create my own speculative opinion out of thin air? (Pigliucci, 2017b)

A good strategy for minimizing confirmation bias is to seek all evidence, both confirming and disconfirming, and use GASPs and CT to critique your own thoughts, conclusions, and behaviors.

Heuristics

Next up is the tragically named **heuristic**. Heuristics are mental shortcuts that we've evolved to reduce the amount of work our brains do, easing the cognitive load of making many decisions or weighing several factors at once. You can think of heuristics as cognitive instincts that make us process and interpret information in specific ways. Sometimes likable people are trustworthy, and sometimes we really do get what we pay for. However, sometimes heuristics lead us astray, such as when we believe likable means trustworthy, popular means worthy, scarce means valuable, expensive means good, authoritative means correct, or confident means competent (Cialdini, 2007; Levine, 2018).

Heuristics include rules of thumb, educated guesses, intuitive judgments, stereotyping, profiling, common sense, assumptions, veridicality, and emotions. Oftentimes, heuristics are adaptive strategies that are good enough, often enough that natural selection never eliminated the occasional error they produced. I'll cover several heuristics so readers can gain an understanding of what they are, how they influence us, and how we might avoid these cognitive traps.

AVAILABILITY ERROR

The *availability error* is the process of reasoning from personal experience. We commit this error by misjudging how often something occurs because of how easy like examples come to mind. If you've thought crime was rampant because it's always on the news or lightning frequently strikes people down, you've committed the availability error. This error is why drug use decreased by half from the 1980s to the 1990s, yet adults rated drug use as the leading danger to young Americans. Likewise, while heart disease was the number one cause of death in men, it received the same amount of media attention as homicide, even though homicide ranked eleventh (Shermer, 2004, pp. 172–73). According to sociologist Barry Glassner, "We waste tens of billions of dollars and person-hours every year on largely mythical hazards like road rage, on prison cells occupied by people who pose little or no danger to others, on programs designed to protect young people from dangers that few of them ever face, on compensation for victims of metaphorical illnesses, and on technology to make airline travel . . . safer still" (Shermer, 2008a, pp. 78–79).

> If someone doesn't value evidence, what evidence are you going to provide to prove that they should value it? If someone doesn't value logic, what logical argument could you provide to show the importance of logic?
> —Sam Harris

The availability error was ingeniously demonstrated in a study conducted by psychologists Amos Tversky and Daniel Kahneman, who pioneered the study of heuristics. In this study (Schwartz, Bless, Strack, Klumpp, Rittenauer-Schatka, & Simons, 1991; Tversky & Kahneman, 1973), participants were asked whether more English words started with a "K" or whether more had a "K" as the third letter. Kahneman and Tversky theorized that the availability heuristic would cause people to guess the former since, as psycholinguists understand, we recall words by their first letter, not their third (Pinker, 2018, p. 42). The majority of participants indeed guessed that more words begin with the letter "K," even though more than twice as many have "K" as the third letter. The same holds true for the letters L, N, R, and V (Kahneman, 2013, p. 7).

The availability error causes us to misjudge the probability of events, like when we misjudge the frequency of shark attacks or plane crashes (Lichtenstein, Slovic, Fischhoff, Layman, & Combs, 1978). Such events are emotionally charged, easy to visualize, and disproportionately covered in the press. The media feed this error by predominantly reporting negative news (Pinker, 2018, pp. 47–51). Emphysema kills more people per year than firearm accidents and murders combined, yet murders make daily headlines while emphysema deaths go unnoticed. Thus, the availability error compromises our ability to assess risk and take measured precautions to protect our health. As skeptics, we must remember that the readily available data are rarely the *only* relevant data. As Kahneman explains, "The world in our heads is not a precise replica of reality; our expectations about the frequency of events are distorted by the prevalence and emotional intensity of the messages to which we are exposed" (Kahneman, 2013, p. 138).

REPRESENTATIVENESS REASONING

In addition to having an awful name, the *representativeness heuristic* is responsible for many biases, including racism and classism. With the representativeness heuristic, we generalize from a stereotypical example, such as assessing the similarity of things and organizing them based on a typical example. In short, this heuristic convinces us that like goes with like and causes should resemble their effects. Occasionally it works out (sometimes the guy nicknamed "Snake" really shouldn't be trusted); other times it doesn't (sometimes the politician is honest and caring).

Demonstrating the representativeness heuristic, Kahneman and Tversky devised an experiment (Kahneman, 2013, pp. 156–65; Tversky & Kahneman, 1983) that's now known as "the Linda problem." Here's my version:

> Jake is a secretive, hot tempered twenty-seven-year-old who has been convicted of burglary, shoplifting, and assault. Jake has been to prison several times and is currently wanted for assault and armed robbery.
>   Based on this description, is it more likely that:
>
> (A) Jake is guilty of the armed robbery or
> (B) Jake is guilty of the armed robbery and assault?

Since the question is one of probabilities, "B" is the wrong answer, since it cannot be more likely that Jake committed armed robbery *and* assault than it is that he simply committed armed robbery. In other words, a combination of two events (committing armed robbery + assault) cannot be more likely than one of those events happening alone. Jake *might* be guilty of both crimes, but probabilistically, it isn't *more likely* that he is.

Answering "B" is a System 2 failure, but don't feel bad. When tested, 89 percent of undergraduates and 85 percent of Stanford Graduate School of Business doctoral students violated the logical rule (Kahneman, 2013, p. 158).

Sometimes the representativeness heuristic is harmless; for example, assuming that a woman wearing running shoes is out for a jog. However, the representativeness heuristic is somewhat responsible for the endangerment of animals, such as when rhinos are killed for their tusks because people believe they resemble virility. For years people believed stress caused ulcers because an ulcer *feels* like being under stress. We only later learned they're caused by bacteria.

Whether it leads to prejudice, animal extinction, or misdiagnoses, the representativeness heuristic is both dangerous and immoral, and can leave us far out of touch with reality. Avoiding these errors begins with an understanding that none of us is perfect and that eliminating prejudice requires significant effort.

ANCHORS AWEIGH

Anchoring is the tendency to "anchor" to the first information we receive about something, basing further judgments about it from that information. For example, if a car is advertised for $15,000, buying it for anything lower than $15,000 may sound like a deal. Here we based what constitutes a good deal on the original price. Similarly, if we learn that Carl is smart, loyal, funny, stubborn, annoying, and aggressive, we'll like him much more than we would Judith, who is violent, irritating, hardheaded, amusing, committed, and intelligent. Of course, these are the same traits, only with Judith we learn about her poor qualities first.

In a set of experiments (Tversky & Kahneman, 1974), Tversky and Kahneman showed just how subtle anchoring is. In the first experiment, they rigged a "wheel of fortune" to land on either 10 or 65, then asked participants to estimate how many African countries are part of the United Nations (UN). On average, participants whose wheel stopped on 10 estimated that 25 percent of UN members were African, whereas the ones whose stopped on 65 estimated that 45 percent were. In the second experiment, high-school students were given five seconds to answer $8 \times 7 \times 6 \times 5 \times 4 \times 3 \times 2 \times 1$. Since five seconds isn't enough time, the students had to guess (the answer is 40,320). This group guessed a median estimate of 2,250. However, a second group was given the same instructions, with the equation inverted ($1 \times 2 \times 3 \times 4 \times 5 \times 6 \times 7 \times 8$). This group's median estimate was 512. In both experiments, the participants were highly influenced simply because they had been unwittingly primed in a particular way.

Marketing experts understand anchoring well (Johnson & Ghuman, 2020). Ads offering "10 for $10.00" encourage us to buy ten (anchoring to the deal) when we might only need one or two, or maybe we don't need any! Anchoring causes us to cling to the first prices we see, the first thing someone says, or the first thing we learn about someone. As such, anchoring impacts our financial lives, our relationships with others, and even how we vote. Between framing (see below) and anchoring, first impressions matter . . . a lot!

PRESENTATION MATTERS

A related heuristic is the *framing effect* (Tversky & Kahneman, 1981), which describes changes in preference based on how information is presented. Framing explains why a $40 dinner looks like the sensible option when it's surrounded by $60 and $70 plates, yet the same dinner looks ridiculous when it's surrounded by $5 and $10 plates. Similarly, people will reject a medical treatment that has a 10 percent chance of causing death but accept one that has a 90 percent chance of survival (Kahneman, 2013, p. 367; McNeil, Pauker, Sox, & Tversky, 1982; Shermer, 2008a, p. 76). Of course, this is the exact same information; it's just been framed differently. Framing matters because, as research shows, "people seem to be more motivated by the thought of losing something than by the thought of gaining something of equal value" (Cialdini, 2007, p. 238).

Although I'm unaware of any research which has looked at framing with respect to criminal sentencing, it is reasonable to suspect that framing influences criminal justice in significant ways. While some judges are bound by sentencing guidelines, they unwittingly become victims of the framing effect by handing out lengthy prison sentences based on how sanctions are framed. When the maximum sentence for murder was twenty-five years, it seemed sensible to give out two- and five-year sentences for drug offenses. Now that the punishment for murder is often life without parole, it seems fitting to give out twenty-five years to drug offenders. By the same token, due to the framing of existing sentencing guidelines—e.g., twenty to forty years for drugs, twenty-five years for burglary, etc.—it now seems absurd to give out *only* twenty-five years for murder. How could a judge sensibly give out twenty years to someone whose had four felony drug possessions and twenty-five years to someone convicted of murder? Framing may even be responsible for driving sentencing guidelines ever upward.

POT COMMITTED

Anyone who's ever played poker is probably familiar with the *escalation of commitment* heuristic, sometimes referred to as the *sunk cost fallacy*. The escalation of commitment heuristic describes our tendency to stick with a decision, regardless of the consequences, because of what we've already invested into that decision. The rational decision is to ignore our investment and look toward the future, yet few of us have the resolve to do so. After making a decision and spending time and energy committing to it, we feel as though justifying that decision requires further support. Our need for consistency (Cialdini, 2007, pp. 59–60; Levine, 2018, pp. 525–26) then leads us to irrational behaviors that align with our original decision.

This heuristic explains why governments continue sending soldiers into a losing war (e.g., Iraq) (Shermer, 2008a, p. 89); why we "throw good money after bad"; why we continue playing a losing poker hand; why cult followers are reluctant to leave; why we stay in jobs we hate; and why we remain in relationships we no longer enjoy.[2] Additionally, this heuristic explains why oftentimes crimes beget more crimes. "I've already robbed a bank; why not shoot it out with the cops?" "I already relapsed; why not go off the deep end?" Awareness and education help us deescalate, preventing us from making mountains out of mole hills. As the billionaire investor Warren Buffet once quipped, "When you find yourself in a hole, the best thing you can do is stop digging."

Other Useful Thinking Tools

What we've covered so far can help identify certain errors in cognition and raise awareness to several cognitive biases that plague our lives. Below I cover a few more helpful tools that assist us in combating our biases.

OCCAM'S RAZOR

First up is *Occam's razor*, a logical device that's used to decide between competing hypotheses when each hypothesis is compatible with the data. The term is named after the English philosopher and theologian William of Ockham (1287–1347), who said that "entities must not be multiplied without necessity" (Novella, 2018, p. 157). Translation: prefer the hypothesis that makes the fewest number of assumptions.

Making assumptions is necessary for any kind of discovery. We assume our surgeon knows how to perform surgery. Scientists assume the world operates according to universal laws. Philosophers use assumptions to probe the ultimate meaning of things. Using assumptions is perfectly fine, provided you understand their limits (chapter 4). However, when a conclusion is offered that rests on assumptions, if multiple explanations fit the data, Occam's razor helps us "slice off" any unnecessary assumptions.

During the coronavirus pandemic, conspiracy theorists wasted no time circulating rumors that philanthropist Bill Gates was responsible for funding the creation of the virus that caused the pandemic (SARS CoV-2) (Polidoro, 2020b). Given the importance of thinking rationally about such issues (Radford, 2020), we'll use this conspiracy theory as an opportunity to apply Occam's razor. Hypothesis 1 claims that Gates funded the development of a novel coronavirus that was subsequently released onto an unsuspecting public. Hypothesis 2 claims that viruses evolve over time and eventually some of them "go viral," infecting millions. Hypothesis 1 assumes, among other things, that SARS CoV-2 was created in a lab, that Gates is interested in the development of deadly viruses, that he has no problem killing millions of innocent people, and that this conspiracy was kept airtight throughout its R&D. Hypothesis 2 assumes that the very successful theory of biological evolution is true; in other words, viruses mutate. Even without weighing the evidence, we can see that Hypothesis 2 is preferable because it rests on fewer and less absurd assumptions.

Occam's razor is a useful device, not an irrefutable principle. Like the empirical sciences, Occam's razor is about probability, not certainty. Sometimes the simplest explanation isn't the right one, and sometimes it is. Accordingly, Occam's razor should be wielded cautiously and intelligently.

REGRESSION PROBLEMS

When we encounter something interesting that we don't understand, our inner child keeps asking, "But why? But why?" Eventually, we realize that at some point explanation must end. To avoid falling into an *infinite regress*, we must be satisfied with the answer we give to the inquisitive child: "Just because!" In his best seller *A Brief History of Time*, late physicist Stephen Hawking retells an example of an infinite regress (Hawking, 1998, p. 1). Philosopher Bertrand Russell was giving a lecture on astronomy,[3] explaining that the Earth rotates around the sun, which in turn rotates around the center of a vast collection of other stars that we call our galaxy. After the lecture, a woman stood up and said: "What you have told us is rubbish. The world is really a flat plate supported on the back of a giant tortoise." Russell gave a superior smile before replying, "Well then, what is the tortoise standing on?" Not giving up easily, the woman replied, "You're very clever, young man, very clever indeed. But it's turtles all the way down!"

We can laugh at the notion of our planet resting on the backs of an infinite number of turtles, but regress problems pop up in more practical conversations, and spotting where an explanation or argument generates a regress is an important CT skill.

DOWN WITH THE PROFOUND

Gravitation toward the "profound" is an intuitive thinking style that allows everyone from marketing experts to con artists, mystics to cult leaders, and other woo-pushing sensationalists to manipulate the gullible (Pennycook, Cheyne,

Barr, Koehler, & Fugelsang, 2015). People are easily exploited by the all-too-human need to believe in what the English philosopher Stephen Law calls *pseudoprofundity*. Put simply, pseudoprofundity is "the art of sounding profound while talking nonsense" (Law, 2007, pp. 214–15; Law, 2011, p. 159).

Law insists that sounding profound is so easy it's stupid. One strategy is to look wise, stating obvious facts v-e-r-y s-l-o-w-l-y. Another is to mix the order of language delivery, a tactic made famous by George Lucas's *Star Wars* character Yoda. Creating a paradox about things that concern nearly everyone—e.g., love, death, freedom, sex, betrayal, power, money—by stringing together wise-sounding verbiage is also especially effective. Pseudoprofundity is so transparent that once you understand how empty it is, you'll wonder how anyone takes it seriously. All that's needed is a topic you want to illuminate—such as one of the categories above—and its opposite—such as hate, life, bondage, impotence, or poverty. Then, don your best wisdom face and string the two together, creating some deep-sounding truism. Imagine it: "Only those who have known hate can truly know love." "Every death must begin with life." Or, to quote Edward Norton's nameless character in *Fight Club*, "It is only after we have lost everything that we are free to do anything."

Technobabble (Smith, 2018, pp. 82–83) is another favored technique used to push pseudoprofundity. In the 1800s, would-be inventor John Keely claimed he invented several perpetual motion machines, giving his devices foolish names like "Vibro-dyne," "Sympathetic Negative Transmitter," and "Hydro-Pneumatic-Pulsating-Vacu-Engine." Referring to his inventions as "positive neutralization" and "spiro-vibrophonic" (Loxton, 2018b), Keely scammed millions from investors who were duped by his pseudoprofound confidence. Keely's legacy survives in books, internet scams, and YouTube videos claiming to have discovered "free energy" or perpetual motion, concepts that violate physicist Lord Kelvin's (1824–1907) first and second laws of thermodynamics. The laws of thermodynamics—the "poster child for iron-clad laws of science" (Novella, 2018, p. 258)—state that in a closed system, you can never get more energy out of the system than you put in because it costs energy for the system to operate. While the U.S. Patent and Trademark Office no longer accepts patent applications for perpetual motion machines, conspiracy theorists, hoaxers, magical thinkers, and all orders of snake oil salespeople still perpetuate the free energy myth, often claiming that "the government" or "Big Oil" conspire to keep us dependent on "their" energy.

In a popular pseudodocumentary titled *What the Bleep Do We Know!?*, filmmakers—not philosophers, neuroscientists, or physicists, to be sure—attempted to explain consciousness by claiming that "the quantum coherent causes neurotransmitters to be released into the synapses between the neurons thus triggering them to fire in a uniform pattern that creates thought and consciousness." A little applied CT quickly dissolves this contention. Quantum effects are orders of magnitude smaller than neurons. Human brains consist of about eighty-four billion neurons (Azevedo et al., 2009; Herculano-Houzel, 2009; Lent, Azevedo, Andrade-Moraes, & Pinto, 2012), each having about 100 trillion atoms each, meaning there's almost 10 trillion trillion atoms in the brain.[4] The neuronal level is gargantuan compared to the quantum level! Thus, physical laws governing the neuronal level differ from those governing the quantum level; in other words, quantum laws don't apply at the macro level. As Nobel Prize–winning physicist Steven Weinberg contends, "There is nothing like intelligence at the level of individual living cells, and nothing like life on the level of atoms and molecules" (Weinberg, 1994, p. 39).

The late particle physicist Victor Stenger explains: "A simple way one can test whether a system needs to be described quantum mechanically is to ask whether the product of the system's typical mass m, speed v, and distance d is of the order of Planck's constant h. If mvd is much greater than h, than [sic] the system can probably be treated classically [i.e., macro]," not quantum mechanically (Stenger, 1995, p. 284). Deepak Chopra is one such spiritualist who uses "quantum jargon as plausible-sounding hocus pocus." Real physicists, like Weinberg, Stenger, and Murray Gell-Mann (2019), refer to Chopra's brand of physics as "quantum flapdoodle" (Shermer, 2011a, p. 150).

Additionally, arguing that quantum effects translate into macro-level events commits the *fallacy of composition*, which states that what's true of the parts is true of the whole. Here are some examples.

> Atoms are invisible. People are made of atoms. Therefore, people are invisible.
>
> That policeman is corrupt. Therefore, the police department is corrupt.
>
> If a runner runs faster, she will win the race. Therefore, if all the runners run faster, they will all win the race.
>
> Standing up in a stadium helps you see better. Therefore, if everyone stands up in a stadium, they can all see better.

Obviously what's true of a part is not always true of the whole, a principle that applies especially to the laws of quantum mechanics.

Once understood, pseudoprofundity is far from impressive and generating your own is simple. However, sometimes understanding is the hard part. When you think you've run into the pseudoprofound, ask for clarification, what Law calls "pseudoprofundity's greatest enemy" (Law, 2011, p. 169).

REASONABLE BELIEF

The schoolyard retort, "You can't prove I'm wrong," is widespread among adults. And, for some reason, we tend to believe that if something isn't conclusively "proven" or "disproven," then both positions are equally reasonable or unreasonable. Proven or not, competing beliefs may still differ radically in their reasonableness and, as we'll see, what is reasonable can vary significantly.

For example, since you're reading this book (as opposed to listening to it) and thinking about what it says (hopefully), it's reasonable to believe that you have eyes and a brain. For the same reason, believing helicopters can fly, even if you've never seen one, is also reasonable. You probably possess tons of evidence that suggests helicopters fly and hardly any evidence that they can't. But just because something is *reasonable* to believe doesn't mean the belief is accurate. You may be schizophrenic—from the Greek words *skhizein* and *phrēn*, meaning "to split the mind"—and have imagined this book, or a massive conspiracy to convince you that helicopters can fly may be afoot. These beliefs are reasonable because they logically derive from available evidence, regardless of whether they're accurate. Since mistaken beliefs can still be reasonable, accuracy is a poor indicator of reasonableness.

Conversely, believing in vampires, witches, Sasquatch, or Atlantis is highly unreasonable since little evidence suggests their existence and hordes of evidence suggest otherwise. While evidence isn't the only criterion by which something is judged reasonable (chapter 4), believing something exists without evidence is highly unreasonable. For instance, while the *Fermi paradox* invites us to question why scientists haven't found evidence of extraterrestrials (Fermi's famous "Where is everybody?" question), the Drake equation shows why believing that intelligent life exists somewhere else in the universe is reasonable, even without

---

Drake equation:

$$N = R_* \cdot f_p \cdot n_e \cdot f_l \cdot f_i \cdot f_c \cdot L$$

R_* = The average rate of star formations in our galaxy

f_p = The fraction of formed stars that have planets

n_e = For stars that have planets, the average number of planets that can potentially support life

f_l = The fraction of those planets that actually develop life

f_i = The fraction of planets bearing life on which intelligent, civilized life has developed

f_c = The fraction of these civilizations that have developed communications

L = The length of time during which such civilizations release detectable signals

direct evidence. Fairies or leprechauns leave no plausible inferences to draw about their existence, so believing they do is unreasonable. However, the existence of extraterrestrial life can be inferred from the available evidence and the vast number of possibilities, so believing they exist is reasonable.

When shown that a cherished belief is false, people sometimes respond with, "You can't *prove* X is false? X might be true!" Sure, but this is loopholism (chapter 4), not reasoning. Moreover, just because something "might" be true doesn't mean its truth and falsity are equiprobable. Additionally, even if the balance of proof was evenly distributed, the most rational position would be to withhold judgment (Elgat, 2019). Bertrand Russell once argued that if he claimed a teapot orbited the sun between Earth and Mars, he couldn't expect anyone to believe him just because he couldn't be proven wrong (more on Russell's teapot in chapter 11). We could invent many incredible claims that cannot be disproven and get exactly nowhere. Nonetheless, believing something simply because it cannot be proven wrong is unreasonable. Additionally, the fact that something is possible doesn't mean it's real.

Notice that we aren't concerned with whether something is true, only that beliefs differ in how *reasonable* they are. It's important to note that when people strongly disagree, that doesn't mean the truth lies somewhere in the middle. As biologist Richard Dawkins explains, "When two opposite points of view are expressed with equal intensity, the truth does not necessarily lie exactly halfway between them. It's possible for one side to be simply wrong" (Shermer, 2013, p. 147). In life, as in court, the burden of proof always lies with the claimant. Just as a prosecutor cannot charge people with crimes and imprison them if they cannot prove they're innocent, I cannot claim that flying dogs exist and shift the burden to you to prove otherwise.

Wrapping Up

Although I've arbitrarily selected my examples here, we've learned a good deal about many brain processes that transpire outside of our awareness. We've also discovered several tools we can utilize for more effective reasoning. Our brains are wired to maximize our own potential and self-preservation, which has a way of leading us away from the truth and toward irrational decisions and personally affects our lives.

Raising awareness about our own fallacious reasoning and personal biases is a big step in the right direction, and mitigating the effects of bad thinking habits takes effort and practice. Furthermore, while we'll never completely avoid fallacious reasoning or biased thinking, making honest, concerted efforts to minimize thinking errors significantly improves the quality of our thinking.

CHAPTER 6

Self-Deception, Part 1—Memory

So far, we've learned about the nature of knowledge, how knowledge is obtained, how it's evaluated, and how those evaluations can go wrong. But before we can know anything, we must be able to remember. Seems like a no-brainer, right? Memory is one of the most familiar cognitive experiences we have, yet while everyone uses memory to function, few see it for the extremely flawed process it is.

Next, we explore how memory works—and how it doesn't. We'll see how memories are stored, how they change over time, their powers and limits, people's beliefs about memory, and the differences between accurate and inaccurate memories. Lastly, we'll discuss the strange phenomenon of false memories and their impact on the justice system.

Building Vocabulary

Metamemory:	False memory:
Knowledge of one's own memory capabilities (and strategies that can aid memory) and the processes involved in memory self-monitoring.	A cognitive error in which someone remembers something that never happened.

Confabulate:	Conflate:
Filling in gaps in one's memory by fabrication.	To erroneously mix or fuse (something) into a single entity.

Introducing Memory

The first culprit in our list of self-deceptions is memory. Memory confers our identities and our sense of self, for without memory, we wouldn't have much in the way of personalities. Our past is but a collection of memories, and our futures are informed by them. Most people store hundreds of thousands of pieces of information in memory, like facts and experiences. And while the term applies to many distinct abilities, when we think of memory, what usually comes to mind is *episodic memory*—our ability to recall experiences—or *semantic memory*—specific memories of facts (Sherry & Schacter, 1987). Recalling what we told our ex at that dinner party last year involves episodic memory, whereas recalling our e-mail password involves semantic memory. Neither episodic nor semantic memory *feels* any different; both just feel like memory. However, their differences are immense. We use semantic memory for rote learning of things that remain relatively constant, such as a phone number, $E = mc^2$, how to spell your name, or what frogs are. We use episodic memory to recall prior autobiographical events, such as what presents we got for our ninth birthday or what happened at our best friend's wedding. Semantic memory can be thought of as knowing, whereas episodic memory can be thought of as remembering. And therein lies the key to the deception: our ability to accurately remember semantic details tricks us into believing we remember episodic details with equal accuracy.

For several reasons discussed below, episodic memory is subject to distortion in ways that semantic memory is not. We might **conflate** the details of a camping trip we took to Yosemite National Park with something we later saw on a nature documentary. To his embarrassment, the former NBC journalist Brian Williams retold a story of landing a military helicopter in Iraq that was hit by an RPG. In previous recantations, Williams claimed that the RPG nearly hit a helicopter flying in front of his and that he had looked down the tube of an RPG that had been fired at the one he was in. Unfortunately, the helicopter in front of him was about thirty minutes ahead, and crew members from his craft disconfirmed Williams's story. Perhaps Williams lied. Perhaps he conflated his own helicopter experience with stories he'd heard from soldiers recalling similar experiences. We'll likely never know whether he made it up or his memory failed him. Nonetheless, considering our understanding of memory, believing that Williams simply conflated his experience with something else isn't too far-fetched.

Before we begin breaking memory down, we need to understand that memory isn't what most people think it is, and it's certainly more than the thing we forget with (Loftus & Ketcham, 1994, p. 38). According to a nationally representative survey of Americans (Chabris & Simons, 2010, pp. 45–46; Simons & Chabris, 2011), what we think we know about memory is more likely wrong than right. Researchers found that the average American believes that:

- our brains keep track of memories like a computer hard drive keeps track of video files
- amnesia causes people to forget their names and identities

- hypnosis helps people accurately recall details of memories
- once a memory is formed, it doesn't change

This survey also revealed that memory experts with more than ten years' experience almost unanimously disagree and that levels of general intelligence and education have little, if anything, to do with misunderstanding memory. Frighteningly, further research shows that although police are usually confident in their understanding of the issues psychological research poses on the law, many endorse myths like those listed above or that people can't have memories of something that didn't happen (Chaplin & Shaw, 2015; Shaw, 2017, pp. 135–36). Given this research, it's no wonder so much of law enforcement wholeheartedly believes the testimony of eyewitnesses over the accused.

Our understanding of memory has been achieved through scientific progress and lots of it! We can better understand memory in several ways, but perhaps the best is engaging the science. Also, arming ourselves with such information can help us develop metamemory. **Metamemory** refers to having an accurate understanding of the capacities and limits of one's memory. Developing your metamemory is like hiring investigators to roam Memory City and learn everything that's going on there. Knowing the limits of our memories and understanding specific problems we have with our memories helps us develop strategies for improving problematic areas. Further, metamemory is about keeping track of what we can remember so that in retrospect, we can decide the likelihood of a memory's accuracy (Shaw, 2017, p. 23).

Drawing on our Street World metaphor, developing metamemory is like building a view overlooking Memory City, where onlookers watch what's going on, who was there, and so on. However, given the laziness of System 2 and our tendency to be cognitive misers (chapter 1), we must consciously activate and exercise metamemory for it to work; even the biggest tourist attractions are closed if no one visits. But before we get ahead of ourselves, we need to understand exactly where and how our memories fail us.

Memory Is Not as It Seems

Skeptic Guy P. Harrison argues that memory works "more like having a little old man who lives inside your skull and *tells you stories* about what you saw or experienced in the past. But every time he tells you one of these stories . . . , he changes it by leaving out some parts and adding other parts that never happened. He might even decide to change the order of events, which means that the storyline in your head won't match reality" (Harrison, 2012, pp. 115–16).[1] As we'll see, Harrison's metaphor is spot on. We'll also see how this little old man is apt to give us both accurate and inaccurate information, and why telling the difference is impossible. Developing metamemory and critically thinking about memory in general helps us figure out whether we can rely on anything our little old man has to offer.

Failing to understand how memory works can affect us in numerous ways. When someone recalls an event differently than we remember, we might think she is lying and get upset or write her off as a fraud. In reality, she may not be lying at all; it may just be her memory (or ours!) that's wrong. Memory failures don't just affect friends swapping stories. Those who don't understand memory—or refuse to accept the findings of science—sit on juries and decide people's future based on the memory of eyewitnesses, a particularly troubling situation given that criminal trials often revisit emotionally charged events; research shows that emotional memories *feel* more memorable, but are actually more susceptible to distortion than emotionally neutral memories (Talarico, LaBar, & Rubin, 2004).

Like the roadways in Street World, memories are constructed by our brains and become entangled with other cognitions, such as beliefs and biases. As psychologists Christopher Chabris and Daniel J. Simons put it, "Memory depends both on what actually happened and on how we made sense of what happened" (Chabris & Simons, 2010, p. 47). Over time, this constant constructing and entangling contaminates memories, slowly and imperceptibly morphing them into something new. Memory creation involves several steps, each one providing an opportunity for something to go awry (McDermott, 2018). Memories are conflated, fade, fuse together—and sometimes they're entirely fabricated.

In fact, the malfeasance of memory led Daniel Schacter to identify the "seven sins of memory": transience, absentmindedness, blocking, misattribution, suggestibility, bias, and persistence (Schacter, 2002, pp. 4–5). According to Schacter, transience occurs as memories are weakened or lost over time; absentmindedness explains our attention deficit, thus our failure to recall information we later need; and blocking is the feeling we get when we attempt,

Schacter's Seven Sins of Memory:

- Transience
- Absentmindedness
- Blocking
- Misattribution
- Suggestibility
- Bias
- Persistence

Being aware of these basic flaws in our memory can help reduce the chances that we embarrass ourselves by overly (and wrongly) relying on flawed memories.

unsuccessfully, to retrieve previously stored information. Schacter calls transience, absentmindedness, and blocking sins of omission, ones we commit when our memories fail to produce desired information.

The remaining four are sins of commission: errors that give rise to incorrect or undesirable memories. Misattribution occurs when we miscredit a memory's source. Suggestibility refers to memories formed as a result of leading questions, comments, or suggestions and involves updating memories with new information and personal biases (Schacter, 2002, p. 9). Due to people's confidence in their memories, coupled with widespread ignorance on the science of memory, the sins of misattribution and suggestibility are especially concerning for the legal system. Schacter's final sin is persistence: those memories, usually painful or unwanted, that refuse to go away, sometimes resulting in flashbacks or post-traumatic stress disorder (PTSD). Schacter's sins crop up every day, and they can be useful guideposts for understanding memory's flawed nature.

Memory's Known Associates

Memories are constructed entirely by our brains (how could it be otherwise?), and if we've learned anything about the brain, it's that the brain is imperfect. Memories are linked to all our thoughts and beliefs, collectively forming our internal models of reality. Thus, memories are associative—they're associated with all the other information in our brains, which is why a smell might remind you of your childhood home or a song brings back memories of a high school sweetheart. Accordingly, memories are also associated with themselves. What researchers call *fuzzy trace theory* (Shaw, 2017, pp. 78–81) suggests that when memories are stored, they're split up into *verbatim* memories (memories of specific details) and *gist* memories (memories of concepts and meanings). Gist and verbatim memories are formed—or "stamped" as it's called—at the same time, although the brain stores each trace separately. So a memory of a single event is split up into several pieces, meaning (1) we recall each trace separately, (2) one trace might be stronger than the other, and (3) in any given situation, one, both, or either trace may be irretrievable. When trying to recall an event, sometimes we remember how we felt (gist), sometimes we remember exactly what happened (verbatim), sometimes we remember both, and we might not remember any of it! You might remember someone who got you out of a bind (gist) but not his name or the bind he got you out of (verbatim).

Gist memories usually remain accurate longer than verbatim memories, although both are just as prone to error as any other memory. Actually, since gist and verbatim memories are recalled separately, they may even be especially error-prone (Brainerd & Reyna, 2002). Gist memories are vague as it is, so when an event is recalled, that imprecision can result in parts or all of the verbatim memory being altered. For example, you may remember *that* someone bailed you out of a jam, and, unable to remember which jam that was, you might remember it happening at a bar when it really happened at a party. If memories weren't associative, making meaningful connections between past events would likely be impossible. However, the associative nature of memories comes with a cost. As Dr. Julia Shaw explains, "Memory illusions are possible because each of our experiences is stored as multiple fragments, and these fragments can be recombined in ways that never actually happened" (Shaw, 2017, p. 81).

The (Un)Reliability of Memory

Problems with memories aren't limited to recall and recombination. Sometimes an event happened so long ago the memory is foggy (transience); sometimes we paid little attention to specifics, thus can't recall important details (absentmindedness); and sometimes recalling a specific memory, such as a name or phone number, is challenging (blocking). These "sins" manifest in statements like, "Don't quote me on this, but . . . " Using such statements declares our uncertainty in our memory, recognizing that our memory cannot be fully trusted.

Researchers have cataloged several types of memories. Some we're familiar with, such as short- and long-term memory. Others we're less familiar with, such as *working memory*—the memory we use to manipulate several pieces of information at once, where those memories are either stored or discarded. Given all the different memory types, there's so much room for error that we shouldn't be naïve or prideful, assuming that what we think we remember is completely accurate.

Take the case of *flashbulb memory*. The phrase "flashbulb memory," coined in 1977 (Brown & Kulik, 1977), describes a typically vivid and long-lasting memory that's formed by an unexpected or emotionally charged event, like being sentenced to prison, witnessing a murder, or being sexually abused. Research shows that flashbulb memories are subject to contamination and degradation just like any other memories (Hirst & Phelps, 2016). Nevertheless, people's confidence in flashbulb memories are oftentimes reinforced by the emotions associated with the event. Furthermore, due to the associative nature of the brain's storage system, people become unjustifiably confident in the accuracy of flashbulb memories.

> What we . . . refer to confidently as memory . . . is really a form of storytelling that goes on continually in the mind and often changes with the telling.
> —William Maxwell

Studying flashbulb memories has led to some helpful and interesting discoveries. The day after the *Challenger* space shuttle exploded, memory researcher Ulric Neisser asked students seven specific questions about where they were and what they were doing when the explosion occurred. Over two years later, Neisser tested the students to see how accurate their memories were. The results: 25 percent scored 0 out of 7, meaning their new accounts didn't match their original accounts at all; 50 percent scored 2 or less; and only 25 percent remembered taking the survey (Neisser & Harsh, 1992)! Less than 10 percent scored a perfect 7 out of 7. Perhaps the most interesting data from Neisser's study revealed that regardless of accuracy, on average, participants rated their confidence in their memories as 4.17 out of 5. Subsequent studies found similar results (Hirst, Phelps et al., 2015; Wright, 1993).

For example, in 2003, psychologists Jennifer Talarico and David Rubin found that accuracy and consistency of memory degrade at the same rate for flashbulb and other types of memories and that participants maintained low confidence in everyday memories but high confidence in flashbulb memories, despite their inaccuracy (Talarico & Rubin, 2003). Echoing both studies, cognitive psychologist Kathy Pezdek studied five hundred students after 9/11, finding that after two years, 73 percent misremembered seeing a video of AA Flight 11 crashing into the North Tower, a video that hadn't aired yet (Pezdek, 2003; and see Smith, Bibi, & Sheard, 2003). Such studies are revealing. Not only are the memories themselves degraded, but we fail to realize *that* they're degraded, leading us to unjustifiable confidence in their accuracy.

Pseudomemories

As far as the brain is concerned, **false memories** are just like real memories. To our conscious selves, false memories are indistinguishable and seem just as real as accurate memories. As you can imagine, this situation gives rise to a number of problems. For one, we don't know when a false memory has formed. For another, we have no way of telling the difference between false and accurate memories. Thus, this inability often leads to false confidence in completely inaccurate memories.

EYE SAW WHAT YOU DID

Eyewitness testimony boils down to what a witness can remember—or what the witness thinks he can remember. With their own eyes, honest and sane eyewitnesses have witnessed Bigfoot, space aliens, mermaids, ghosts, and other nonexistent creatures. Many such witnesses can provide detailed and compelling descriptions of what they saw. Is the fact that they really believe they saw what they think they saw compelling evidence? Should we believe them? As it turns out, the answer is a resounding no. In a study on the ability to accurately recall details (Palmer, Brewer, Weber, & Nagesh, 2013; Shaw, 2017, pp. 150–51), a researcher on the street asked people to fill out a consent form agreeing to participate in a study. Once the form was signed, the researcher asked the participant to look at a second researcher who would step into view for a few moments before stepping away. Half of the participants were asked to immediately identify the second researcher in a photo lineup, while the other half were asked to do so a week later. The participants who were given the photo lineup immediately identified the second researcher 60 percent of the time, while the participants who waited a week identified the second researcher 54 percent of the time. So just over half the participants identified the person they just saw, and that number begins to drop within a week: a dreadful outcome that gets even worse.

When it comes to identifying crime suspects, people are race, age, and gender biased, meaning we suck at accurately identifying "others"—other races, other genders, and people in other age groups (Shaw, 2017, pp. 151–54). If a white person perpetrates a crime against an Asian person, the Asian person will be worse at identifying the white culprit than if the culprit were Asian. The same goes for any combination of race, age, or gender—more bad news for a legal system that relies heavily on eyewitness testimony. Given the evidence, strong reliance on the memory of eyewitnesses, combined with a general lack of knowledge about the fallibility of memory, is a recipe for wrongful convictions.

Given the science on the reliability of memory, if memory is a poor source of accuracy, its usage should be curtailed by the courts. In fact, psychologist James E. Alcock argues, "The most common cause of wrongful convictions is faulty memory in the form of an erroneous eyewitness report" (Alcock, 2018, p. 93). While laypeople remain blissfully uninformed, many convicts understand that eyewitness testimony is notoriously unreliable. Convicts may not know, however, that sometimes even honest and well-intended eyewitnesses innocently misremember the facts. Worse yet, sometimes memories are fabricated on the spot by something as simple as a three-letter word. In 1974, psychologist Elizabeth Loftus showed students short clips of car accidents (Loftus & Palmer, 1974). After the viewing, some were asked, "About how fast were the cars going when they hit each other?" Others were asked the same question, but rather than "hit," the verb was changed to "smashed," "collided," "bumped," or "contacted." A week later, the students' estimates depended on which word was used to describe the collisions. Those who were primed with the verb "hit" estimated an average of 40.8 mph, followed by "collided" (39.3 mph), "bumped" (38.1 mph), and "contacted" (31.8 mph). Interestingly, those who were primed with the verb "smashed" were more than twice as likely to report having seen broken glass than those who were primed with the verb "hit," despite that there was no broken glass.

Loftus's experiment illustrates that leading questions can alter both *how* a witness remembers an event and *what specific details* the witness remembers. Eyewitnesses rely on memories that can fade, change, or warp into partially or completely fictitious events. That shady testimony that couldn't be explained by anything other than a conspiracy between the witness and the prosecutor may be nothing more than the (mis)workings of memory.

Confabulation is the brain's way of filling in blank spots in memory, sort of like how the brain fills in tiny bits of missing information whenever we blink. When we go to recall an event, if the brain is missing pieces of information, it simply confabulates details that align with its internal narrative. The memories that fill the gaps are based on beliefs, expectations of what should be remembered, information we've gathered since the memory was encoded, our internal models of reality, and a slew of other factors. Since the brain confabulates memories outside of our conscious awareness, controlling this process or knowing it happened is impossible.

False Memories

False memories are memories of events that never happened. They comprise an array of memory distortions (Wade, Sharman, Garry et al., 2006) and can have deadly consequences. In 2002 (Whitlock & White, 2002),[2] for example, a series of shootings that became known as the D.C. sniper attacks resulted in the deaths of seventeen people from Maryland, Virginia, and Washington, D.C. In the rush to identify the culprits, eyewitnesses falsely identified the shooters' vehicle as a white box truck or van, leading investigators to search for white trucks and vans while the killers continued on. (The killers' vehicle was a blue 1990 Chevrolet Caprice.)

False memories can even contaminate judicial proceedings long before they begin.[3] In 2015, researchers showed that adults generated false memories of committing crimes after just three hours of police interrogations (Shaw & Potter, 2015). So not only do suspects contend with the errors of perception and the false memories of eyewitnesses, they must also worry about how their *own* memories came about![4] Criminal defense lawyers often breathe a sigh of relief when they learn that the only evidence against their client is the memory of eyewitnesses. By contrast, jurors are often clueless that memory is plagued by errors. In 2002, legal scholars understood "the general failure of the law to reflect virtually any of the insights of modern research on the characteristics of human perception, cognition, memory, inference or decision under uncertainty, either in the structure of the rules of evidence themselves, or the ways in which judges are trained or instructed to administer them" (Tavris & Aronson, 2019, p. 198). However, the science on memory fallibility has pushed some criminal courts to recognize the unreliability of eyewitness testimony (Wells, Malpass, Lindsay, Fisher, Turtle, & Fulero, 2000).[5] Laws requiring courts to instruct juries on the suspect nature of eyewitness testimony are slowly being passed (Weiser, 2012).

The noted French author La Rochefoucauld once observed, "Everybody complains of his memory, but nobody of his judgment" (Taylor, 2017, p. 180). We shouldn't be surprised, considering that since most of us have no idea memory is so flawed—a process we don't control—why would we doubt our own judgment—a process we directly control? Our inability to tell the difference between real and false memories (Schacter, 2002, p. 114)[6] leads to unjustifiable confidence in the accuracy of our memories, especially when we can remember specific details. Concerning this inability, Loftus concluded that "manufactured memories are indistinguishable from factual memories" (Loftus, 1993). Additionally, false memories traumatize people just as real memories do. False memories of past lives, being abducted by aliens, or satanic sexual abuse (Goertzel, 1994b; Spanos, Burgess, & Burgess, 1994; Spanos, Cross, Dickenson, & DuBreuil, 1993; Spence, 1996) can seem just as real as memories of events that actually occurred.

Some memories are certainly false. For example, some claim to remember being born, being in the hospital shortly after their birth, and even their very early childhood. Such memories must be false because long-term memories are stored in parts of the brain that don't even begin to develop until about eight months of age (Shaw, 2017, pp. 3, 11–15),[7] while others aren't developed until around age three (McKee & Squire, 1992). Moreover, we develop a sense of self around age two, and children under two don't encode the world in language and schemas as adults do. Accordingly, few (if any) people remember anything from before this period (Wade & Tavris, 2017, pp. 372–73). How, then, can people have these impossible memories?

In an ingenious experiment that used the power of suggestion, Schacter created false memories by simply showing people photographs (Schacter, 2002, pp. 112–13; Schacter, Koutstaal, Johnson, Gross, & Angell, 1997). Schacter took participants to a park where they watched a young couple—confederates of the study—enjoy a picnic. The couple engaged in typical picnic behavior, such as applying sunscreen, drinking, taking pictures, and eating a sandwich. Two days later, Schacter questioned the participants about pictures taken of the young couple during the picnic. However, some of the photos were taken of things the participants didn't witness. A few minutes later, Schacter read out objects and actions, asking the participants to say yes when he read one they remember from the picnic. Schacter warned them to be careful because some of the actions and objects could only be found in the photos they just saw and didn't actually happen at the picnic. As predicted, the participants began saying yes when Schacter described objects and actions that were in the photos but didn't occur at the picnic. Schacter created false visual memories in his participants' minds with just a few moments and a little suggestion.

Affirming Schacter's sin suggestibility, Loftus has mastered the art of creating false visual memories simply by asking leading questions (Loftus, 1975; Loftus, Miller, & Burns, 1978). Moreover, Loftus has shown that implanting false memories ("memory-hacking" as Shaw calls it), even in emotionally and psychologically healthy people, can be achieved just by repeating a suggestion until it becomes a memory (Loftus, 1997). This tactic is often abused by overzealous prosecutors in child abuse cases. Thus, Loftus concludes, "Memories are like a Wikipedia page; you can change it, but so can other people" (Loftus, 2018).

Suggestibility even distorts flashbulb memories. In 1992, after a plane in the Netherlands crashed into an eleven-story apartment building, researchers asked people, "Did you see the television film of the moment the plane hit the apartment building?" Fifty-five percent of respondents answered yes. In a follow-up study, two-thirds also said yes. Despite participants' ability to recall details of the crash, such as the speed of the plane and the angle it hit the building, footage of the plane hitting the building doesn't exist (Crombag, Wagenaar, & Van Koppen, 1996; Schacter, 2002, p. 112). Further research reveals that sleep deprivation can impair real memories and increase one's chances of creating false memories (Frenda, Patihis, Loftus, Lewis, & Fenn, 2014)—a grim outcome, especially considering that around 30 percent of Americans report being sleep deprived every day and between fifty and seventy million Americans have chronic sleep disorders (Centers for Disease Control and Prevention, 2017).

Causes of false memories:

- Imagination
- Dream interpretation
- Hypnosis
- Sleeping disorders
- Absorption

- Leading questions
- Fantasy role-playing
- Overconfidence
- Brain biochemistry
- Exposure to false information

Multiple factors can lead to the complete fabrication of false memories.

CREATING MEMORIES

Simply observing someone else perform some act can generate a false memory of having been the one to perform the act (Lindner et al., 2010). Furthermore, researchers have successfully implanted rich false memories of spilling punch on a bride's wedding dress, being attacked by a vicious dog (Hyman, Husband, & Billings, 1995; Porter, Yuille, & Lehman, 1999; Shaw, 2017, p. 174), and getting one's finger caught in a mouse trap (Bruck & Hembrooke, 1997). If manipulating memory in a lab is this easy, what about outside the lab? Unfortunately, the research is nothing less than a devastating indictment.

The forming and strengthening of a memory can be compared to building a new road in Street World (Schacter, 2002, pp. 32–33). When an event happens, a new road (neural pathway) is built. The importance of the event determines the road's popularity. Thus, popular roads are more active than unpopular roads, just as important events garner more attention than unimportant events. A road's popularity determines how many additional routes are built to access it. Likewise, the importance of an event dictates how many neurons are associated with it. Additionally, the greater the demand to use the road, the greater the demand for expansion and upkeep. Similarly, the more we think about something, the more we strengthen neuronal connections. "Neurons that fire together wire together" as the saying goes.

Memories are built, maintained, and reinforced by importance and repetition. Does this mean that once a memory is formed, it'll always exist? Experience says no. For the same reason arterial roads in Street World can be closed, memories can be completely lost over time. However, we must be careful not to push our metaphor too far. If a road is closed and deteriorates over time, a substantial amount of construction is needed to restore it to working order. Memories are different. A memory lost to time can sometimes be brought back by cues that remind us of the original event. A friend tells you a story you would've never thought about again, but its telling brings the event back in vivid detail. Or something happens during the day that reminds you of a dream you had last night that would otherwise have never been remembered. Memory's ability to lie dormant indefinitely can cause us to glow with euphoric nostalgia, but it's also a source of one of the biggest misunderstandings about memory, which led to the so-called repressed memory movement.

Repressed Memories

False memories shouldn't be confused with so-called *repressed memories*, memories that people supposedly block because of the trauma of an event. Although the repressed memory movement is rooted in outdated Freudian psychology (Shaw, 2017, pp. 229–33; Stanovich, 2013, pp. 1–2),[8] it survives nonetheless (Tavris, 2019). This movement succeeded in creating an epidemic of false memories and hardships, including multiple prison sentences (Hagen, 2018) where the only evidence was repressed memories. Proponents believed that victims suffered not from recurring memories of traumatic events but from *not* remembering traumatic events. Repressed memory therapists employed many "techniques" to "recover" repressed memories, such as hypnotherapy, dream interpretation, or fantasizing about what could've happened. Failing to remember a traumatic experience, such as being sexually abused, was interpreted as evidence that abuse occurred. Accordingly, failing to remember being abused meant you were abused, and remembering abuse also meant you were abused (readers of chapter 3 will immediately recognize this as a completely unfalsifiable hypothesis). Put simply, the argument for repressed memory suggested that everyone was abused.

Repressed memory therapists argued that accessing repressed memories assisted victims through the healing process by allowing them to identify the source of their problems and process those events through the comfort and safety of therapy. Thus, therapists were

> Memories are like flagstones; time and distance work upon them like drops of acid.
> —Ugo Betti

healing their patients while providing a platform for protecting others from abuse. By contrast, skeptics pointed to a lack of empirical evidence supporting repressed memories, arguing that "recovered" memories could be false memories created by suggestion and influence.

The case for repressed memories is doubtful at best. Research doesn't support the idea that traumatic memories are repressed or forgotten (Engelhard, McNally, & van Schie, 2019). In fact, the opposite is true. Memories of trauma are typically recalled more accurately over a longer period than any other type of memory (Shaw, 2017, pp. 164–65). Additionally, researchers haven't found a single case of anyone who forgot being in a concentration camp, fighting in combat, surviving an earthquake, or experiencing any other traumatic event (Wade & Tavris, 2017, pp. 370–71). In cases of trauma, victims suffer from an inability to *forget* the event, not an inability to remember it (McNally, 2003), because the release of adrenaline during the traumatic event results in the memory being stamped into the brain. In the end, skeptics won the day. In case after case, repressed memories were actually caused by "suggestive questioning, guided visualization, age regression, hypnosis, body-memory interpretations, dream analysis, art therapy, rage and grief work, and group therapy" (Loftus & Ketcham, 1994, p. 141).

After studies on false memories came to public light, one project dubbed "Eyewitness Misidentification" by the Innocence Project (Innocence Project, 2009; Laney & Loftus, 2018) gathered information on more than three hundred people who were convicted of crimes they didn't commit, crimes often of a devious sexual nature. Out of the three hundred–plus innocent people, at least three quarters (75 percent) of them were convicted based on false memories! And these are only the few U.S. cases in which DNA exonerated the innocent. Another project conducted by professor of law Brandon Garret showed that out of 250 people who were exonerated by DNA evidence, seventeen faced a death sentence, about a third were sentenced to life, and two-thirds of the convictions were based on the testimony of eyewitnesses who initially had trouble identifying the culprit (Alcock, 2018, p. 93).[9] Given that every year in the United States more than 75,000 criminal trials are decided on the basis of eyewitness testimony (Ross, Ceci, Dunning, & Toglia, 1994), the need for the law to catch up with the science has never been more apparent.

MEMORY AND THE LAW

False memories have serious consequences within the justice system. False incriminating evidence has induced false confessions (Kassin & Kiechel, 1996)[10] that, based entirely on false memories, have also led to numerous bogus convictions (Schacter, 2002, pp. 119–23). Additionally, research supports the notion that, intelligence notwithstanding, people—suggestible people, in particular (Alcock, 2018, p. 228; Gudjonsson, 1991)—are often coerced into false confessions (Alcock, 2018, p. 228; Gudjonsson & MacKeith, 1990; Ofshe, 1989). In one study, Shaw succeeded in planting false memories in her participants' minds, 70 percent of whom later became convinced that they had committed crimes that never occurred (Shaw & Potter, 2015). Moreover, Shaw's participants conveyed specific details about their supposed "crimes," oftentimes expressing guilt about having committed them.

Furthermore, memories are susceptible to what's known as the *misinformation effect*—the contamination of memories due to information obtained after the fact (Alcock, 2018, p. 84; Loftus, 2005; Loftus & Hoffman, 1989). In the first study on the misinformation effect (Loftus, Miller, & Burns, 1978), Loftus and others showed participants slides of a car stopping at a stop sign. Participants were then asked to answer a questionnaire about the slides they just saw. Some of the questions contained misinformation (the car stopping at a yield sign), while others contained accurate information (the car stopping at a stop sign). On a recall test, participants who were given the misinformation were significantly more likely to report having seen the car stop at a yield sign than those who were given the correct information.

Given the prevalence of misinformation and the ease by which it spreads (Helfand, 2017; Innes, 2020), the misinformation effect has serious implications for both jurors and eyewitnesses. A common and deceptive tactic used by lawyers is exposing juries to evidence the judge will surely strike from the record. Everyone suspects that juries don't just forget evidence because it was ruled out on a technicality, and research confirms these suspicions. In cases where prosecutors used this tactic, jurors exposed to damning evidence that was later stricken were significantly more likely to convict than jurors who weren't (Hastie, Schkade, & Payne, 1999; Hawkins & Hastie, 1990).

Putting the misinformation effect into context, let's say a witness is questioned after a crime by a detective using leading questions and insinuating details about what happened. Later, while testifying on the stand, the witness "recalls" those insinuated details, forgetting that they were only suggestions or possibilities offered to help "jar" the witness's memory (recall how Schacter's participants "remembered" false details from the picnic and Loftus's participants "remembered" the car stopping at a yield sign). Naturally, the prosecutor persists with this line of questioning, pressuring the witness to provide convincing details. We're left with a theme that may be completely false, stated under oath by an eyewitness who believes she's honestly and accurately recalling important information. Believing a conspiracy was hatched between the prosecutor and the witness may seem perfectly plausible, but the flaws of memory show that a conspiracy isn't necessary.

Remember Forgetting

While memory improves substantially from the time we're infants, it doesn't improve as much as we believe. Everyone has memories they've forgotten, and when we've forgotten something, our brains are wont to confabulate. Accordingly, unjustifiably defending a false memory is fairly common. In 1999 (Lie & Newcombe, 1999), researchers showed eleven-year-olds pictures of three- and four-year-old children, some of whom they went to school with seven years earlier. Few of the eleven-year-olds could identify which pictures were of their former classmates. Considering that by age nine human brains are at about 95 percent volume and reach 100 percent by age thirteen (Shaw, 2017, pp. 15–16),[11] if eleven-year-olds have this much trouble remembering something from just a few years prior, should we really expect to remember anything from twenty, thirty, or even fifty years ago?

> Any event, no matter how important, emotional, or traumatic it may seem, can be forgotten, misremembered, or even be entirely fictitious.
> —Julia Shaw

Memory is also riddled with a phenomenon related to misattribution: *source amnesia*—our tendency to forget when, where, or how information was acquired, while retaining details about the information itself (causing us to misinterpret confidence for accuracy). While we've probably all heard something like, "I can't remember where I heard 'X' but I know 'X' is true," skeptics would say, "If I can't remember the source, how can I know whether that source, or the information it offered, is reliable?"

Truth amnesia—the inability to remember whether a claim is true or false—is another example of forgetfulness. A 2005 study (Skurnik, Yoon, Park, & Schwartz, 2005) showed that over 25 percent of young adults misremembered a false statement as being true only three days after they heard it, and 40 percent of older adults misremembered a false statement as being true. Participants usually remembered having heard the statement yet couldn't recall that it was false. Truth amnesia is particularly dangerous when it comes to, say, our children's health and safety. For example, people might remember hearing something about vaccines and autism yet forget important details such as there's no causal link between them. As a result, parents may skip important vaccinations, harming the community and possibly their own children.

A related phenomenon is the *illusory-knowledge effect*, which occurs when we mistake information learned after the fact for information we "knew all along" (Begg, Robertson, Gruppuso, Anas, & Needham, 1996). For instance, if a juror recalls information about the case from media reports before trial, the juror runs the risk of convicting the defendant based on that recalled information—which may be false—rather than the facts presented during trial.

Our Ever-Changing Memories

Memories clearly change over time. Since we've all experienced vague or lost memories, and we've probably all disagreed with people who remember an event differently than we do, the changing nature of memory shouldn't surprise anyone. When it comes to memories, we often defend our own version of events against all objections, not realizing that memories don't just blur and disappear; they can become completely new memories that seem identical to the original one. The content of our memories even morphs into new content, conflating with other memories (or beliefs or other people's stories or movies or dreams, etc.) without our knowing or being able to tell the difference. Actually, pseudo-memories might even seem *more* real than accurate memories!

As we saw earlier, when our brains create memories, they're attached to emotions and themes associated with the event from which they were created. Once a memory is stamped, whenever it's accessed, our brains re-create it (Shaw, 2017, pp. 64–65), updating it (Budson & Price, 2005; Burton, 2008, p. 82) with what we might call the latest software—our latest biases, memories, information, beliefs, perspectives, opinions, and so on—so as not to upset our existing thematic narratives (Ross, 1989; Schacter, 2002, p. 139). Returning to our Street World metaphor, accessing a memory isn't like selecting a route and driving to a destination; it's more like visualizing the route and literally constructing the entire route from the bottom up. Whenever we rebuild a road (access a memory), it (the memory) changes in small ways, from which type of pavement is used (e.g., who was there) to how big it is or where it's built (e.g., where we were when it happened). Each change depends on what new technologies or materials have become available since we last built it, much the same as how our brains incorporate details into our memories that we're exposed to after the original memory is formed; the past becomes contaminated with all that post-event information.

Strange Memories

False memories have led honest people to sincerely believe that their own family sexually abused them when they were children (Laney & Loftus, 2018, pp. 235–36; Loftus, 1995; Loftus & Ketcham, 1994). Accordingly, the astronomer Carl Sagan wondered: "If some people can with great passion and conviction be led to falsely remember being abused by their own parents, might not others, with comparable passion and conviction, be led to falsely remember being abused by aliens?" (Sagan, 1996, p. 158). In 2002, Harvard professor of psychology Richard McNally et al. sought to answer Sagan's question. When his team asked whether people who reported being abducted by aliens were prone to developing false memories, they found that "adults who reported either repressed or recovered memories of alien abduction exhibited heightened propensity to false memories in the laboratory" (Clancy, McNally, Schacter, Lenzenweger, & Pitman, 2002). It makes sense, then, that alien "abductees" will point to the vividness and detail of "memories" of alien abduction as compelling evidence that the experience happened (McNally, 2012).

McNally asserts that factors that contribute to such false memories include: "New Age" beliefs, familiarity with cultural narratives of alien abduction, elevated fantasy proneness (absorption), sleep paralysis and hypnopompic (the state of consciousness between sleeping and waking up) hallucinations, and quasi-hypnotic memory recovery sessions. His theory suggests that memory formation has a cultural component where memories are influenced by expectations that are based on prior beliefs. As McNally and his team explain, "Individuals who are more prone to develop false memories in the laboratory are more likely to develop false memories of experiences that were only suggested or imagined" (Clancy, McNally, Schacter, Lenzenweger, & Pitman, 2002, p. 459). McNally's theory accords with much of Loftus's research on memory recovery therapy, research on hypnotic susceptibility (Shermer, 1997, pp. 108–13; Wegner, 2002, pp. 283–85; Weitzenhoffer & Hilgard, 1959; Wilson & Barber, 1981), and other research showing that people claiming they were abducted by aliens have fantasy-prone personalities (Basterfield & Bartholomew, 1988) and are especially susceptible to false memories (Clark & Loftus, 1996; Orne, Witehouse, Orne, & Dinges, 1996; Wilson & French, 2006).

The hypothesis that memories are influenced by our expectations (suggestibility) was tested in an ingenious experiment (Brewer & Treyens, 1981) where participants were taken to a grad student's office and instructed to wait while researchers checked to see if a previous participant was done. Seconds later, the researcher would take the participant to another room where they were unexpectedly asked to make a list of everything they could remember seeing in the office they just left. The office was typical in that it contained a desk, shelves, chairs, and so on. However, the office didn't have a filing cabinet or any books, items one would expect in a grad student's office. Almost every participant recalled the chairs, desk, and shelves. However, 30 percent of participants misremembered books, and 10 percent even misremembered a filing cabinet. Just seconds after leaving the office, the participants developed actual false visual memories based on what they expected to remember rather than what was really there.

So memories are not literal recordings of reality that can be replayed like a movie. Memories are edited versions of reality that our brains re-create whenever the memory is recalled. Based on our brain's expectations—these expectations themselves based on all the information the brain has logged since the memory was formed—it reconstructs a version of reality that aligns with its own narrative rather than with external reality.

For decades, researchers have recognized that humans are highly suggestible and susceptible to false memories. Police lead witnesses to misidentify suspects;[12] therapists lead patients to believe dreams are repressed memories (Mazzoni & Loftus, 1998); interviewers lead children to wrongly believe they've been physically or sexually abused (Loftus & Ketcham, 1994). Suggestibility is why anyone soliciting information cautiously pursues their task. Everyone is susceptible to suggestibility. "Who you are does not matter on this point," explains skeptic Guy P. Harrison. "Even master skeptics and committed critical thinkers face great dangers when the stakes are high" (Harrison, 2015, p. 153). Thus, memories do not deserve the uncritical trust we typically allow.

Two Distortions Do Not Equal Clarity

Later on we explore the relationship between beliefs and memory. In fact, memory is responsible for any long-term belief, as believing something we couldn't remember would be difficult at best. Beliefs and memory both are imperfect, and sometimes combining the two results in completely distorted beliefs and entirely fabricated memories. Recalling a memory generates a belief about that thing, and due to memory distortion, we can be certain that any beliefs associated with memories are likewise error-ridden. This effect can be particularly troubling when it comes to *autobiographical memories*—memories associated with self-identity and self-esteem. If we see ourselves as honest, for example, yet we do something that was less than honest, over time, the memory of the event becomes consistent with the narrative that we are honest (Alcock, 2018, p. 94). Without any conscious effort, our brains construct consistent narratives without regard for truth, explaining one way people come to beliefs that are almost certainly false.

Imagine you have a clear memory of a childhood event that's backed up with photos, only to have someone tell you it never happened! Well, when beliefs and memory conspire with what appears as good evidence, we're especially vulnerable to generating false memories and erroneous beliefs. Research shows that memories can be manipulated and outright generated simply by being shown doctored photos. In one project (Shaw, 2017, pp. 21–22; Strange, Sutherland, & Garry, 2006), six- and ten-year-olds were shown two photos of real events and two doctored photos of events that never happened. During follow-up interviews, a large percentage of the children (31 percent of six-year-olds, 10 percent of ten-year-olds) not only recalled the false events as real, some even provided elaborate details.

Wrapping Up

We can never exercise too much caution when critically analyzing our memories. Memories are created by our brains and combine with all the other information it maintains, including our beliefs, perceptions, emotions, internal narratives, and constructed realities. Memories fade away, change over time, and are fabricated by exploitation or accident. Our confidence in, the details of, and our emotional attachment to memories are poor indicators of accuracy. As Michael Shermer understands, we rehearse "the memory and in the process change it a bit, depending on emotions, previous memories, subsequent events and memories, and so on. This process recurs thousands of times over the years, to the point where we must ask whether we have memories or just memories of memories of memories" (Shermer, 1997, p. 270). Although memory is a powerful tool that helps us navigate our lives, we'd be wise to Think Straight before over-relying on this subjective and flawed process.

CHAPTER 7

Self-Deception, Part 2—Belief

We now turn our attention to one of the most important yet misunderstood components of our cognitive makeup: belief. A key attribute of CT is understanding the limits, flaws, features, and influence of this fundamental aspect of our mental lives. We are at the mercy of our beliefs, and we form new ones all the time. We also become emotionally wedded to some of our beliefs, making it difficult for us to change or abandon them.

No belief should be off limits to skepticism, regardless of how deeply ingrained or sincerely held. Keeping this in mind, I'll explore the naïveté in assuming our beliefs are reliable without having any external, corroborating evidence. I will also endeavor to evaluate and question the reliability of our beliefs. Living rationally, valuing evidence, and being sensitive to the truth all require an understanding of some of our most basic cognitive functions.

Building Vocabulary

Cognitive dissonance:	Fundamental attribution error:
Mental discomfort or tension resulting from conflicting cognitions (e.g., ideas, beliefs, opinions) or when one's beliefs differ from one's behavior.	Emphasizing disposition to explain the behavior of others, while emphasizing the situation to explain our own behavior.

Essentialism:	Dogmatism:
The belief that inanimate objects have "essential" properties that make them what they are or give them a certain character.	Absolute certainty that one's belief or opinion is correct; passionate closed-mindedness.

Beliefs

We enter the world without any beliefs. Over time, we accumulate beliefs that range from what we think about toasters to ideological positions on gun control. To a large degree, our beliefs define our lives by informing our actions and guiding our decisions. Many of our choices are based on beliefs about the likelihood of events, such as whether a suspect is guilty, whether our surgeon knows what he's doing, or whether our car will start in the morning. Accordingly, if we care about making the right choices, basing our beliefs on reality, or understanding how probability influences our beliefs, then the accuracy of our beliefs makes a huge difference. Some of our beliefs lead us to success, and some lead us to failure; some are right, and some are wrong. Our question is: when is a belief worth believing, and when should we abandon a worthless belief? That is, how do we assess the quality of our beliefs?

> Beliefs are only as reliable as the evidence and information upon which they are based.
> —James E. Alcock

The word "belief" means many things, and many things make up our beliefs. For instance, we might believe so-and-so got what she deserved or that our "soulmate" awaits us on Match.com. Such beliefs are matters of taste and speculation. When we profess a belief, we're making a statement about something we believe to be the case. Belief statements vary depending on their context. Saying "I believe in evolution" is not the same thing as saying "I believe in restorative justice." Evolution is either true or false, but whether restorative justice is the best approach to dealing with crime is subject to the opinions and goals of those making the argument. Thus, belief is multilayered, and understanding the context in which a belief is being expressed is key.

The novelist Ursula K. Le Guin observed that "belief is the wound that knowledge heals" (Le Guin, 2000). Typically, belief is our veridical, System 1, default mode. Nobody asks, "How strong is the evidence for evolution?" or "How much evidence shows that God exists?" or "What evidence do you have for alien visitation?" Instead, the questions are asked as "Do you *believe* in evolution?" or "Do you *believe* in God?" or "Do you *believe* aliens have visited Earth?" demonstrating that we're much quicker to believe than to weigh evidence.

Those who already believe in ghosts, Bigfoot, and aliens, for example, are the same ones who report seeing ghosts, Bigfoot, and aliens. If such unlikely beings exist, how come skeptics never see them? How come those who already believe in them are the only ones who ever encounter them? Why don't demons ever possess people who don't believe in demonic possession? Why don't ghosts or aliens reveal themselves at skeptic conventions? By contrast, if unlikely things don't exist, why do people everywhere believe they've seen them? Can we explain away millions of ghost sightings or alien abductions? Is it fair to say that millions of people have just been mistaken? The countless counterfeit fallacy says yes, but let's delve a little deeper into the mechanisms of belief.

Reasonable Belief

The prolific author Isaac Asimov was once pressed about his skeptical beliefs. Specifically, he was asked if he believed in the Bermuda triangle, life after death, telepathy, ancient astronauts, flying saucers, and other incredible claims. Asimov, ever the skeptic, answered no on all counts. "But don't you believe in *anything*?" insisted his interviewer. His response epitomized what it means to maintain reasonable belief. "Yes," he said. "I believe in evidence. I believe in observation, measurement, and reasoning, confirmed by independent observers. I'll believe anything, no matter how wild and ridiculous, if there is evidence for it. The wilder and more ridiculous something is, however, the firmer and more solid the evidence will have to be" (Asimov, 1997, p. 349).

To believe something, two conditions must be met: One, we must have knowledge of the belief, since we cannot believe something we aren't aware of, and two, we must believe it, since it makes no sense to say we believe something we don't believe (Gilbert, 1991). Now, we believe the things we believe because we think our beliefs are correct, yet insisting that our beliefs are all impeccable would be pretty arrogant. Addressing this very problem, British philosopher Julian Baggini advises that "we should believe what it is most rational to believe because to reject the idea that we ought to have reasons for what we believe would be to say that we can believe what we like without reason. That takes away any requirement to justify our own beliefs or criticize those of others" (Baggini, 2017, p. 214). Baggini's argument gets to the bottom of exactly what being reasonable means, explaining that beliefs that aren't based on reason are essentially built on sand and that grounding our beliefs on reason provides the foundation needed to support our beliefs and evaluate others'. Accordingly, it's reasonable to believe what we have good reasons to believe, and it's unreasonable to believe what we have good reasons to doubt. If we want to believe what's most likely true, we must base our beliefs on the best evidence available. Likewise, basing beliefs on inferior evidence, or, worse, no evidence at all, is unreasonable and absurd. To this end, philosopher Lewis Vaughn laid out the following seven principles (Vaughn, 2019, pp. 112–21) for evaluating information:

1. If a claim conflicts with other claims we have good reason to accept, we have good grounds for doubting it.

2. If a claim conflicts with our background information, we have good reason to doubt it.
3. We should proportion our beliefs to the evidence.
4. It is not reasonable to believe a claim when there is no good reason for doing so.
5. If a claim conflicts with expert opinion, we have good reason to doubt it.
6. When the experts disagree about a claim, we have good reason to doubt it.
7. It's reasonable to accept the evidence provided by personal experience only if there's no good reason to doubt it.

Following Vaughn's principles ensures that our beliefs are reasonable and maximizes the chances that our beliefs are true. Even so, many of us maintain our beliefs simply because we want to, although generally people widely reject this idea. Instead, people insist that their beliefs are held for rational reasons. In reality, we rarely arrive at our beliefs for rational reasons, which is one reason we're so reluctant to give one up after being presented with a rational reason to do so. However, hesitating to abandon our beliefs isn't always a bad thing. Our coffee-drinking habits don't change with every new study on the effects of caffeine, and we wouldn't stop believing light allows sight because someone on the bus said so.

Nonetheless, since many beliefs are resistant to logic, evidence, and proof (Alcock, 2018, p. 454), maintaining accurate beliefs isn't simply a matter of being exposed to accurate information. Even the highly intelligent and well-informed believe outright falsehoods. Furthermore, culture has a lot to do with one's beliefs. For example, one hundred years ago people believed that evil spirits visited them in the night causing panic and terror. Nowadays, people attribute the same experiences to alien abduction. Psychologists attribute these experiences to hypnopompic and hypnogogic hallucinations, rather than succubae or aliens. But, although specific beliefs are difficult to accurately predict, the psychological tendency to believe incredible claims can be predicted by measuring other character traits, such as whether someone is more of an intuitive or analytical thinker (Shermer, 2013, pp. 111–18). Additionally, while research currently suggests that false beliefs don't necessarily imply low intelligence, the question is far from settled (Wiseman & Watt, 2006). Smarter people are probably better at rationalizing false beliefs, which is a bit of a double-edged sword (Shermer, 2011a, pp. 35–36). According to Guy P. Harrison, "Those who believe they could never fall for a silly belief already have" (Harrison, 2017, p. 13). What's more, our beliefs influence how we think about the world, and how we think about the world influences our beliefs. Intuitive thinkers are more likely to interpret strange events as paranormal, while more analytical thinkers tend to be satisfied with worldly explanations (Bouvet & Bonnefon, 2015); a concerning prospect given that paranormal beliefs not only underlie delusional thinking (Irwin, Dagnall, & Drinkwater, 2012), they're also correlated with failing to understand how the world really works (Svedholm, Lindeman, & Lipsanen, 2010). Thus, thinking analytically gives us a layer of protection against falling for false beliefs (Alcock, 2018, p. 454; Pennycook, Cheyne, Seli, Koehler, & Fugelsang, 2012).

Critically thinking about our beliefs demands cognitive responsibility. As humans, we often mistake the fact of having a belief for believing that what we have is a fact—a clear error in cognition, since people obviously maintain inaccurate beliefs while fully believing those same beliefs are true. Let's say you believe using cocaine is harmless, so you decide to try it. You discover that all the negative things you've heard about cocaine were wrong: you don't feel any cravings, you didn't miss any work or end up in jail, and your health didn't deteriorate. You may be inclined to continue believing that using cocaine is harmless, and so continue using it until you're unemployed and addicted and fifteen of your teeth are missing. Clearly our beliefs can be wrong, so maintaining accurate beliefs is important for maintaining freedom, health, and well-being.

> Men are apt to mistake the strength of their feeling for the strength of their argument.
> —William E. Gladstone

Origins of Beliefs

Most of us have a sense of what beliefs are, though when pressed, pinning down the word "belief" can be challenging. We're also rather possessive with our beliefs, a sentiment expressed in the language we use to talk about our beliefs (Gilovich, 1991, pp. 85–87). An idea "catches on," or we "hold" or "cling to" a belief. We "abandon" a conviction, we "change" our opinion, we "chew on" an idea, or we become "entrenched" in a worldview. We use phrases such as "I don't *buy* that" or "I *take* that back." This possessiveness means that when someone presents us with information that contradicts something we believe deeply, we get defensive because we feel as if that person is trying to take something that's ours.

Physiologically, beliefs are little more than the conscious experience we have when electro-chemical signals travel across neurons (Biswas-Diener, 2018; Taylor, 2017).[1] Consequently, we cannot open a brain, locate a belief, and physically remove it. In practice, beliefs motivate us to act and directly influence our behavior. Beliefs are the reason why we commit crimes (or don't), why we eat healthy foods (or junk), why we care for our children (or not), why we visit a physician (or a homeopath), and why we do just about anything. Additionally, there's a widespread assumption that we choose what we believe. However, reality is a bit more complicated. If you think you choose your beliefs, try believing that the sun is cold, that you walk on your back, or that you don't have a brain.

Modern psychology tells us that we usually arrive at our beliefs for emotional reasons and only later rationalize them. Psychologist Daniel Kahneman states, "If we think that we have reasons for what we believe, that is a mistake. Our beliefs and our wishes and our hopes are not always anchored in reasons" (Macdonald, 2014). Michael Shermer agrees, explaining that "the human brain is a 'belief engine.' We form our beliefs for a variety of subjective, personal, emotional, and psychological reasons in the context of environments created by family, friends, colleagues, culture, and society at large; after forming our beliefs we then defend, justify, and rationalize them with a host of intellectual reasons, cogent arguments, and rational explanations" (Shermer, 2011a, p. 5; Shermer, 2013, pp. 2–3). Shermer's central argument says that "beliefs come first; explanations for beliefs follow," which is the opposite of developing a belief from the available evidence, then modifying the belief according to the evidence. Because our first response to our environment is largely emotional, becoming slaves to our emotions is easy and leads us to defend our beliefs even without a reason.

Beliefs come in an almost endless array, including religious, political, and moral beliefs and beliefs about science, health, family values, and numerous other topics. Since the term "belief" has so many layers and possible interpretations, I need to level the playing field. As such, for our purposes, I define a belief as *a sure conviction that someone or something exists or something is true or false.* Nothing about a belief confirms (or even suggests) that it's true (or false); as far as beliefs go, we have only feelings or opinions. Whether a belief *is* true is another matter entirely.

Most of the time beliefs are arrived at through exposure to events or information, not through our own ponderings. Teachers, parents, friends, newspapers, TV, books, magazines, documentaries, the internet, and other sources of information help shape our beliefs. Thus, those who believe in grand conspiracies come to those beliefs the same way skeptics come to reject them: through word of mouth, not firsthand observation (Shtulman, 2013). Accordingly, understanding how to evaluate sources and separating expert consensus from non-expert opinion is perhaps the most important skill a skeptic can possess.

Evaluating the Source

Throughout the coronavirus pandemic, many people believed a wide variety of incredible claims that contradicted what I believed. I thought that masks reduced the spread of the virus, that hydroxychloroquine and ivermectin were ineffective for treating COVID-19, that mRNA vaccines are safe and effective, and that the death toll was actually higher than was being reported. Others thought just the opposite: that masks didn't make a difference, that hydroxychloroquine and ivermectin worked to treat COVID-19, that mRNA vaccines are dangerous, and that the death toll was much lower than was being reported. On each of these issues, since the issues involve matters of fact rather than matters of judgment, either I was right or I was wrong. The question for this section, then, is: how do we know who's right?

Like most human endeavors, evaluating sources can have more to do with prior beliefs than discovering the truth, and amending beliefs isn't something everyone does eagerly (Lobato & Zimmerman, 2018, pp. 28–29). In court, lawyers tout "experts" to support their case. The prosecution presents "experts," only to have the defense present "experts" that say just the opposite. Jurors are left with the impression that experts disagree, judging the accused based on who was more persuasive rather than on the facts. Sometimes expert opinion is indeed split. However, there's usually a consensus among experts within their specialized fields. So what's an expert, and how do we know?

Experts are individuals with specialized training and knowledge in a particular, usually narrow field. We trust experts specifically because their training and knowledge means they're more likely to be right than we are. Their training provides access to more and better information than we have, along with the ability to better evaluate that information. The number of people who understand quantum mechanics is tiny. Yet the number of people who doubt the existence of electrons is also tiny because most people believe the experts when they say electrons exist. Expert opinion is generally accepted *unless that opinion conflicts with one's beliefs or interests*. For example, experts agree that human activity is causing global warming, vaccines are safe and effective, evolution happened, and genetically engineered crops are just as safe as any other foodstuff (chapter 13). Unfortunately, special interests and humans' misunderstanding of cause and effect lead otherwise rational, intelligent people to reject these very facts. Additionally, the internet has convinced many that Googling something and reading a few articles gives them a license to weigh in against experts. However, most people lack the necessary training to conduct adequate research or evaluate the information they find (Helfand, 2017). For example, although most Google-goers can competently locate information on any given topic, only a fraction can tell the difference between a pilot study and a double-blind, randomized, placebo-controlled trial, and even fewer are aware that there *is* a difference. Moreover, evaluating the quality, methodology, and confounding variables of a study is not something non-experts do competently, if at all. Worse still, the *Dunning–Kruger effect* (chapter 8) convinces many that they're learning all about a subject when all they've really done is ingest false information (Vyse, 2018b).

Before forming a belief, we can take steps to determine the validity of a source, such as seeing whether experts agree, looking for disconfirming information, and asking critical questions. Is the source reliable? Do they have relevant credentials? Are they qualified in this particular field? Do unbiased experts agree? Whom we trust, whom we choose to believe, and what sources we find credible speak toward the accuracy of our beliefs. Just as information from the *National Inquirer* is predictably less reliable than information from Reuters (Vyse, 2019c), information from the History Channel's *Ancient Aliens* has a lesser chance of being accurate than information from PBS's series *NOVA*. Importantly, experts aren't always right, but that doesn't mean they're always wrong, that we shouldn't listen to them, or that our opinion carries more weight than theirs.

The Expert Non-expert

In recognition of the vast amount of available information, physicist and skeptic Victor Stenger observed that "much of that information is untrustworthy, and it takes a trained thinker to filter out the good from the bad. Magical thinking," Stenger continues, "and blind faith are the worst mental systems we can apply under these circumstances. They allow the most outrageous lies to be accepted as facts" (Stenger, 2012, p. 322). Indeed, while the internet and media are treasure troves of accurate information, it's no secret that both abound with completely fraudulent claims, sometimes from seemingly reliable sources. For example, should Deepak Chopra, an author and endocrinologist, be explaining quantum mechanics? The answer is no, for the same reason you wouldn't get your car fixed by a dentist or visit a psychologist for a broken leg. Chopra has no education in physics, chemistry, quantum mechanics, or any other relevant field. Experts in each of those fields (as well as others) totally disagree with Chopra's incredible claims about "quantum consciousness"; the information he relies on is outdated, misunderstood, or irrelevant; and he doesn't have any accomplishments in any field related to quantum mechanics. Therefore, trusting Chopra over experts is highly unreasonable. Although this example puts Chopra in the spotlight, it isn't limited to authors or quantum mechanics and should be generalized to all claims made by all people. On this very point, Bertrand Russell argued that:

> (1) when the experts are agreed, the opposite opinion cannot be held to be certain; (2) when they are not agreed, no opinion can be regarded as certain by a non-expert; and (3) when they all hold that no sufficient grounds for a positive opinion exists, the ordinary [individual] would do well to suspend [. . .] judgment. (Schick & Vaughn, 2005, p. 124)

Let's examine Russell's argument more closely. Russell's first point is that if experts claim to know that something is the case, laypeople cannot claim to *know* the experts are wrong. For example, if you went to one hundred mechanics and all one hundred of them told you your alignment was off, you couldn't reasonably claim to *know* the problem was with your suspension. Russell's second point is similar: if experts disagree, non-experts cannot reasonably claim to *know* how to solve the disagreement. That is, if those with the best background knowledge and training haven't found the answer, it's extremely unlikely that you have. Lastly, if experts are unsure about something, non-experts cannot reasonably claim to be certain about that thing. If no dermatologist on the planet could identify the cause of your rash, a chiropractor cannot claim it's caused by spinal misalignment.

So how do we know whether someone is an expert? If an electrician is giving psychological advice or a carpenter is explaining the formation of stars or a physicist is weighing in on how the heart circulates blood, doubting what they say is reasonable. Not because they cannot be right; they certainly can. However, we cannot *know* they're right because they're speaking on matters outside their field of expertise. To be considered an expert, one should have:

- education and training from a relevant and up-to-date program or school
- experience making reliable judgments in their field of expertise
- specialized professional accomplishments
- current expertise (outdated expertise may be irrelevant)
- respect among their peers and other experts in their field (Smith, 2018, p. 53; Vaughn, 2019, pp. 117–18)

Thus, we are justified in believing experts who speak on matters within their field of expertise, provided we don't have a good reason to doubt them. And we have good reasons to doubt experts if they are motivated by something other than the search for the truth. We should also doubt expert opinion if their findings contain factual errors, if their claims conflict with what we have good reasons to believe, if their claims aren't supported with current and reliable evidence, if their writings contain logical errors or contradict GASPs, or if most experts in their specialized field disagree (Vaughn, 2019, pp. 118–19). Furthermore, an expert in one field is not an expert in another. Moreover, just because one expert makes a claim doesn't mean the claim is settled. Outliers exist in every field, and it would be folly to assume that since an expert said X, X is therefore true. Wrong. Expert opinion is to be judged by the consensus of knowledge shared by the majority of experts. And even then, falling for nonsense still happens. Enter predatory publications.

Predatory Publications

Sometimes a source's (un)reliability is obvious, such as the *National Inquirer*. The sheer preposterousness of the claims touted by the *National Inquirer* is enough for most of us to discount them as cheap entertainment. But sometimes the reliability of a source is hidden in deceptive language, such as the *American Journal of Modern Physics*, the *International Archives of Medicine*, *Pure and Applied Mathematics Journal*, the *Cancer Research Journal*, the *American Journal of Immunology*, or *Progress in Physics* (Beall, 2019). A name like the *Cancer Research Journal* sounds legit but masks its pseudoscientific nature. Such journals, called *predatory journals* or *predatory publishers*, are publications that charge fees to publish papers but without the services—such as peer review and editorial critique—typically offered by legitimate journals. Predatory journals are financially interested in publishing any article its editors receive. As such, many operate under the pretext of a legitimate peer-reviewed journal; they "perform a fake peer review, . . . fail to perform any peer review but claim to, or [go] through the motions of peer review, but ignore 'reject' recommendations from peer reviewers" (Beall, 2019, p. 285).

Source credibility is further complicated when well-known and even respected sources publish unscientific and even anti-scientific literature (Benson, 2019) or when the World Health Organization pushes ancient and anti-scientific "remedies" as legitimate medicine (Renckens & Dorlo, 2019).

Evaluating Beliefs

Many people believe that if something has been around for a long time, there must be something to it. For example, some believe that thousands of years of traditional Chinese medicine (TCM) history means it must work. Such faulty reasoning is a logical fallacy called the *appeal to tradition*, since believing that because something has been around for a long time it must work makes no sense. Murder and genocide are extremely old, but few people believe either is useful. Believing ideas based on how long they've been around makes

> The most common of all follies is to believe passionately in the palpably not true. It is the chief occupation of mankind.
> —H. L. Mencken

as much sense as believing that Zeus causes lightning or that earthquakes result from Poseidon striking the ground with his trident. Thus, it isn't just our beliefs that must be evaluated, but also the reasons we believe them.

Oftentimes we naïvely assume that we arrive at our beliefs through reliable processes. Sometimes we do, so we don't readily abandon them. However, much like memories (chapter 6), beliefs can be distorted and grounded on fictions, such as the belief that "I can handle my liquor." In fact, beliefs are tied to memories, which raises the following dilemma: we're equally as convinced by a distorted belief as we are by clear (but false) memories. Memories build confidence in our beliefs, which are then reinforced by association with emotions. Psychologist Daniel Kahneman argues that confidence in beliefs is related to *cognitive ease* and *cognitive coherence* (Kahneman, 2013, pp. 87–88). That is, we hold our beliefs confidently "when the story we tell ourselves comes easily to mind, with no contradiction and no competing scenario" (Kahneman, 2013, pp. 239–40). Evidence counts for very little or, at least, not as much as our subjective confidence does, and we often fail to consider that we might lack the evidence needed to make accurate judgments.

Since our beliefs are closely tied to our emotions, how we formulate our beliefs has a lot to do with our emotional makeup, and our brains tend to follow a path of least resistance. It's emotionally easier to say, "It's not my fault," than admitting we've done something immoral or shameful. Unfortunately, the road to belief isn't paved on reason and evidence; it's paved with cognitive biases, thinking errors, logical fallacies, and strong emotions, which are extremely unreliable routes to accurate beliefs. Emotions help us make quick decisions, and this is an adaptive, evolved strategy. Instead of wasting time contemplating whether a snake is dangerous, we experience fear. Instead of investigating the pros and cons of rotten meat, we experience disgust. Instead of deciding whether our mates should fraternize with potential rivals, we experience jealousy (Shermer, 2008a, pp. 158–59). Thus, "emotions are proxies for getting us to act in ways that lead to an increase in reproductive success" (Shermer, 2008a, p. 158). Such *heuristics* (chapter 5) prevent us from wasting time calculating the risks of, say, confronting a predator versus fleeing, eating something that might make us sick, or spending precious resources rearing a child that isn't ours. Throughout evolutionary history, experiencing an emotion and reacting swiftly led to more descendants than receiving the message after it was too late.

Evolutionary psychologists think emotions evolved to override our thought processes in emergencies by motivating us to behave in particular ways when survival or death hangs in the balance. This sort of motivated reasoning is described as *hot cognition*, where our thinking is influenced by our emotional state (Hess & Pickett, 2018). Feeling good puts us in a state of cognitive ease and lets our guards down. In this state, we're more prone to like what we see, believe what we hear, and trust our intuitions (Kahneman, 2013, p. 60). At times of *cognitive strain*, we're more suspicious, less comfortable, or more focused on our surroundings. Thus, we're more likely to accept something veridically while sitting around a campfire and more apt to see an enemy in the forest while out on a hunt. Our ability to score accurately on a test is also influenced by our emotional state. Researchers have shown that putting people in a good mood by making them think happy thoughts before taking a test more than doubles their accuracy, whereas people placed in unhappy moods were less capable of producing accurate answers (Kahneman, 2013, pp. 68–69). Much about how we live is based on subtle cues from our environment and how we react to those cues emotionally.

The Highways and Byways of Belief

We maintain beliefs and reluctantly replace them with new ones because of what psychologists call "belief stability," or the stability of beliefs over time. In his book *Belief*, psychologist James E. Alcock outlines the following eight factors that keep beliefs stable: importance, primitiveness, value, interconnectedness, self-esteem, public pronouncement, general acceptance, and group influence (Alcock, 2018, pp. 176–77). When outlined this way, we easily see why beliefs are resistant to change. Conspiratorial beliefs could fall into each category, so let's use it as an example. Oftentimes, conspiratorial beliefs are personally significant (important), fundamental and foundational (primitive), morally cherished (value-laden), tied to many other beliefs (interconnected, monological), and part of the personal identities one shares with the world (self-esteem, public pronouncement, general acceptance, and group influence). Any one of Alcock's eight factors explains why belief change is uncommon. Accordingly, the more factors that apply to a belief, the more resistant to change we should expect that belief to be. Conspiratorial beliefs typically incorporate several of Alcock's eight factors, so we should expect such beliefs to change less frequently than weakly

Alcock's Model of Belief Stability:

- Importance
- Primitiveness
- Value-laden
- Interconnectedness

- Self-esteem
- Public pronouncement
- General acceptance
- Group influence

Alcock's model of belief stability provides a framework that explains why some people are close-minded about certain beliefs. The more factors a belief is associated with, the less susceptible the belief is to change.

held beliefs, which is precisely why conspiracy theorists typically remain conspiracy theorists, despite the poverty of their beliefs.

Interconnectedness is particularly important when it comes to maintaining strongly held beliefs. To demonstrate how important interconnectedness is, let's take a trip to Street World and head for Belief City. Changing a belief requires changing other beliefs connected to it. A new and conflicting belief increases cognitive dissonance (see below), making us resistant to the new belief. Let's say you reject the idea that the mind is simply the product of brain activity (known as monism), and instead you strongly believe in dualism, Descartes's idea that the mind exists independent of the body. If you're comfortable living on Dualism Blvd., moving to Monism Lane would seem less appealing, simply because if monism were true, other connected beliefs could be jeopardized, such as belief in an afterlife or belief in the soul, which in turn might call into question certain religious beliefs or beliefs about morality. Staying on Dualism Blvd. is cognitively easier than constructing detours or entirely new neighborhoods just to make sense of the new belief.

We can push our Street World metaphor even further. If we want to upend or replace existing beliefs, demolition and reconstruction is the name of the game. Given our propensity to resist change, this task isn't easy. We don't want our old beliefs demolished, just as we prefer not to use resources constructing new ones. Building new roads inevitably prevents access to old roads and removes others from the rest of Street World. If construction crews come along and replace a dirt road with a paved residential street, we're less likely to resist that change than if an entire interstate were demolished and replaced by a superhighway. The longer we've held a belief, the more roads that are connected to it, the more resources it takes to upend it, and the less interested we are in changing it.

For example, getting to Destiny Drive requires taking Karma Boulevard to Jinx Avenue, then from Jinx Avenue to Law of Attraction Lane, and from there it's a straight shot to Destiny Drive. If we demolished Jinx Avenue, access to Destiny Drive would be compromised. And, as we saw in chapter 1, demolishing a road doesn't mean travel becomes unnecessary. If we demolish the route to Destiny Drive, we'd have to replace Jinx Avenue with Random Road, replace Karma Boulevard with Coincidence Street, replace Law of Attraction Lane with Chance Avenue, and replace Destiny Drive with Life Just Happens Street until travel can continue. Needless to say, it's a lot easier to continue believing in karma than it is to question whether karma, jinxing, the law of attraction, and destiny are all illusions and then replace those beliefs with unsatisfying concepts such as randomness and chance. Since strong beliefs got that way by reinforcement or trauma (the more a road is used, the higher the demand for it to become a highway or interstate), it's much easier to continue believing than it is to tear down an entire network of beliefs *and* replace it with an unfamiliar belief structure.

The Roadways of Belief

When a deeply held belief is challenged by contradictory evidence, provided the evidence is reliable, we like to think we'd change the belief to conform to the evidence. However, research shows that contradictory information usually causes us to double-down on our beliefs rather than abandoning the one that's wrong. We believe we're moral, sensible people, despite the severity or frequency of our moral and intellectual transgressions. **Cognitive dissonance** plays a vital role in making it cognitively and emotionally easier to rationalize why we maintain a belief (adding a new road) rather than discarding it altogether (demolishing a highway). For example, a heavy smoker confronted by data linking smoking to cancer reduces dissonance by telling himself he really doesn't smoke that much or finding some way to discredit the evidence.

> When our beliefs are contradicted by evidence in the world, it is better to adjust our beliefs than to deny the evidence and cling to dysfunctional ideas.—Keith E. Stanovich

Cognitive dissonance has been explained using Aesop's "The Fox and the Grapes" fable. The fox spots some delicious-looking grapes just out of reach and begins salivating. Realizing he can't reach them, the fox adjusts his beliefs, convincing himself the grapes are sour so he no longer wants them. Another example is the dissonance we feel when we commit some immoral act. We like to tell ourselves we're good people, and when we do something not so good, we experience dissonance: a conflict between the belief that we're good and the fact that we did something bad.

Cognitive dissonance explains why committed anti-vaxxers rarely change their position (Nyhan & Reifler, 2015; Nyhan, Reifler, Richey, & Freed, 2014) or admit that the entire anti-vaccine movement is based on misinformation and logical fallacies (Howard & Reiss, 2019). When we work hard or suffer in pursuit of a goal, we try to convince ourselves that we value the goal, even if we really don't (Aronson & Mills, 1959). Psychologists call this the *justification of effort effect*, which I'll explain using a rather colorful example. Let's say you've spent years committed to a street gang. You got jumped in, you spent countless hours reinforcing the benefits of gang life, you are covered in gang tattoos, you dedicated your wardrobe to your gang's colors, and no doubt you believe that your gang is righteous above all others. Your image, reputation, and even your income are all tied to your gang, and your family and peers support your involvement. Now, after all this effort, evidence of how wrongheaded you've been will more likely convince you to double-down and remain in the gang rather than admit you've been wrong for years and have caused some harm in the process.

Cognitive dissonance motivates us to resolve conflicting beliefs, even if false beliefs are the result. To reduce cognitive dissonance, some reject the conflicting belief, some change the behavior, some deny the evidence, and some engage in special pleading (Wade & Tavris, 2017, p. 314). Another dissonance reducer is keeping conflicting beliefs separated, as if prohibiting access between cognitive roads. For example, in one study (Russell & Jones, 1980), researchers gave skeptics and paranormal believers articles to read that either supported the existence of ESP or claimed ESP didn't exist. Believers remembered the articles that supported their beliefs well but couldn't accurately remember the ones that didn't. Moreover, over 15 percent actually *mis*remembered an unfavorable article as favorable. By contrast, skeptics remembered both the articles that supported their beliefs *and* those that didn't. Thus, the authors concluded that believers are less willing than skeptics to accept new evidence that contradicts their beliefs and that skeptics are better at recalling the evidence for incredible claims they disagree with.

Beliefs based on first impressions are highly resistant to change, even when we're presented with evidence that completely discredits them (Alcock, 2018, pp. 181–84; Ross, Lepper, & Hubbard, 1975). Moreover, the longer we've maintained a belief, the more we've reinforced it and the more resistant it is to change. Fortunately, part of the brain keeps us in touch with reality, and when counterevidence becomes overwhelming, prior beliefs sometimes give way, allowing us to reevaluate the original belief and accept the new evidence (Redlawsk, Civettini, & Emmerson, 2010). Even deeply entrenched beliefs can be undermined if the evidence is presented clearly and concisely (Nyhan & Reifler, 2018).

We put a ton of effort into building our worldviews that make up Street World. It takes a ton of effort and resources to demolish a superhighway. Likewise, it takes a ton of effort to tear down beliefs we've spent years reinforcing. The more effort we spend

> It is better to see a few patterns that are not actually there than miss one that is.—Richard Wiseman

reinforcing a belief, the less willing we are to participate in its demolition. This dilemma is similar to the escalation of commitment fallacy (chapter 5). Rather than expending effort reconstructing our cognitive highways, we work conflicting information into our existing narratives or wall them off completely.

Making Connections . . . Real or Imagined

Research shows that people who believe incredible claims are more likely than skeptics to identify illusory patterns (Mohr, Landis, & Brugger, 2006; Reed, Wakefield, Harris, Parry, Cella, & Tsakanikos, 2008; Shermer, 2011a, p. 120). More surprising, however, is learning that seeking out and identifying patterns is human nature. Who believes incredible claims and why they believe them are complicated questions. Nonetheless, many factors contribute to incredible beliefs, including one's biology, environment, mental health, socio-cultural elements, and even expectancy biases.

Shermer argues that beliefs are evolutionarily adaptive and reflect our pattern-seeking tendencies (Shermer, 2003a, pp. 38–39; Shermer, 2011a, p. 59). We recognize a pattern (e.g., standing upwind of game animals is bad for the hunt) and believe the pattern reflects reality. Unfortunately, figuring out the difference between meaning*ful* and meaning*less* patterns isn't part of the deal. According to Shermer, meaningless patterns (e.g., painting animals on a cave wall before a hunt) are usually harmless and may even reduce anxiety. Thus, evolution hardwired the following two thinking errors into our brains: *Type 1 Errors—Believing Falsehoods*—and *Type 2 Errors—Rejecting Truths* (I'll return to Shermer's theory on patternicity in chapter 8). Believing a falsehood could be something like, "Big Pharma is conspiring with all the world's governments." Likewise, we might reject a truth such as "Conspiring with all the world's governments would be virtually impossible."

Evolution would only favor this arrangement if it provided some advantage. Supporting his theory, Shermer argues that *Type 1* and *Type 2* errors exist not only because they're usually harmless but because they're part of the hardware that lets us make *Type 1* and *Type 2 Hits: Not Believing a Falsehood* ("It's not likely that Big Pharma is conspiring with all the world's governments") and *Believing a Truth* ("Big Pharma is neither one entity, nor does it have the resources to get the world's governments to engage in their conspiracy"). Accordingly, *Type 1* and *Type 2* "*Hits*" serve valuable evolutionary functions. And, while we may make the occasional *Error*, if those *Errors* are harmful less often than the *Hits* are helpful, natural selection will favor such hardware as an evolutionarily advantageous tradeoff. Thus, we can begin to understand false beliefs as a consequence of false pattern recognition, which, in turn, helps us realize that people believe incredible claims because evolution favored believing credible ones.

THE ROOTS OF BELIEF

Our environment, culture, and personal experience contribute immensely to our beliefs. This too is adaptive. As children, believing adults is a successful survival strategy. For example, burning our hand on a hot stove after Mom's forewarnings leads us to believe what she says. When tribal elders told children not to wander off or they might be eaten by wolves, those children that wandered off anyway did not, on average, survive long enough to pass on descendants, so listening to authority was hardwired in. It follows that the more reasons we have to support our beliefs, the less likely we are to abandon them. If Mom says that vaccines are dangerous and we witness the neighbor get sick after being vaccinated, we're more likely to believe vaccines are dangerous than if the neighbor didn't get sick. Keep in mind, however, that none of this means our beliefs are true, only that *sometimes* we have reasons to believe them. After all, the neighbor could've caught the flu from someone at the doctor's office.

Humans have many needs that, when satisfied, cause our brains to release a pleasure-inducing chemical called dopamine. Dopamine is the brain's way of telling us to repeat a behavior. One of the reasons drugs are so addictive is because they release huge amounts of dopamine, so our brains tell us to "do it again." This reward system isn't without an evolutionary basis. When our ancestors did something that increased their chances of survival or reproduction, they felt good—that is, their brains released dopamine. Likewise, when exercising a belief satisfies a need, the brain rewards us with dopamine. Just as the brain cannot tell the difference between overeating that leads to diabetes or drug use that leads to addiction, it cannot tell the difference between a true or false belief when each one satisfies some human need. When exercising a belief satisfies a need, that behavior is reinforced with dopamine, thus motivating us to maintain the belief, or "repeat that behavior."

Thus, sharing our beliefs is biologically valuable. For example, although overwhelming evidence shows that acupuncture and sham acupuncture are equally as ineffective at relieving lower back pain (Cherkin et al., 2009), believing in the positive effects of acupuncture and exercising this belief by telling others about its benefits, visiting an acupuncturist, or talking to other believers, satisfies many of our socio-cultural needs, such as the need to belong, the need to feel important, or the need to help others. We may not be helping anyone by recommending an acupuncturist, but *believing* we're helping is enough to trigger the release of dopamine. As such, dopamine likely plays a huge role in why people believe incredible claims. Because of its ability to induce symptoms of psychosis, such as hallucinations and delusions, psychologist Bruce Hood believes that "if there is a smoking gun for the biological basis [for our tendency to believe incredible claims], it seems to be firmly held by the hand of dopamine" (Hood, 2009, p. 251). In fact, dopamine is so powerful that in a study of both skeptics and believers, skeptics whose dopamine levels were increased showed a heightened propensity to identify illusory patterns (Mohr, Landis, & Brugger, 2006), leading Shermer to dub dopamine "the belief drug" (Shermer, 2011a, p. 117).

Other needs:

- Simplicity
- Self-esteem
- Justice
- Control

- Meaning
- Consistency
- Belonging

Our needs motivate us to formulate our beliefs and behave in ways that maximize our well-being.

But not everything that feels good is good. It may feel good to shoot heroin or drink liquor, but the resulting addiction and health damage, coupled with violations of parole, the law, or prior commitments, are not what we'd call good by any objective standard. It may feel good to express anger through retaliation or eat every sugary treat we find, yet if we measure what's good by the consequences of our actions,[2] we'd have to conclude that these feelings are counterproductive and don't serve the same purpose today as they did when they first evolved.

Mere-Exposure Effect

Another marketing favorite, the *mere-exposure effect* explains our tendency to favor things we're more familiar with. Take branding. Branding is a process of using names and symbols to communicate the qualities of a particular product. Brands are designed to signal uniform quality—customers who try and like a product can return to it by remembering its name or logo. Branding tells consumers what quality to expect and ensures a quality consumers know (Ebert & Griffin, 2013). In fact, a product's quality doesn't make it "name brand" at all; its popularity does.

The fact that we like something that everybody else likes is cause for skepticism. As Bertrand Russell once observed, "The fact that an opinion has been widely held is no evidence whatever that it is not utterly absurd; indeed in view of the silliness of the majority of mankind, a widespread belief is more likely to be foolish than sensible" (Russell, 1929, p. 58). The mere-exposure effect even enters our romantic lives. Research shows that increased exposure to someone is significantly correlated with how that person was rated for attraction (Moreland & Beach, 1992). However, as anyone who's ever had a roommate knows, the rule doesn't always apply. Over time, people's repeated habits can drive us crazy and make us avoid them at all costs (Cunningham, Shamblen, Barbee, & Ault, 2005).

Continued Influence Effect

The *continued influence effect* also sheds light on just how complicated beliefs are. This effect occurs when we originally judge information as true and later discover that information is false, yet continue being influenced despite its falsity (Alcock, 2018, pp. 217–19; Ecker, Lewandowsky, Swire, & Chang, 2011). When a judge instructs a jury to ignore a particular piece of evidence, jurors don't simply forget it. Likewise, when someone is charged with a heinous crime but found not guilty, how many people believe the person is innocent (think of O. J. Simpson's or Jody Arias's murder trials)? Similarly, when scientific studies are printed and later retracted, many people continue being influenced by the original debunked paper.

This effect seems to exist because we don't like incomplete explanations. If an initial bit of information fully explains something we find important, we have a difficult time discounting that information, even if we later discover it's false, unless, of course, a more complete explanation is offered in its place. One study showed that criminal defendants were found not guilty more often when an alternative suspect was named than when the defense simply argued that the defendant was not guilty (Tenney, Cleary, & Spellman, 2009). As psychologist James E. Alcock reflects, "We often construct a neat little package of events in our heads, and then if told later that part of the package is not true, this creates a gap. . . . And so, in the absence of another explanation, it is not easy to put the original false explanation totally out of mind" (Alcock, 2018, p. 217).

To Control or Be Controlled

Our physical and psychological needs motivate our behavior in specific ways. Understanding where these motivations originate can reduce how often we commit *Type 1 errors*. For example, decision making and beliefs are two basic needs. Since the feeling of not being in control is uncomfortable, the desire for control motivates us to believe we're in control even when we're not (Blackmore & Troscianko, 1985). Consequently, sometimes randomness doesn't appear random at all (Shermer, 2011a, pp. 77–84; Whitson & Galinsky, 2008). In addition, our cognitive makeup creates what's been dubbed the *illusion of control*: believing we can influence chance events (Matute, Yarritu, & Vadillo, 2011). For example, the outcome of tossing a fair coin is purely random. Nonetheless, under the right conditions, we'll convince ourselves that we can skillfully predict the outcome (Langer, 1975; Langer & Roth, 1975). The first study on the illusion of control (Jenkins & Ward, 1965) exposed participants to two switches and a light, allowing them to control when the switches connected to the light. Then, unbeknownst to the participants, researchers put the light on a random timer, yet the participants continued believing they controlled when it turned on or off. Furthermore, lacking control also enhances our tendency "to perceive a variety of illusory patterns . . . perceiving conspiracies, and develop other superstitions" (Whitson & Galinsky, 2008, p. 75).[3]

Feeling in control is an important psychological need, so powerful that people can tolerate more pain if they think they control when it ends (Hood, 2009, p. 32; Salomons, Johnstone, Backonja, & Davidson, 2004). Think about waiting around to hear about test results, the feeling you have after a boss tells you "something serious" happened at work, or the uncertainty of not knowing how some important future event will unfold. We anticipate something dreadful, yet we want it sooner rather than later. That feeling of uncertainty and helplessness you experience before the news is delivered is tremendously uncomfortable, thus we become motivated to satisfy our need for resolution by any means we can.

Superstitions (chapter 11) provide an excellent example of the illusion of control. Research suggests that superstitions nurture such illusions (Blackmore & Troscianko, 1985). People believe that behaving a certain way increases the chances of some desired outcome. When behaviors or actions "work" (we wear a certain hat and our team wins the game), we confirm our sense of control; when those same behaviors or actions don't "work" (we wear the hat and our team loses), we simply forget or ignore it. We count the hit because it becomes an event, and we ignore the miss because it's a non-event—something we're unlikely to remember. Thus, superstitions make us feel like we exert control over completely random events, even though the feeling is illusory.

Stereotyping is another way cognitions motivate behavior; it's System 1's way of saving energy by simplifying complexity (Macrae & Bodenhausen, 2000; Macrae, Milne, & Bodenhausen, 1994). A stereotype "is an unwarranted conclusion or generalization about an entire group of people" (Vaughn, 2019, p. 175). Grouping others into "categories" helps us predict their actions based on what we've seen from others in that "category" (Macrae, Milne, & Boden-hausen, 1994). Negative stereotypes make us feel closer to others that are like us by reinforcing that "they" are different in some undesirable way (Tavris & Aronson, 2019, p. 75). Accordingly, stereotyping is helpful if "we understand that the rule is just a schematic, or an oversimplified representation of a much more complicated reality" (Novella, 2012, p. 14). If you don't get a job be-cause you have tattoos, you've been negatively stereotyped, and perhaps your potential employer lost out on the best applicant because he stereotyped you. Therefore, by exaggerating differences between groups, encouraging selective perception, and ignoring differences within groups, stereotyping distorts reality in ways we're frequently unaware of (Judd & Park, 1993; Judd, Park, Ryan et al., 1995). Failing to understand that stereotyping is little more than a cognitive shortcut can lead to bigoted mindsets if we fail to realize its limits. Sometimes the rich white guy driving a Mercedes is an arrogant prick, and sometimes he's a generous humanitarian. Stereotyping can inform us based on prior experience and give us an idea of what we *might*

> A great many people think they are thinking when they are merely rearranging their prejudices.—William James

expect, but it cannot tell the difference between the prick and the humanitarian. That task is one for System 2.

The need for meaning is another strong motivator of behavior. Professor of neurology Steven Novella argues that we desire meaning because we "want there to be an overarching meaning to our existence because it gives us a sense of purpose" (Novella, 2012, p. 14). Novella's argument makes sense, as we tend to reject any notion that our lives lack purpose.

For fear of social embarrassment, we have trouble admitting we're wrong. This reluctance motivates us to avoid social embarrassment, which in turn interferes with our ability to reach objective conclusions. By holding others to different standards than we hold ourselves, blaming our own mistakes on external (situational) causes while unfairly blaming internal (dispositional) factors for the behavior of others, we commit what social psychologists call the **fundamental attribution error** (FAE). We see Jody talking to the cops and immediately assume she's snitching; when *we* talk to the cops, it's because "My car was stolen" or "I was getting served divorce papers" or whatever.

The FAE also works the other way, as when we attribute our own good fortune to how hard we work and how smart we are, while attributing the same good fortune in others to luck or circumstance (Davis & Stephan, 1980). For example, if we receive a promotion at work, we attribute it to our talent and good work ethic, while if someone else gets the promotion, we attribute it to their dad being friends with the boss. In addition to situational and dispositional attributions, the FAE also describes our tendency to interpret our own behaviors and beliefs as rationally driven (intellectual attribution) and our tendency to interpret others' beliefs and behaviors as emotionally driven (emotional attribution) (Shermer, 2013, p. 31).

Research repeatedly shows that situational factors typically have more influence on behavior than individual characteristics (Alcock, 2018, p. 34; Zimbardo, 2008). In one study, researchers told participants that they would receive a vitamin shot to investigate how vitamins affected vision. Unbeknownst to them, the participants received either a placebo or an adrenaline shot. Adrenaline affects the body in several ways: increased arousal, sweaty palms, increased heart rate, accelerated breathing, and, importantly, making us more keen to our surroundings. After receiving the shot, participants were joined by a confederate who pretended to be either euphoric or angered. Interestingly, the participants who received adrenaline reported feeling either happier or more irritated depending on the confederate's behavior (Schachter & Singer, 1962).

Moreover, the FAE affects everyone from the rich and privileged to the poor and underprivileged. Wealthy people rationalize their position as being justified and earned by their moral commitment to others, while the underprivileged believe the rich are undeserving, that they got rich by accident or by morally corrupt means (Shermer, 2008a, p. 68). One antidote to the FAE is Hanlon's razor—a cousin of Occam's razor—which advises us to "never attribute to malice that which can be adequately explained by stupidity." The FAE can be toxic to a society that doesn't Think Straight about such tendencies. This cognitive error drives us to see ourselves as smart and rational, while viewing "others" who disagree with us as dumb and irrational. Moreover, if we attribute all our mistakes and failures to external factors, we're less likely to grow as individuals. After all, those who already know everything have no reason to learn anything.

Unchallengeable Beliefs

In chapter 1, I discussed open-mindedness—the willingness to change your mind as the evidence demands. Some beliefs are so entrenched that no amount of evidence or logical argument can dislodge them. Such beliefs can be classified as *delusions*. While some delusions are not beliefs (Vyse, 2018a), a delusional belief is defined as a strongly held false belief that is maintained in the face of contradictory evidence (Alcock, 2018, p. 189; Barch, 2018). Those suffering from delusions are typically unwilling to consider the opinions of others (Mullen & Linscott, 2010), believing instead that what *they* know is the truth. In his 1913 book *General Psychopathology*, the philosopher and psychiatrist Karl Jaspers laid down three criteria which define delusional beliefs (Jaspers, 1913/1997, pp. 95–96). First, a belief must be held with unwavering surety of conviction. Second, the belief must be unchangeable and strongly resistant to counterargument or contradictory evidence. Lastly, the belief must be strange, bizarre, absurd, or patently false. While no guidelines are perfect, a belief that is (a) held with absolute certainty, (b) immune to contradictory evidence, and (c) patently absurd certainly carries the markings of a bona fide delusion (Alcock, 2018, p. 191).

Delusions differ from beliefs that are wrong in that false beliefs are amenable to change. As philosopher Peter Boghossian explains, "The difference between misconstruing reality and being delusional is the willingness to revise a belief" when the evidence demands (Boghossian, 2013, p. 82). Psychologists have catalogued several types of delusions and, importantly, discovered that delusions are heterogeneous, meaning there's a spectrum of delusional thinking and behavior (Barch, 2018, p. 426). Delusion sufferers cannot distinguish truth from fiction and tend to suffer from dysfunction in perception, in turn rendering them incapable of recognizing the irrationality of their delusions (Alcock, 2018, pp. 190–91).

A **dogmatic** belief is held with absolute certainty and cannot be changed. Dogmatic individuals typically go to great lengths to avoid subjecting their beliefs to contradictory evidence or logical challenges. Thus, any evidence or argument that conflicts with their belief, regardless of its truth value, stands little chance of being incorporated into their body of knowledge. Dogmatic beliefs develop for several reasons, including childhood indoctrination, insecurity, self-identity, a need to be "right," the need for certainty, and emotional significance (Alcock, 2018, pp. 186–87). Dogmatic beliefs are the most stable form of belief, thus where we find dogmatic beliefs we should expect to find most (if not all) of Alcock's eight factors concerning belief stability (see above).

Therefore, it shouldn't surprise us that high scores on tests of dogmatism correlate with resistance and hostility to new ideas. In contrast to their skeptic counterparts, dogmatic individuals exhibit anxieties about the future, low tolerance for ambiguity, less willingness to change their mind, and less flexibility in their problem-solving behavior (Rokeach, 1960). Conversely, skeptics understand the changing nature of evidence and argument and thus willingly change their minds as the evidence demands.

The "Essence" of Belief

People also believe that inanimate objects carry an "essence" that makes it unique. **Essentialism** is like the childhood belief in cooties, and continues into adulthood (Gelman, 2004). Essentialism even drives racist prison politics that dictate what sinks or showers prisoners use, whom they eat after, and even where they stand on the prison yard. Harvard psychologist Bruce Hood argues that essentialism is a product of how human brains are designed, tracing the roots of essentialism to how children interpret the world (Hood, 2009). Hood calls this tendency to believe in the paranormal and supernatural a *supersense*. A supersense is similar to common sense, but where common sense is guided by reason, supersense is guided by intuition (Hood, 2009, pp. 6–7).

To demonstrate this phenomenon, during public lectures, Hood asks audience members if they will wear a sweater he brought along (Hood, 2009, pp. 36–37). The sweater is old and a bit tattered but is otherwise clean. Usually about a third of the audience is willing to try it on. Hood then ups the ante, offering a prize for wearing the sweater, at which point several more volunteer to try it on. After describing details of several gruesome murders committed by a serial killer named Fred West, Hood informs his audience that the sweater belonged to West. At this point, the volunteers change their minds, fearing that the sweater carries the "essence" of Fred West and has somehow been magically contaminated.

> The only force I fear more than human irrationality is irrationality armed with passion.—Leo Rosten

In another of Hood's research projects, participants were asked to rate the faces of people for attractiveness, intelligence, and how happy they'd be to accept a heart transplant from the people in the photos. Once the participants rated twenty pictures, researchers told them half the pictures were of convicted murderers and half were volunteers. The participants were then asked to rerate the pictures for attractiveness, intelligence, and willingness to accept their hearts. Predictably, the attractiveness and intelligence of the "murderers" dropped. However, more surprising was that the biggest drop was the willingness to accept a heart transplant. Hood concluded that the unwillingness was due to the belief that an "essence of evil" would be transmitted through the heart (Hood, 2009, p. 196).

It isn't just the "essence of evil" that people believe in, as demonstrated by the ridiculous prices paid for memorabilia. People purchase clothes from their favorite celebrities and absolutely refuse to have them dry cleaned before picking them up (Hood, 2009, pp. 4–5). Essentialism is also shown in research on how children feel about certain toys. For instance, the way children feel about a toy depends on whether they imbue it with essence (Mikkelson & Mikkelson, n.d.). Children wouldn't have a problem accepting a replacement toy unless they believed there was something essential about the original that would be missing if it were replaced.

Neurologist Steven Novella contends that essentialism makes sense from an evolutionary perspective, arguing that parents would not likely accept a replacement of their child and would prefer their own child over a replacement (Novella, 2012, pp. 41–42). Shermer agrees, insisting that "evolutionary reasons for . . . *essentialism* [are] rooted in fears about diseases and contagions that contain all-too-natural essences that can be deadly (and hence should be avoided), and thus there was a natural selection for those who avoided deadly diseases by following their instincts about essence avoidance" (Shermer, 2011a, p. 88). Hood goes further, arguing that essentialism comes with our ability to intuit and is necessary for human societies to maintain sacred values (Hood, 2009, pp. 264–67).

An evolutionary origin of essentialism would explain why even ardent skeptics are essentialists. Even skeptics value an original painting, an autographed book, or genuine memorabilia over replicas. Imagine your child hitting a game-winning home run at the World Series and a kind fan bringing you the home-run ball. You happily place the ball on your trophy mantel to keep the memory alive. Would you trade the ball for an exact replica, or would you feel like by trading you would have lost something essential?

Wrapping Up

Beliefs are the foundation of everything we do. We believe wearing seatbelts will better protect us, so we wear seatbelts. We believe that eating vegetables is healthy, so we eat vegetables. We believe that freedom beats incarceration, so we obey the law. Our beliefs are based on multiple factors, some of which align with reality, some of which merely reinforce our constructed models of reality. The fact that our beliefs seem to confirm what we think we know about reality and define who we are makes them enormously influential on our decisions and behaviors. Since we're often wrong about what we believe, Thinking Straight demands a willingness to question our beliefs rather than blindly following them into the dreary realms of false realities.

CHAPTER 8

Self-Deception, Part 3— Perception and Pattern Recognition

What we believe and what we know come to us by way of our senses. We see rainfall and learn where rain comes from. We taste pizza and learn that we love it. We hear our mother call us by our legal name and learn we're in trouble. We touch a cactus and learn that spiky things hurt. Most of the time, we have little reason to doubt that what we see, touch, hear, taste, and smell are exactly what they seem. But research shows that our senses bring us inaccurate information often enough that they can corrupt our entire worldview. Moreover, all sensory information goes to the brain, where it gets processed and a completely subjective model of reality is constructed (Hawkins, 2021).

This chapter is about how we process information, or, more accurately, how unreliable the brain is at processing sensory information. We'll often defend something to the end because we heard it ourselves, or we saw it with our own eyes. It turns out that we have much more reason to doubt what we see and hear than we realize, and Thinking Straight about those matters is necessary for anyone who prefers reality over fantasy.

Building Vocabulary

Inattentional blindness:	Illusion:
An inability to perceive something that is in plain sight.	A distortion of sensory information in which a thing appears as something it is not.

Open-mindedness:	Pareidolia:
The willingness to change your mind according to evidence; applying an equal standard of evidence and proof to all claims.	The tendency to see illusory resemblances in random noise and interpret those resemblances as meaningful patterns.

Perception

Perception, the process by which our brains organize and interpret sensory information (Wade & Tavris, 2017, p. 180), can be extremely misleading. After all, for millennia people could clearly *see* the Earth was flat and the sun, moon, and stars all orbited around it. From the ground, our planet looks immense; from beyond Pluto, it is but a pale blue dot suspended in a vast ocean of empty space (Sagan, 1994). Beginning with Eratosthenes's investigations more than two thousand years ago (Sagan, 1980, pp. 14–17; Upton, 2014), scientific investigations repeatedly verified the claim that Earth is spherical and not at the center of the universe. Despite the overwhelming evidence, to this day not everyone is convinced (Foster, 2018; Loxton, 2019; Nazé, 2018).

In his book *Against Method*, philosopher of science Paul Feyerabend explained that information brought in by our senses doesn't accurately reflect reality because "sensory information is distorted in many ways" (Feyerabend, 1975/2010, p. 245). More recently, professor of neurology and psychology Terence Hines said that "perception is a function not only of the actual sensory stimulus that is picked up by the eye or the ear, but also of what we know and believe about the world, even if that knowledge and belief are wrong" (Hines, 2003, p. 238). Throughout this chapter, we'll learn how our senses construct a narrow and subjective model of reality, beginning with an examination of the limits of visual information as Feyerabend alluded.

Looking at a distant star may seem like a rather simple observation, and it is. But what lies beneath the surface of this seemingly mundane activity is much more interesting. The light from the star began its journey to your eyes thousands of years ago and trillions of miles away before finding its way to you. Since it takes time for light to travel, the star you're seeing might no longer exist; it could've gone supernova years ago, and you wouldn't know because the light from its explosion hasn't had time to reach you. Though the star appears fixed and motionless, this is an illusion. So long as you're observing it with the unaided eye, the star you see is traveling at over 500,000 miles per hour as it orbits a supermassive black hole at the center of the Milky Way galaxy, itself speeding across the universe at galactic speeds. Once the star's light reaches Earth, that light is diffracted (broken up and reformed) and refracted (bent) by Earth's atmosphere, which is why stars appear to twinkle. Finally, the light reaches photoreceptors (cones and rods) in your eyes and then confronts any limitations or shortcomings in vision you might have. But seeing happens in an area of the back of the brain called the visual cortex, not the eyes. So the light (photons) reaches the eyes, where photons knock electrons in the atoms of your photoreceptors into higher energy levels, or out of the atom completely, creating electrochemical signals that pass through the thalamus before being interpreted by the visual cortex, creating what we experience as vision (Stenger, 2012, p. 201).

Human vision suffers from a number of design flaws (Wade & Tavris, 2017, pp. 186–89). For starters, we all have a blind spot where the optic nerve—the nerve that brings visual information from the retina to the brain—connects to the retina—the part of the eye that allows sight. Additionally, the images we see are only clear in the fovea, a tiny area of photoreceptors within the retina that lets us see clearly an area about the size of your thumb held at arm's length. The visual information outside of the fovea is shrouded in a fog, meaning most of our visual field is blurrier than pictures of Bigfoot.

Making matters worse, our eyes will fatigue if we stare too long, which causes small involuntary eye movements that can make it appear as though what we're looking at is moving. This phenomenon is known as the *autokinetic effect* and is likely responsible for many supposed paranormal experiences and UFO sightings (Hines, 2003, p. 242; Smith, 2018, pp. 190–91). Research on the autokinetic effect shows that we're also influenced by what others claim to see. Someone claiming she sees a light moving in a particular direction will prompt others to see it as well (Otani & Dixon, 1976). Each stage in this seemingly simple process of stargazing presents an opportunity for reality to be misrepresented. This is our perceptual experience.

One phenomenon demonstrates the lengths the brain goes to confabulate details to maintain a coherent model of reality. What's known as *apparent motion* or the *color phi phenomenon* is a powerful illusion where the brain projects an image of what it believes *should have appeared in the past*! The experiment is easy to create.[1] Turn off all the lights in a room, carefully ensuring the room is as dark as possible. On the left side of a screen, flash a light of one color, say red. In close succession to the red flash, flash a blue light on the right side of the screen (Kolers & von Grünau, 1976).

If you guessed what people would see in this experiment, you'd likely say that they'd see a red light flash on the left side of the screen, followed by a blue light flash on the right side. What people *actually* see is a light shooting across the screen, turning from red to blue at the halfway point. Accordingly, what we see is the blue light flash *before it's turned on*. Researchers aren't sure exactly how the brain does it. Amazingly, the brain projects the image of the blue light backward in time, so long as the discrepancy between "brain time" and "external time" is about 120 milliseconds or fewer (Damasio, 2002).

Our perceptions of the world and reactions to events hugely impact our lives. If someone says something in jest, and we perceive it as being said with malice, we might react with violence or aggression rather than humor and laughter. In one of his dialogues, Plato has Phaedrus say, "Things are not always what they seem," encouraging readers to take a skeptical approach, but what's this say about perception? At the very least, it suggests that perception depends on the individual and the situation. If I ask Glen, "Are lawyers helpful?" Glen will think of instances in which lawyers were helpful. If I ask Glen, "Are lawyers corrupt?" he will search his memory for examples of corrupt lawyers. With either question, Glen's brain automatically searches for instances that confirm the question, ignoring any instances that don't (Kahneman, 2013, p. 81). This brain function isn't limited to Glen or to questions about lawyers, and it gets even more interesting. If I ask if you're outgoing, provided you view being outgoing as a positive trait, you'll search your memories for instances that demonstrate how outgoing you are, while ignoring any memories where you weren't (Sanitioso, Kunda, & Fong, 1990).

Some aspects of perception are learned and develop over time, some beginning in early infancy. For instance, the term *naïve realist* has been used to describe how people assume the world is exactly as it appears and how they can't distinguish between themselves and the world around them. Researcher Rheta DeVries looked at naïve realism in an ingenious study where she allowed young children to play with an even-tempered housecat (Maynard). DeVries wanted to know whether the children could tell the difference between objective reality and what they see (subjective perception). The younger children, around three years old, understood that Maynard was a cat. However, when DeVries put a mask of a dog on Maynard's head, some children began thinking that Maynard was now a dog. Older children, around six years old, understood that Maynard was a cat and that he remained a cat even when he was wearing the mask (Siegler, 2018).

Naïve realism follows some into adulthood (Hood, 2009; Pronin, Gilovich, & Ross, 2004), even when perceptions lead to false realities. Naturally, we believe we're better than average at almost everything we do and we perceive things as they really are, and we're excellent at identifying the biases in everyone except ourselves (Pronin, Lin, & Ross, 2002; Tavris & Aronson, 2019, pp. 54–55). Since we know we're rational and we see things clearly, we assume other rational people see things as we do. This outlook lures us into a deceptive logical trap: rational, fair, and open-minded people ought to agree with a rational opinion; since I'm rational, fair, and open-minded, any opinion that I hold is rational, fair, and open-minded; therefore, anyone that doesn't agree with me is irrational, unfair, or closed-minded (Ross & Ward, 1996). This self-justification sounds a lot like logic, but the argument fails because rational, fair, open-minded people can be irrational, unfair, and closed-minded. They can also be honest but mistaken, they can disagree, and they can behave irrationally. Accordingly, even if we are rational, fair, and open-minded, it doesn't follow that anyone who disagrees with us is not.

How Our Brains Construct Our Realities

If you're a hippo, you run through water as easily as people run through air. If you're a bacterium, you *dig* your way through water, much like people dig through mud (Dawkins, 2006, p. 157). So while the physical properties of water remain constant, its viscosity (density) is reduced to a matter of perception. The hippo and the bacterium could argue endlessly about the *real* viscosity of water without ever settling the score.

Based on a small portion of the information it receives, our brains are constantly constructing a consistent narrative of what's happening, constructing our internal models of reality. In his book *The Illusion of Conscious Will*, Daniel Wegner asserts, "We can't possibly know (let alone keep track of) the tremendous number of mechanical influences on our behavior because we inhabit an extraordinarily complicated machine" (Wegner, 2002, p. 27). Thus, some people cannot simultaneously pat their head and rub their stomach or focus on a song and simultaneously absorb information from a book. Multitasking, put simply, is largely an illusion (Shaw, 2017, pp. 188 95). You may be able to read and listen to music at the same time, but you cannot pay serious attention to both. So our perception is based on the small amounts of information we're focusing on or the limited amounts of information our brains decide are important. What we experience as "our" perception is really an illusion that's constructed by our brains without conscious awareness.

Tricking the Senses

The word **illusion** comes from a Latin word that means "to play, mock, or trick." Illusions are extremely convincing, especially when we cannot imagine how we could be being tricked or deceived. Thus, illusions convince us that some very unlikely things are true or that some very true things are rather unlikely. In fact, everything in the entire universe can be thought of as part of an illusion played on the human mind. The French astronomer Camille Flammarion (1842–1925) captured this best:

> We see the sun, the moon and the stars revolving, as it seems to us, round us. That is all false. We feel that the earth is motionless. That is false, too. . . . We touch what we think is a solid body. There is no such thing. We hear harmonious sounds; but the air has only brought us silently undulations that are silent themselves. . . . Sensation and reality are two different things. (Alcock, 2018, p. 47)

Illusions fool us easily and often and not just visual illusions. For example, the *illusory truth effect* describes how hearing something repeatedly makes us believe it's true (Bacon, 1979). Good examples include the belief that a full moon causes people to act in bizarre ways (Rotton & Kelly, 1985) or that humans only use 10 percent of their brains (Boyd, 2008; Radford, 1999). The illusory truth effect explains why we believe something simply because we've heard it; it's also the reason courts sequester juries. In one study (Hasher, Goldstein, & Toppino, 1977), participants were given sixty plausible-sounding statements, some true and some false, and later asked to rate their truth values. The participants returned two weeks later and were given another set of statements that included some of the statements from the previous visit. Two weeks later, the participants were again asked to rate whether they believed a series of statements were true. This third visit showed that participants more readily believed the statements that had been repeated, regardless of whether they were actually true. This effect is so powerful that hearing something only once is enough for us to misremember it as true (Fazio, Brashier, Payne, & Marsh, 2015).

We're also easily fooled by *optical illusions*. The brain constantly makes assumptions about incoming information, and optical illusions force the brain to make assumptions based on what's likely true. Probably the most famous optical illusion is the Necker Cube.

When you first look at a Necker Cube, your brain constructs an image of the cube. After staring at it for a bit, the image switches and the cube appears to shift positions and face a different direction. As you continue staring, the cube shifts back and forth between the two positions. The cube isn't physically chang-

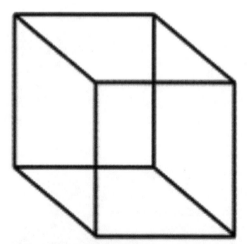

Figure 8.1. Necker Cube.
Courtesy of the author.

ing from one position to another; it's just that your brain is constantly constructing new images of the cube based on the information it has.

Although accurate most of the time, sensory input is often misleading, warranting skepticism about our perceptions. Thinking Straight demands seeking corroborating evidence that validates reality, rather than relying solely on our subjective experiences. Stage magicians have perfected the art of exploiting the many shortcomings of our brains by understanding that the limits of our perception allow them to do things right in front of our eyes and we're none the wiser.

Just as the ovals at the intersections of Figure 8.2 are neither black nor blinking, there's nothing literally different about plain old water, whether you're a bacterium or you're a hippo. The difference lies in the perception of the beholder.

Figure 8.2.   Disappearing black dot illusion. *https://commons.m.wikime dia.org/wiki/Optical_illusion.*

The Lengths Our Brains Go to Trick Us

Sayings like "Look at it from her perspective" and "Appearances can be deceiving" caution us against taking our powers of perception for granted and suggest that personal perception cannot validate reality. The brain warps reality in powerful ways. To use an extreme example, in Fatima, Portugal, in 1917, after three children prophesied that the Virgin Mary would appear to perform miracles, a whopping seventy thousand people allegedly witnessed the sun zigzag, dance, enlarge, and hurtle toward the earth (Dawkins, 2006, pp. 116–17). Given that such a claim is highly incredible, could the corroborating eyewitness accounts of so many people be wrong? How does one explain away eyewitness testimonies of seventy thousand people? By using GASPs, and beginning with simpler explanations.

You'll remember from chapter 1 that a Straight Thinker asks, "What else must be true if X is true?" If the Miracle at Fatima really happened, obviously the laws of physics must've been suspended—stars don't typically zigzag or dance; the earth wasn't vaporized by the sun's heat or sucked in by its gravity. Yet as far as we know, nothing can suspend the laws of physics, and no instance of a violation of the laws of physics has ever been discovered (Weinberg, 1994, pp. 129–30). (In fact, the very definition of a miracle is a violation of the laws of physics!) Given that the laws of nature are far more reliable than eyewitness testimony, any testimony that suggests a violation of those laws is an incredible claim. What else must be true? Well, if the sun began moving erratically, everyone on the planet would have seen or felt the effects, yet people just outside of Fatima (and even some just outside the large gathering *in* Fatima) failed to witness the alleged event. If the sun suddenly rushed toward the planet, we should expect people attempting to flee or at least turning around to protect themselves from the heat. Yet photos of the "event" depict a typical day at the beach.[2]

Writing *Of Miracles* in his classic 1748 treatise *An Enquiry Concerning Human Understanding*, renowned Scottish philosopher David Hume (1711–1776)

Maxims of CT:

- Unexplained does not mean unexplainable.
- Incredible claims require credible evidence.
- Absence of evidence is not proof of absence.
- Correlation does not mean causation.
- What can be asserted without evidence can be dismissed without evidence.

These maxims are guideposts for weighing evidence and evaluating incredible claims.

SELF-DECEPTION, PART 3   175

argued that "no testimony is sufficient to establish a miracle, unless the testimony be of such a kind, that its falsehood would be more miraculous, than the fact, which it endeavors to establish." In other words, we shouldn't take someone's word that an incredible claim is true unless it'd be even crazier if it were untrue. Instead, when confronted with incredible claims, we should ask, "What's more likely?" As for the miracle at Fatima, what's more likely: that the laws of nature were violated, or that fallible humans were mistaken? Since many attendees of the event never saw anything unusual, yet their statements are unknown to or were discounted by those who believe in miracles, we can say the latter is more likely. But we can also apply this test to other incredible claims. For example, what's more likely: 99 percent of scientists are mistaken or wrong about anthropogenic climate change, or non-scientists don't understand or refuse to accept the data? What's more likely: hundreds of thousands of scientists conspire to publish fake research claiming that vaccines and genetically modified foods are safe, or people reject those findings for lack of understanding, exposure to misinformation, or political or ideological reasons? If ninety-nine out of one hundred dentists say we should floss our teeth once a day, that doesn't mean that one dentist is onto something; it means that one dentist missed something.

Trusting Skepticism

In his parable of the Indian prince and the ice (Baggini, 2006, pp. 7–9), Hume illustrates the need for skepticism when approaching incredible claims, reminding us how limited our perception really is. As the story goes, an Indian prince leaves his desert home for an epic voyage to the mountains. Returning two years later, the prince tells his cousin Michonne about lands so cold that water ceases to move, instantly becoming solid, like a rock.

Michonne, not wanting to embarrass the prince publicly, pulls him aside to express her disbelief. The desert is full of tall tales, and Michonne rightly refuses to believe all the marvelous stories about fire-breathing dragons and grotesque giants. She refused to believe this marvelous tale about rock-hard water, even if it did come from her own trusted cousin. The question is, was Michonne wrong to discount something we all understand as a universal fact? Michonne thought she couldn't be fooled by a seemingly incredible tale. What the prince told her flew in the face of all her experiences, so her skepticism is healthy. Yet she was wrong. So how can we know our own skepticism will not lead us down a similar path to false conclusions?

The key here is understanding Michonne's—and by extension our own—limited experience and the skeptic's demand for quality evidence. Skeptics raise their objections and present their case, while keeping an **open mind** about the possibility of being wrong. Thus, Michonne's error wasn't to apply skepticism, but to dismiss the prince's claim before evaluating the evidence. As for our own CT, Hume's parable illuminates our duty to examine the evidence before reaching a conclusion (recall Sagan's "postjudice" principle from chapter 1) and instructs us that even strong beliefs are based on subjective experience. We can be so certain of something yet so completely wrong. The takeaway: just because you think something, that doesn't make it so, regardless of how certain you are.

What You See Is NOT What You Get

The argument that X knows something because he saw it with his own eyes is a terrible but common argument. For instance, these days nobody seriously doubts the existence of electrons or black holes, even though no one's ever "seen one with his own eyes." I've never seen Japan, but that's not a very good reason to think it doesn't exist. Similarly, people have seen many things that simply weren't there. Whether something has been seen tells us nothing about whether it's real or whether its reality exists solely in the mind of the seer.

Psychologists have long since discovered that our eyes deceive us on a regular basis. In a 1949 experiment (Bruner & Postman, 1949), psychologists asked participants to identify playing cards. Some of the cards were ordinary, while others were deceptive, such as a red six of spades or a black four of hearts. The participants identified the cards with ease and even identified the deceptive cards as normal. For example, a participant might identify the black four of hearts as the four of hearts, without noticing that something was off. When the experimenters increased the number of exposures to the deceptive cards, participants began hesitating, making comments such as "That's the six of spades, but something's wrong with it." Upon further exposures, the participants began noticing the deception and, once identified, usually had no problem "seeing" the deceptive cards for what they were. This experiment showed that sometimes, even something that's right before our eyes might as well be invisible. Recent research has revealed even more.

> We don't see things as they are; we see them as we are.—Anonymous

In what's become one of the most famous experiments in psychology (Chabris & Simons, 2010, pp. 5–8; Simons & Chabris, 1999), Christopher Chabris and Daniel J. Simons showcased what they call the "everyday illusion"[3] of attention. The illusion of attention leads us to think we pay more attention to our surroundings than we really do. In their now-famous experiment, Chabris and Simons instructed participants to watch a twenty-five-second film of six people (three in white T-shirts, three in black) weaving about, passing a basketball back and forth. Participants were told to pay close attention to detail and to count the total number of times the ball is passed by people in white shirts. The players move about quickly, making the participants' task difficult. As soon as the film ends, a series of questions are asked including "How many of you saw the gorilla?"

A few seconds into the film, a woman wearing a gorilla suit casually strolls onto the scene, faces the camera, pounds her chest, and then walks off nonchalantly. The "gorilla" was in full camera view for nine seconds, which is over a third of the film! Nonetheless, around 60 percent of participants never noticed the "gorilla" and would likely testify in court that nobody in a gorilla suit was in the video. Even more shockingly, experiments using eye movement tracking showed that the same participants who didn't notice the gorilla spent an average of a full second looking directly at it (Memmert, 2006), which was the same amount of time the people who did see it looked at it! Further research on the illusion of attention found that better working memory was strongly correlated with having witnessed the gorilla (Seegmiller, Watson, & Strayer, 2011), suggesting that people who see the gorilla have a larger attention budget (see below) than people who don't. In a follow-up experiment, participants were told about, but not shown, the invisible gorilla video and were asked if they thought they would have noticed the gorilla. Although only 40 percent actually *did* notice it, 90 percent of people predicted that they would've seen the gorilla (Levin & Angelone, 2008).

The Miracle of Fatima and the invisible gorilla experiment tell us that we see things that definitely aren't there, we miss quite obvious things that are right before our eyes, our perceptions are not reality, and our level of confidence in our perceptions is often unjustified. Misunderstanding the limits of our perceptions can come with some serious consequences. When we wrongly think we see more than we do, our confidence in our abilities unjustifiably increases, and we engage in dangerous behavior such as texting while driving. Most of us wrongly think we notice most things in our experience when in fact we notice only a tiny fraction, and even then, we're only consciously aware of a tiny fraction of *that*. Think of how many noises you're tuning out while reading this book. You may be blocking out the sounds of the A/C, doors opening and shutting, people's conversations, or your roommate snoring. The same way you ignore irrelevant sounds, your brain is constantly tuning out sensory information that *it* finds irrelevant.

EXPERIENCING THE WORLD

With an understanding of how our perceptions deceive us, we're ready to ask the question, "What does it mean to see something?" Or for that matter, "What does it mean to smell, taste, or touch something?" We indirectly experience the world through our senses. For example, we see objects because they either emit or reflect light particles (photons) before entering our field of vision. Depending on the object, when photons bounce away, they do so at a specific *wave frequency* of the electromagnetic spectrum that our brains translate into different colors. That green grass isn't *really* green; green is simply a mental construct of how brains interpret a part of the electromagnetic spectrum that we *experience* as green. To be sure, the wavelength itself isn't green or any other color. The only difference between, say, green and purple is the distance between the light's waves. Try to imagine the color of the grass before any photons hit it. Before any photons hit the grass, no electromagnetic waves are leaving the blades. Thus, without light and a brain to "translate" that light into color, the grass is completely colorless.

So despite *appearances*, the color of grass (and everything else) doesn't even exist outside of brains. Our brains are simply assigning different wavelengths their own signatures, which they then compare with past experiences and other sensory information to create an image in the visual cortex. The image we *see* in front of us is really inside of our heads! Even more, since it takes time for light to travel from the grass to our eyes, we're actually seeing the grass in the past. The time it takes for light to travel is why we can look at the night sky and see stars that no longer exist; the light from their destruction is still on its way to us. What's true for light and sight is equally as true for our other senses.

For instance, just as light travels from objects to the eyes and is then translated into colors and images by the brain, sound works similarly. Sound begins when an object or disturbance creates ripples across air molecules, which then propagate outward, much like ripples in a pond propagate away from the pebble that caused them. When those ripples reach our ears, they vibrate tiny bones in our ears in specific ways. Those vibrations are then translated by the brain as different "sounds." If a deaf person claps in a soundproof room, the clap will still cause ripples in the air molecules, but if the person's ears and brain cannot translate those ripples, there is no sound. Those rippling molecules will eventually settle down, leaving no trace of anything resembling a "sound." And just as it is with light, sound takes time to travel from its source to our ears, so what we hear actually happened in the past: think of sitting in a baseball stadium and seeing a batter hit a ball long before hearing the sound (a phenomenon we'll return to shortly).

The Illusion of Confidence

Responding to charges that the origins of humans could never be known, the famed English naturalist Charles Darwin (1809–1882) said in *The Descent of Man* (1871) that "ignorance more frequently begets confidence than does knowledge: it is those who know little, and not those who know much, who so positively assert that this or that problem will never be solved by science." Long before psychology demonstrated it empirically, Darwin was illuminating the illusion of confidence.

Just like any other brain process, certainty is a feeling (Burton, 2008),[4] one that happens *to* us rather than something we *do*. When we make deliberate decisions that cause *un*certainty, the amygdala, a part of the brain associated with feelings, is activated (De Martino et al., 2006), providing motivation to seek certainty over uncertainty. Feeling certain results from our brains telling us we're right, regardless of whether we are. The brain does this by selectively assigning undue weight to the confidence of ourselves and others. For instance, in a study on overconfidence (Wagenaar & Keren, 1986), researchers found that overrating one's own abilities undermines seat belt safety because most people wrongly believe they can drive better than everyone else (De Craen, Twisk, Hagenzieker, Elffers, & Brookhuis, 2011).

The authors of *The Invisible Gorilla* identify confidence as another of their everyday illusions (Chabris & Simons, 2010, pp. 80–115). Actually, evidence shows that confidence is strongly influenced by genetics. People who are confident in one area of their lives are usually confident in others, despite knowledge or ability (Chabris & Simons, 2010, pp. 100–101). In chapter 1, we learned that people's confidence in themselves often exceeds their abilities. This fact applies equally to predictions we make about ourselves and others (Vallone, Griffin, Lin, & Ross, 1990). Such overconfidence can and does lead to inaccurate and distorted perceptions of reality. Since at least 1977, researchers have studied one of the keys to feeling certain: the illusion of confidence (Lichtenstein & Fischhoff, 1977). Researchers believe that overconfidence is "one of the most consistent, powerful, and widespread" psychological biases of all (Johnson & Fowler, 2011). The illusion of confidence is so powerful that people are regularly tricked out of their hard-earned money by confident salespeople or stockbrokers. In fact, the word "con" in "con man," "con artist," and "con game" comes from the word "confidence."

> What a man sees depends both upon what he looks at and also upon what his previous visual-conceptual experience has taught him to see.—Thomas Kuhn

The illusion of confidence is twofold. "First . . . it causes us to overestimate our own qualities, especially our abilities relative to other people. Second . . . it causes us to interpret the confidence—or lack thereof—that other people express as a valid signal of their own abilities, of the extent of their knowledge, and of the accuracy of their memories" (Chabris & Simons, 2010, p. 85). We not only overrate ourselves; we also overestimate the abilities of others based on how confident they are.

Beginning with the first part of this illusion, people who are the least skilled are more likely to overrate their abilities than others. The philosopher Bertrand Russell captured this elegantly when he said, "The whole problem with the world is that fools and fanatics are always so certain of themselves, but wiser people so full of doubts." What's now understood as the *Dunning–Kruger effect* was first developed by psychologists Justin Kruger and David Dunning in a paper aptly titled "Unskilled and Unaware of It" (Ehrlinger, Johnson, Banner, Dunning, & Kruger, 2008; Kruger & Dunning, 1999). The Dunning–Kruger effect describes our dual inability to perform competently and to recognize that very inability. Recall the term "metacognition," introduced in chapter 1. Metacognition is basically thinking about thinking and accurately assessing our performance. But when it comes to rating our abilities, especially abilities we believe we have, all bets are off. As Dunning asserts, "Incompetent people do not recognize—scratch that, cannot recognize—just how incompetent they are," adding that "the incompetent are often blessed with an inappropriate confidence, buoyed by *something* that feels to them like knowledge" (Novella, 2018, pp. 45–46). In short, the Dunning–Kruger effect says that we're incompetent, that that very incompetence prevents us from realizing we're incompetent, and consequently we drastically overestimate our own abilities.

This illusion is easy to find, in others or in ourselves. Asking why the world is full of incompetent, deluded people, Professor Steven Novella answers that "we all are these people" (Novella, 2018, p. 48). Novella drives his point home with the following invitation:

> Think about some area in which you have a great deal of knowledge, at the expert to mastery level (or maybe a special interest in which your knowledge is above average). Now, think about how much the average person knows about your area of specialty. Not only do they know comparatively little, they likely have no idea how little they know and how much specialized knowledge even exists. Furthermore, most of what they think they know is likely wrong or misleading.
>
> Here comes the critical part: Now realize that you are as ignorant as the average person is in every other area of knowledge in which you are not expert. The Dunning–Kruger effect is not just about dumb people not realizing how dumb they are. It is about basic human psychology and cognitive biases. Dunning–Kruger applies to everyone. (Novella, 2018, p. 49)

No doubt we all suffer from the illusion of confidence in multiple areas of our lives. System 1 heuristics equate confidence with competence, even though confidence rarely reflects competence; confidence is a terrible measure of one's ability, and intelligence doesn't guarantee accuracy. As psychologist Daniel Kahneman explains, "Declarations of high confidence mainly tell you that an individual has constructed a coherent story in his mind, not necessarily that the story is true" (Kahneman, 2013, p. 212).

Also, it's easier to identify negative qualities in others than identifying them in ourselves (recall the fundamental attribution error). Fortunately, when we expect more of ourselves, we try harder. If you expect to get an A on a test, you'll try harder to make sure you do. Unfortunately, by misjudging our own incompetence, we're less likely to try to improve and thus more likely to remain incompetent. Believing you already know the answers to the test, you're less likely to study. Believing you already know how to Think Straight, you're not likely to read a book about critical thinking. This attitude traps us in a revolving door of ignorance and incompetence.

The second part of this illusion is mistaking one's confidence for competence, an error in judgment that can have serious consequences. Many have had lawyers who believed they were great attorneys, and their overconfidence convinced clients to trust them far beyond what was warranted. Even meaningless details, such as how someone dresses, can hugely impact how we judge their competence. For example, even though a terrible doctor is just as capable of dressing up as a good doctor is, we tend to believe a doctor that dresses in slacks and a T-shirt is less capable than one wearing a white lab coat over a suit (Cha, Hecht, Nelson, & Hopkins, 2004; McKinstry & Wang, 1991; Rehman, Nietert, Cope, & Kilpatrick, 2005). Over 37 percent of people believe that the confident testimony of an eyewitness is enough to convict a criminal defendant of a crime (Simons & Chabris, 2011), even though experts agree that "an eyewitnesses [*sic*] confidence is *not* a good predictor of his or her identification accuracy" (Kassin, Ellsworth, & Smith, 1989), that highly confident witnesses are only accurate 70 percent of the time, and that low-confidence witnesses are only accurate 30 percent of the time (Sporer, Penrod, Read, & Cutler, 1995). What's more, something as simple as offering an encouraging "OK" can influence witnesses to misidentify a crime suspect (Schacter, 2002, p. 116; Wells & Bradfield, 1998; Wells, Malpass, Lindsay, Fisher, Turtle, & Fulero, 2000). Whether it's lawyers, doctors, eyewitnesses, or ourselves, a good rule of thumb is to approach confidence with a healthy dose of skepticism.

More Errors of Perception

INATTENTIONAL BLINDNESS

The invisible gorilla experiment was conducted to study what psychologists call inattentional blindness: our total blindness to things we're not focusing on. That

Eyewitness testimony:

- Suggestion
- False confidence
- Confabulation
- Biases

One of the most unreliable forms of evidence is eyewitness testimony.

driver that seemed to "pop out of nowhere" didn't really pop out of nowhere. He was just somewhere you weren't paying any attention to.

Attention is a limited resource. If our brains processed all the information collected by our senses, we'd be completely overwhelmed. We use phrases like "pay attention" to convey that attention takes from the brain's limited attention budget. We might think of the brain as an economist that tries to reduce costs. Attention costs energy and thus is an expense. Since the brain has a fixed amount of energy, attention must be conserved. As a result, we miss a lot of what happens around us. Think about how much goes on around you when you're sucked into a good movie or focused on an art project. Our brains devote much of their attention budget to those tasks, while ignoring other sensory information it doesn't find relevant to the moment.

In an interesting experiment on inattentional blindness, a mock purse-snatching was shown on a television show. The thief's face was clearly visible, and so viewers were shown a six-person lineup and asked to call in and identify the perpetrator. More than 2,000 viewers called in, and more than 1,800 identified the wrong perp (Buckhout, 1975; Wiseman, 2011, p. 131).

Throughout our lives, we miss so much sensory information that if Chabris and Simon's gorilla represented all the information we encounter, we attend to only about one hair's worth. As a point of fact, "we experience far less of our visual world than we think we do" (Chabris & Simons, 2010, p. 7). It'd be impossible to pay heed to all that information—or even a significant fraction of it—so the brain ignores the information it regards as irrelevant (the rest of the gorilla) and processes only what it deems is important (that single hair). Being the cognitive misers we are, bound by veridical vending and System 1 thinking, we often discover that the brain's goals, such as conservation of energy or maintaining consistent internal narratives, are frequently at odds with our own goals of analytical reasoning or truth discovery. This disconnect is one of the many reasons Thinking Straight is such an important intellectual endeavor.

CHANGE BLINDNESS

A closely related phenomenon is called *change blindness*. Change blindness occurs when we fail to notice dramatic changes to our immediate environment.

Experiments show that when interacting with an unfamiliar person, the person can change outside of our view, and most of the time we won't notice that we're talking to an entirely different person. In one study, college students approached a counter to speak with a clerk. The clerk would bend beneath the counter to grab something, and a second clerk, sometimes looking nothing like the first clerk, would swap places with the first clerk (Simons & Ambinder, 2005). In another study, college students were stopped on campus by a bystander and asked for directions. In the middle of their conversation, confederates of the study carrying a door would walk between the college student and bystander and swap the bystander for another person (Simons & Levin, 1998). Again, the students wouldn't notice the switch.

EXPECTANCY ILLUSIONS

Psychologists also recognize what they call *perceptual constancies* (Wade & Tavris, 2017, pp. 194–96): how the brain interprets sensory input based on expectations. For example, we expect that when we turn our eyes, the objects in the world remain put (location constancy). We expect a plate to remain circular, regardless of which angle we view it from (shape constancy). In one experiment measuring *color constancy*, participants looked at cutouts of trees and donkeys that were illuminated by red light. Even though the cutouts were made from the same green material, the participants said the trees were green and the donkeys were gray (Duncker, 1939). Their brains expected gray donkeys, so they manipulated the data so that it matched their internal model of reality.

Size constancy—perceiving the size of everyday objects as constant regardless of their distance—is also of interest to psychologists, since it appears learned, not innate.[5] We understand that when something looks smaller as it gets further away, it's not literally shrinking. Nonetheless, this is how our brains interpret the world. Anthropologist Colin Turnbull, who was studying an African tribe that live in such dense forest that they never see anything far away, noticed that the tribespeople had never learned size constancy. Turnbull was shocked when he took a tribesman to an open plain where a herd of buffalo could be seen grazing in the distance, and the tribesman asked what kind of insects they were. After being told they were not insects but were buffalo twice the size of any he'd ever seen, the tribesman didn't believe him. As they neared the buffalo, the tribesman realized the "insects" were indeed buffalo and became frightened and confused: he wasn't sure whether the "insects" had turned into buffalo or if Turnbull was a witch (Turnbull, 1961).

The history of the phrase "flying saucer" provides an excellent example of both visual constancy and expectancy illusions. Aside from a few isolated reports of cigar-shaped "UFOs" between 1896 and 1897 (Hines, 2003, p. 235), and a wave of reports of flying machines leading up to the Wright brothers' invention of the airplane (Novella, 2018, p. 8), people didn't report seeing "flying saucers" until 1947, when a civilian pilot named Kenneth Arnold reported seeing nine shiny flying objects that he originally said flew like geese in formation (Shermer, 2013, p. 71), later saying the objects were shaped like the heel of a boot that moved across the sky "like that of stones or saucers skipping across a pond" (Prothero & Callahan, 2017, p. 65). Arnold never said the objects *looked* like saucers; he said they moved erratically, like a saucer skipping across the water, and the reporter misquoted him. Arnold later clarified, telling the journalist Edward R. Murrow that the press "did not quote me properly. . . . When I described how they flew, I said that they flew like they take a saucer and throw it across the water. Most of the newspapers misunderstood and misquoted that too. They said that I said that they were saucer-like; I said that they flew in a saucer-like fashion" (Prothero & Callahan, 2017, p. 318, and see Sagan, 1996, p. 70). Having been misused by the media and sensationalized by Hollywood, the term went to fixation and people have reported seeing flying saucers ever since.

It should be no surprise, then, that since the UFO myth began, the popularity of UFO sightings has increased proportionally to the rate at which they were portrayed by Hollywood and the media, as well as the rate at which someone could get their name in the papers by reporting a UFO sighting (Sagan, 2007). Political scientist Michael Barkun pointed out that within months of Arnold's report, 90 percent of the population had heard of flying saucers. By 1966, 46 percent of Americans believed in flying saucers, and in 1978 that number increased to 57 percent (Barkun, 2013, pp. 81–82). Even more shocking, Barkun points to a poll of Americans suggesting that 2 percent of Americas believe they were abducted by aliens (Barkun, 2013, p. 82)! Putting this into perspective, if that figure were cut in half, aliens would have abducted some 3.25 million Americans. A more generous figure says that if only one out of a hundred reported abductions left a single witness—e.g., pilots, neighbors, family members, amateur astronomers, sky watchers, military personnel, people getting a midnight snack, air traffic controllers—that would still leave more than 30,000 people who've witnessed someone get abducted by space aliens. Those witnesses would surely include some skeptics and other disbelievers, and those testimonials would be touted by believers. However, witness reports are predominantly those of believers.

Scientists also recognize what they call *temporal synchronization*. When we see something that also involves sound, the light reaches our eyes long before the sound reaches our ears. To compensate, so long as both lines of information reach us within 80 milliseconds of each other (about 90 feet distance), the brain will stitch together a seamless event as if there were no lapse between sensory inputs, thus constructing an alternative reality. We usually only recognize this when the sensory stimuli don't align, such as when we hear a baseball being hit shortly after we see the batter hit the ball. Temporal synchronization may also explains how echo-locating bats can zero in on moving prey that will be in a different place from when the sound wave left it (Simmons, 1979; Wiegrebe, 2008). Temporal synchronization is clearly adaptive, as without it we couldn't perform even basic functions like dodging a spear or avoiding a falling tree.

Baseball players can hit fastballs not because they can "see" the ball over the plate. They couldn't possibly "see" that happen, since that event is still in the future when they begin swinging the bat. They know when to swing the bat and where to aim it because the brain projects (confabulates) the information needed to perform the task. Because of temporal synchronization, the brain projects an image of where a moving object should be, and as a result, every game of catch doesn't end in injury, and professional baseball players strike out *only* 70 percent of the time.

Since the brain can only process a fraction of the information it receives, it makes sense of everything by tapping into prior beliefs and expectations and by confabulating details to create a consistent narrative. You see the same symbol differently depending on its context (see figure 8.3) because of information that's already stored in the brain. Based on prior expectations, System 1 constructs the letter B in the top image and constructs the number 13 in the bottom image.

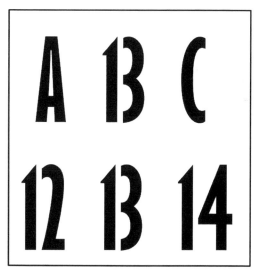

Figure 8.3.   Referential illusion.
Courtesy of the author.

Pattern Recognition

Have you ever bought new shoes and then started seeing the same ones everywhere? A friend gets a new car and it seems like the same model is everywhere you look? You learn a new word and everyone seems to be saying it? A loved one dies in Chicago, and suddenly you're hearing about Chicago every day? This phenomenon is a cognitive bias known as *frequency illusions*, or the *Baader–Meinhof effect*. It's caused by heightened awareness of the object of awareness and can lead us to jump to the conclusion that the universe is trying to bring our attention to whatever that thing is. "This must be a sign" or "The universe is trying to tell me something" are common phrases when it comes to recognizing patterns.

Pattern recognition is one of the most useful tools humans ever evolved. But as the saying goes, appearances can be deceiving. If the saying is true, are the patterns we see ever an accurate guide to reality? To be sure, several adaptive advantages to recognizing patterns exist. For instance, recognizing patterns led to the discovery of agriculture, and helped our ancestors avoid sickness, outsmart predators, and find the foods they needed to survive. Because pattern recognition was so useful throughout our evolutionary past, our brains are hardwired to seek them out. Yet, as we've seen, the brain is an error-prone mess, and sometimes it finds illusory patterns that we don't realize are illusory. Harrison puts it thus: "On one hand it makes sense for us to see some patterns of things that aren't really there in order to be good at seeing real ones that matter. On the other hand, we need to be aware of this phenomenon because it can lead to a confident belief in things that are not real or true" (Harrison, 2013, p. 64). We have an obvious interest in believing what is true, which is why learning about how our brains construct a completely subjective reality is a vital step toward Thinking Straight.

Patterns Galore

According to neuroscientist Steven Novella, "brain processing is based largely on pattern recognition" (Novella, 2012, p. 36). Seeking connections, our brains detect patterns in everything from numbers and words to events and objects. Evolution sculpted our brains to constantly seek patterns, whether real or illusory. We evolved to take patterns seriously because sometimes recognizing a pattern meant life or death; for example, when a spider with a red hourglass on its belly bites, the person bitten gets deathly sick. Accordingly, our brains attach meaning and emotion to the patterns it finds (red hourglass rouses fear), convincing us that all patterns are real and meaningful (hence arachnophobia), even when they're not (some spiders are harmless).

The tendency to seek patterns and assign them meaning is what psychologists call the everyday *illusion of cause*. The illusion of cause "arise[s] from the fact that our minds are built to detect meaning in patterns, to infer causal relationships from coincidences, and to believe that earlier events cause later ones" (Chabris & Simons, 2010, p. 153). Since we know System 1 is terrible at testing reality, Thinking Straight demands that we take a skeptical look at what our brains do on autopilot.

In fact, the part of the brain that tests reality is the same part that filters and processes pattern recognition (Novella, 2012, p. 36). When the brain detects a pattern, it runs the pattern against its internal model of reality to determine whether the pattern is consistent with the existing narrative. If any inconsistencies arise, the brain "fudges the data" rather than throwing it away. Note that nothing written into this programming tells the brain whether it's making accurate deductions: System 1 is just doing its thing. More often than not, the process works correctly, and since an occasional misfire doesn't usually kill us, evolution didn't select against this software (it's better to fear *all* spiders than to not fear a deadly one).

While dreaming, our reality testing software isn't as active as when we're fully awake—hence hypnopompic (the state of consciousness leading out of sleep) and hypnogogic (the state of consciousness leading into sleep) hallucinations. We hardly need research to tell us this, as we've all experienced bizarre dreams that seem perfectly sensible, like surviving death or breathing under water. When awake, if we remember our dreams at all, oftentimes they don't make any sense; that is, they don't accord with reality.

A well-known psychological condition known as *psychosis* is characterized by an inability to test the patterns one sees against reality (Novella, 2012, p. 37). Psychosis doesn't prevent pattern detection, only the ability to test them against reality. Consequently, those suffering from psychosis tend to interpret illusory patterns as real and compelling. People with psychosis become so convinced by illusory patterns that it's difficult to understand why others don't see them as well. Moreover, psychosis isn't an all-or-nothing phenomenon. From fully diagnosable to hardly recognizable, psychosis covers a broad spectrum. At nearly every stage, the sufferer is usually clueless about their delusions and convinced about their reality.

Another powerful pattern-recognizing phenomenon is **pareidolia**—pronounced "pehr-uh-dowl-ee-uh"—the unwavering compulsion to see patterns in everything, along with the inability to *un*see patterns once they're detected. Seeing human faces in rocks, animals in the clouds, or Elvis's face on a piece of toast are examples of pareidolia.

Figure 8.4. Happy pareidolia. *Shermer, M. (2010, February).* **The pattern behind self-deception.** *TED. https://www.ted.com/talks/michael_shermer _the_pattern_behind_self_deception (accessed July 7, 2020).*

VISUAL PAREIDOLIA

Our brains have an area called the *fusiform gyrus*—specifically, the *fusiform face area* (FFA)—which recognizes, remembers, and identifies human faces (Harris, 2011, pp. 238–39; Sergent, Ohta, & MacDonald, 1992). Humans can distinguish between thousands of faces, even when the differences between them are slight. Our lifelong fascination with faces begins at birth, which is why babies stare at the pattern on the top (figure 8.5) much longer than they stare at the one on the bottom (Hood, 2009, pp. 123–24). The moment the brain sees something with face-like features, the FFA treats it like it's a face, processing the information differently than it would other objects. Studies show that it takes the brain only a fifth of a second to distinguish faces from ordinary objects like chairs, although objects or images that resemble faces—e.g., ambiguous pictures—still activate the FFA. Fortunately, it takes the brain only slightly longer to distinguish a face from something merely resembling a face (Hadjikhani, Kveraga, Naik, & Ahlfors, 2009). Artificially stimulating the FFA can even cause

Figure 8.5.   Seeing faces.
Courtesy of the author.

people to see faces spontaneously (Shermer, 2011a, p. 130). Unfortunately, damage to the FFA can cause a disorder known as *prosopagnosia*, where the ability to recognize faces is impaired but other cognitive functions are not (Hood, 2009, pp. 123–29). Imagine being unable to recognize your own face or the face of someone you love!

The first example of pareidolia on Wikipedia is the so-called Face on Mars, a picture of an area on Mars that slightly resembles a human face (or a Jason mask). Initial photographs were very low-resolution (worse than 250 m/pixel [820 ft/pixel]), and the image resembles a human face, especially after having been primed to see one. The image was captured with antiquated technology, half of it is in shadow, and what looks like a nostril is actually just a missing pixel of data (Sagan, 1996, p. 55). Twenty years later, high-resolution pictures (at a resolution of 14 m/pixel [46 ft/pixel] or better) of the same structure showed a normal geological formation that looks nothing like a face (Posner, 2000).

When our brains create memories and construct reality, they're also trying to create patterns that resemble the ones they're familiar with. Once it finds one, the pattern is usually locked in and we can't unsee it. *Cognitive priming*—another form of pareidolia—occurs when the brain has primed the visual cortex to see something specific. For instance, in a familiar image that depicts

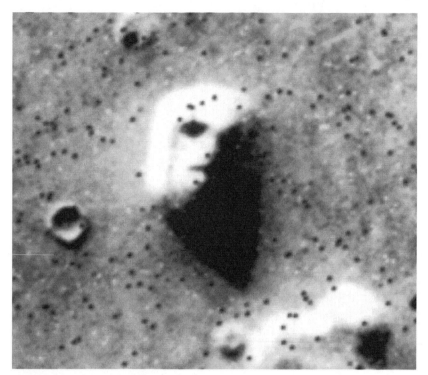

Figure 8.6.   Face on Mars. *Wikipedia, https://en.wikipedia.org/wiki/Cydonia_(Mars)#%22Face_on_Mars%22.*

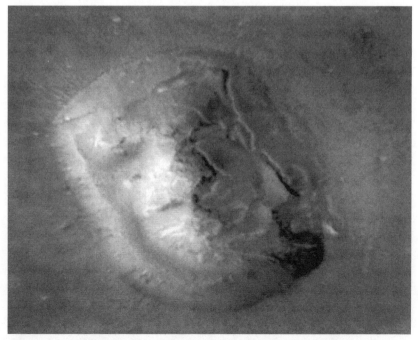

Figure 8.7.   Science reveals the face on Mars. *Wikimedia.org.*

an old woman and a young woman, depending on how you look at it, about 95 percent of the time people see the young woman in the image if they're primed with pictures of young women, whereas they see the old woman if they're primed with pictures of old women (Leeper, 1935). Other research shows that people who believe in the supernatural or paranormal see faces in inanimate objects more often than skeptics (Riekki, 2012). Additionally, people suffering from specific phobias are usually prone to having multiple phobias (Barlow & Ellard, 2018; Hofmann, Lehman, & Barlow, 1997).

Pictures of ghosts or supposed portals to other dimensions are good examples of pareidolia or delusions. A flash might bounce off moisture or dust in the air or even end up on the photo paper in the development room, creating a so-called orb or some other anomalous effect. A finger or strand of hair obstructing the lens or over and double exposure can also corrupt a photo, caus ing our brains to see shapes and objects that look familiar but aren't really there (Nickell & Biddle, 2020).

AUDIO PAREIDOLIA

We don't just recognize patterns in visual stimuli; we also hear them. For example, it was once widely rumored that demonic lyrics could be heard when Led Zeppelin's "Stairway to Heaven" was played backward (Shermer, 2006). The alleged reason behind "backmasking"—as it's called—hidden messages was to influence people to behave in certain, perhaps sinister ways. Fortunately, backmasked songs are nothing more than pareidolia, and influencing human behavior by backmasking hidden messages isn't possible (Vokey & Read, 1985).

The tendency to hear patterns in randomness was described in a set of experiments in 1936 by famed psychologist B. F. Skinner (Skinner, 1936). In Skinner's experiments, participants listened to nonsense syllables—such as only vowel sounds—and later reported hearing sounds similar to normal speech, including consonant sounds they couldn't have heard. Psychologist Diana Deutsch made similar findings. Deutsch took a two-syllable nonsense word and played it on a repeated loop. After several minutes, her participants reported hearing a variety of words that weren't present in the recording, sort of like hearing the phone ring when you're in the shower, even though it never rang. The words they heard also changed meaning over time and tended to reflect the participants' personal interests (Deutsch, 2008; Deutsch, Henthorn, & Lapidis, 2011). Electronic Voice Phenomena (aka EVP) are prime examples of phenomena people genuinely believe exist that are based entirely on audio pareidolia.

Another phenomenon associated with auditory stimuli is a *mishearing*. Mishearings occur when we hear something unclearly and our brain makes out what it thinks it heard, like hearing "super salad" when what was actually said was "soup or salad." Our brain takes the distorted information and instantly taps into its database to formulate a pattern (Burton, 2008, pp. 75–76).[6] *Paracusia*, having an auditory hallucination despite there being no auditory stimuli, is another type of pareidolia that demonstrates the unreliability of the things we think we hear.

The Prevalence of Pattern Seeking and Pattern Recognition

To generate hypotheses or gather information, researchers sometimes do something called *data mining* (dishonest researchers also engage in what's known as "p-hacking," which is like data mining's evil twin). Data mining involves looking for information in a dataset to find potential patterns. In some cases it's a good way to form hypotheses. However, good researchers don't *rely* on data mining as confirmatory evidence because they understand that *the law of large numbers*—a mathematical theorem that explains why improbable events must occur regularly—says patterns should be found in random data anyway. Data mining is just a guidepost. Good researchers attempt to confirm that any patterns found while data mining are real by independently testing them against new sets of data. Using new datasets decreases the chances that faulty data are carried into future analyses, and independent datasets won't likely carry the same random patterns, if indeed they're random (Smith, 2019).

Because of the brain's pattern-seeking habits, data mining also happens subconsciously, which is very different from doing it deliberately. Subconsciously data mining can be dangerous, as it can cause people to see patterns without seeing all the data the pattern is buried in. Some drugs, such as amphetamines or hallucinogens, can amplify this experience (Hood, 2009, p. 251) and increase our tendency to draw conclusions that are far removed from reality. For example, Melisa might wrongly think (i.e., hallucinate) that more cops are on patrol than usual, concluding that the cops are looking for her. Perhaps because she's hallucinating, Melisa doesn't realize that her brain constantly searches for patterns. However, the patterns it finds aren't necessarily real or meaningful.

The law of large numbers and the shortcomings of data mining are why using more than one dataset when forming a hypothesis is so important. If Melisa investigated a completely independent dataset—such as the cops' patrol logs—she might discover that the usual number of cops were out, or perhaps there were more but for reasons that had nothing to do with her. We can infer that seeing agency where it doesn't exist leads to conspiracy thinking. I'll return to conspiracy thinking in chapter 13. For now, I'll simply point out that sometimes paranoia is little more than an inability to explain something or come up with alternative explanations.

Similar to data mining is establishing what's known as an *arbitrary stop point* (Smith, 2018, pp. 128–29). Let's say some sugar from a prison's pantry came up missing and the guards believe someone stole it. They might begin their investigation by reviewing camera footage that shows who went in and out of the pantry. Now, if the guards watched the footage up until they saw someone go into the pantry and concluded that that person stole the sugar, they'd be establishing an arbitrary stop point; that is, stopping their investigation as soon as they found any evidence that confirms their suspicions. It could be that the first person to enter the pantry did steal the sugar, but it could be someone who entered the pantry hours later, or it could be that someone just used too much sugar in their cookie dough. To properly read data, investigators must review all the data rather than just the part of the data that confirms a hypothesis.

Also hardwired into our brains is a phenomenon called *hyperactive agency detection*, or *apophenia*. Apophenia is the tendency to perceive agency and meaning in random patterns. Michael Shermer calls this "patternicity," which he defines as *the tendency to find meaningful patterns in both meaningful and meaningless data* (Shermer, 2011a, p. 5). Shermer argues that selection pressures urged our ancestors to assume that all patterns are real and to treat all "patternicities" as real and important. Committing *Type 1 Errors* became evolutionarily advantageous because "there would have been a beneficial selection for believing that most patterns are real" (Shermer, 2008b; Shermer, 2011a, pp. 60–62).

According to Shermer, patternicity occurs whenever the cost of making a *Type 1 Error* is less than the cost of making a *Type 2 Error*, so the default is to assume all patterns are real (Shermer, 2013, p. 46). Accordingly, detecting false or illusory patterns is the low-cost tradeoff for the ability to make *Type 1* (not believing a falsehood) and *Type 2* (believing a truth) *Hits*. When committing a *Type 1 Error* (believing a falsehood; finding an illusory pattern) by, say, fleeing from the sound of a rustling bush when it was only the wind, the cost is relatively low: a few wasted calories. But when committing a *Type 2 Error* (rejecting a truth; ignoring real patterns) by, say, not fleeing from the sound of a rustling bush that hides a dangerous predator, the consequences range from serious injury to certain death (Nesse, 2001). Such hardware, argues Shermer, laid the basis for superstition and magical thinking.

Notably, pattern recognition isn't unique to humans. In a set of classic experiments, the behaviorist B. F. Skinner created "superstitious" pigeons by placing them in a "Skinner box" and delivering food every fifteen seconds (Skinner, 1948; and see Wright, 1962). It didn't take long before several pigeons were ritualistically repeating whatever they were doing just before the food was released. Further, Skinner found that randomly assigning rewards had the same effect, producing dramatic displays of superstitious behavior such as spinning counterclockwise, bobbing their head, pecking the ground, and flapping a left wing. The pigeon's brains were seeking patterns where none existed.[7]

Another way we misperceive patterns is committing the *gambler's fallacy*, which can be illustrated with the following problems (Stanovich, 2013, p. 161).

Problem A: In a trial in which a fair coin is flipped, the coin has just landed on heads five times in a row. For the sixth toss, is it more likely that (1) tails will come up; (2) heads will come up; or (3) heads and tails are equally likely to come up?

Problem B: People playing slot machines win something about one out of every ten tries. Patrick just won on his first five tries. What are the chances that Patrick will win on his sixth try?

In Problem A, the answer is 3, since the chance of getting heads or tails is exactly 50 percent with each flip. Although intuition suggests that it's now more likely for tails to show up because we've seen heads five times in a row, each flip is statistically independent, and the improbability of the first five flips doesn't carry over into future flips. As skeptic and author Guy P. Harrison explains, "Coins do not have an internal memory system that is linked to a guidance control mechanism that allows them to adjust their landing positions based on previous flips" (Harrison, 2012, p. 84). The same reasoning applies to Problem B; no mechanism exists by which information from prior tries carries over into future tries. On any given flip of the coin or pull of the lever, we cannot predict the outcome any better than the odds suggest. Nonetheless, we can say that over the course of many trials, the *average* over those trials is statistically predicted.

Despite the common feeling that we're "due" for something to happen to us, the universe doesn't keep track of our behavior, doling out punishment and reward. Moreover, there's no known mechanism by which the universe even could. The moment we lose focus, all the errors of our brains that we've been discussing—e.g., pattern recognition, agency detection, belief in essence, data mining—conspire to create an illusory picture of the world that is nearly impossible for us to notice. Illusions may be emotionally compelling, so we need CT skills to systematically test apparent patterns to truly know if what we see is an illusion or completely real.

We also interpret the world based on how we believe the universe works. To this effect, psychologists recognize what they call a *locus of control*, referring to the "general expectation about whether the results of your actions are under your own control (internal locus) or beyond your control (external locus)" (Wade & Tavris, 2017, p. 401). People with an internal locus of control attribute what happens to them to their own actions. For example, when Rebecca scores high on a test, she'll praise her own intelligence and study habits. On the other hand, people with an external locus of control attribute what happens to them to other sources, such as luck, fate, or other people. If Rebecca's test scores tanked, she might blame the results on bad teaching or missing information in the textbook. Research by Susan Blackmore showed that believers in ESP tend to interpret data and see evidence of the paranormal, whereas skeptics came to the opposite conclusion. Blackmore also discovered that, although believers recognized more patterns than skeptics, they also made more *Type 1 Errors* (false positive) than skeptics (Blackmore & Moore, 1994). Thus, how we view the world impacts how we perform basic tasks, such as interpreting straightforward data.

Wrapping Up

In this chapter, I covered some complex aspects of our mental lives. While we've learned how unreliable some of our mental processes are, we've also equipped ourselves with tools for understanding those processes, preparing ourselves to deal with cognitive flaws without being overly invested in something that might be wrong. Our senses bring us the information we rely on to navigate our lives. Most of the time, we can trust that information, though not unconditionally. Certain circumstances and human attributes contribute to a wide range of perceptual errors that oftentimes lead us astray. Thus, we can trust our senses unless we have good grounds not to, and we have good grounds not to if we cannot corroborate the information they bring us with external, reliable mechanisms. Each of our senses serves a unique purpose in continuously shaping our internal models of reality. However, if we want our perceptions to accord with reality, we must recognize their fallibility and always err on the side of skepticism.

> If you would be a real seeker after truth, you must at least once in your life doubt, as far as possible, all things.—René Descartes

CHAPTER 9

Innumeracy and the Deception of Coincidence

Some things are just hard for our brains to grasp, and numbers are certainly one of them. With anything involving huge scales, such as the age of the Earth, the size of the universe, the number of stars in the sky, or the inevitability of coincidence, we have great difficulty wrapping our heads around how far, big, or long they are. Why? Because our ancestors evolved to think in terms relative to their existence. In terms of distance, they were concerned with miles (at most). In terms of time, they were concerned with decades (at most). Since there was never any survival or reproduction advantage to understanding vast scales, there was never any selection pressure against misunderstanding large scales. Consequently, understanding extreme scales doesn't come naturally, which creates problems in advanced societies.

This chapter is about probability, exploring the reasons why we have trouble dealing with extreme scales, and why we seek meaning in random coincidences. By the end of the chapter, readers should have a good understanding of why extreme scales are so hard to grasp, and we might even develop a greater appreciation of the immensity of those large-scale units, despite how hard it is to wrap our heads around them.

Building Vocabulary

Cold reading:	Barnum effect:
The technique of using high-probability guesses and assumptions to give the impression of having information about a person or event.	The tendency to give high-accuracy ratings of a vague and generalized description of one's personality.

Confederate:	Negligible:
An aide of an experiment whose behavior is determined prior to the study being carried out.	Too small or unimportant to be of concern.

Innumeracy

Although less familiar, the word "innumeracy" is similar to "illiteracy." While illiteracy refers to words, innumeracy refers to numbers. Innumeracy describes our general misunderstanding of large numbers and their consequences. Humans are driven largely by folk numeracy, which is the tendency to misperceive probabilities, think anecdotally rather than statistically, and focus on and remember short-term trends and small-number runs (Shermer, 2013, p. 38). Although both illiteracy and innumeracy can be attributed to poor education (Lefevre, 2000), the roots of innumeracy span from psychological holdups to misunderstandings of mathematics (Paulos, 2001, pp. 96–132). Fortunately, the cure for innumeracy is knowledge and skills, such as mathematics and CT.

> They should have known better. The probability of a train derailment was infinitesimal. That meant it was only a matter of time.—N. K. Jemisin

Research shows that numerate people are less susceptible to framing effects (chapter 5) and better understand numerical risk (Weller, Dieckmann, Tusler, Mertz, Burns, & Peters, 2013). Thus, innumeracy explains why we break phone numbers up into sections and why we opt for treatments with a 90 percent survival rate but reject one with a 10 percent death rate (Kahneman, 2013, p. 367; McNeil, Pauker, Sox, & Tversky, 1982). Statistical information permeates our lives (Lusty, 2006), and to make sense of the complexities of the modern world, we need a basic understanding of numbers and statistics; the problem is most of us lack these basic skills. For example, it may come as a surprise that fully half of all Americans are below-average intelligence, though this is a fact about averages, not a lament about Americans. IQ is generated by collecting information on many people, then drawing a line in the middle, which is considered the "average." Anyone below the line is below average; anyone above the line is above average. The same reasoning applies to height, age, and so on, yet astonishes many who learn it for the first time.

Consequently, even averages are oftentimes deceptive. The average American earns about $54,000 per year (Dobelli, 2014, p. 164). Now, imagine you're at a party with fifty average Americans and Bill Gates shows up. What's the average wealth among party attendants? Bill Gates is worth around $126 billion, which skyrockets the average wealth of party attendants over two million percent. The new average is meaningless because of one outlier. When you hear the word "average," try thinking about the distribution. The average American city has between 100,000 and 300,000 people, but this average is just as meaningless as our Bill Gates example because extremes are found at both ends.

While people are good at recognizing patterns—sometimes too good, as we saw in chapter 8—our intuitive understanding of numbers, probability, statistics, and coincidence is far from impressive. For instance, guess how long a million seconds is. The answer is about 11½ days. While that answer isn't exactly shocking, if you guessed how long a billion seconds is, you wouldn't likely have said nearly 32 years! Even this number pales in comparison to the time it takes for a trillion seconds to pass: nearly 32,000 years. And the age of the universe in seconds is about 10^{18} (a one followed by 18 zeros) seconds.

One example that highlights the unreliability of intuition is the bat and ball problem (Kahneman, 2013, p. 44). Off the top of your head, what does intuition tell you the answer to the following question is?

> A bat and ball cost $1.10.
> The bat costs one dollar more than the ball.
> How much does the ball cost?

If you're like most people, the number that came to mind is $0.10. However, this intuitive answer is incorrect, since the total cost would be $1.20 if the ball cost $0.10. If the bat costs one dollar more than the ball and the total is $1.10, the ball must cost $0.05. That is, $0.05 + $1.05 = $1.10. System 2 thinking is required to figure out what the ball costs. Even simple numerical problems can easily throw us off. When numbers get bigger and the resolutions are more subtle, all bets are off.

Coincidence Demystified

The word "coincidence" "refers to the chance alignment of two variables or events that seem to be independent, especially if it seems as if the occurrence defies the odds" (Novella, 2018, p. 134; Stanovich, 2013, p. 173). Lacking training to the contrary, almost everyone seeks meaning in coincidence. When rare events occur, we look for meaningful explanations. Add in the emotional investment of, say, picking lottery numbers of your kids' birthdays then winning the lotto and a "hidden meaning" appears. We search for meaning because, as Sagan explains, "We're significance junkies" (Sagan, 1996, p. 372). Although clinging to significance can be comforting, it misses the forest for the trees. Searching for hidden meaning in a random event means we aren't appreciating the fact that coincidences happen at all. Even if some meaning were offered for a coincidental occurrence—for example, "The universe meant for 'X' to happen"—we'd never know that the offered explanation is *really* the meaning behind the event. Since forcing a coincidence isn't possible, the claim is untestable and therefore is unknowable.

For example, no one finds it odd that thousands of people dream about plane crashes every day or that planes crash on occasion or that people die every day. The number of dreams, plane crashes, and deaths that occur daily represent a dataset. When someone dreams (very common) about a plane crash (fairly common) and the next day knows someone involved in a plane crash (a statistical certainty given the numbers), the tendency is to ignore the size of the dataset and look for meaning in this otherwise random string of events. By all accounts, due to the sheer numbers involved here, the chances of someone having a dream about a crash, followed by knowing someone who was involved in a crash, is a statistical certainty (Shermer, 2013, pp. 11–12). Many explanations could be offered to "explain" this coincidence; for example, "everything happens for a reason" or "it was all part of a cosmic plan." Whatever you come up with, it's not likely you'll find an explanation that is ultimately testable, and without a testable hypothesis, there's no reason to think this event is anything other than a random coincidence that's statistically demanded.

Seeing the world as though coincidences have some hidden meaning completely ignores all those events that don't line up, which are way more numerous than those that do; imagine all the dreams about plane crashes that don't align with real-world events. Such thinking makes coincidences appear more meaningful than they really are. The "meaning" behind a coincidence is basically an illusion. Playing spades, hearts, bridge, or pinochle, if we ever saw a so-called perfect hand dealt, where every player has all cards of one suit, we're likely to decry how statistically unlikely this hand is, not realizing that every possible arrangement of hands is equally as unlikely.

In an excellent article (Molé, 1999) that takes the popular book *The Celestine Prophecy* to task, Phil Molé explains three mistakes we make when determining the likelihood of coincidences:

1. *We often fail to realize that many different combinations of events can result in the same coincidence.* In other words, we ask the wrong question. For example, rather than asking "What are the odds that I would dream about a plane crash and find out that so-and-so died in a plane crash the next day?" we should ask, "What are the chances that anyone would dream about a plane crash and learn about a friend's death in a plane crash the next day?"[1]

2. *Even if we know the correct probability for an event, we often misunderstand the information it contains.* To use Molé's example, "If a particular event (or intersection of events) has a 0.05% chance of happening during a given time interval, the odds of it happening to you or me are quite low. But if it does happen to either of us, we should not conclude that there was only a 0.05%

chance of the event arising from chance, and a corresponding 99.95% chance that it happened through some other (more mysterious) means."

3. *We have a natural tendency to force order on the world around us, and this tendency can lead us to misinterpret the significance of our experiences.* This tendency compels us to make irrational conclusions regarding the events that occur, often in the face of evidence that it's completely random. Nobody wants to think their lives are so insignificant that even the most meaningful thing that can happen to them—like a death of a loved one or winning the lotto—was purposeless. However, as Molé sagely observes, "personal feelings are not a measurement of truth, no matter how passionate they are."

That's So Random

Molé's observations reflect a commonly overlooked shortcoming about human nature that most of us live blissfully unaware of: our difficulty understanding randomness and coincidences. For example, which of the following four sets of numbers isn't random?

Set 1: 6840;   **Set 2:** 8403;   **Set 3:** 7777;   **Set 4:** 5387.

Of course, this is a trick question, playing on what's called a *clustering illusion*: seeing clusters of patterns in randomness. Most people say 7777 isn't random, and when asked to write a random sequence of numbers, few create one that resembles Set 3. People don't see four consecutive digits as random; they see it as a non-random pattern. Psychologists dubbed this common tendency *repetition avoidance*. A good example is seeing patterns in the arrangement of stars, like stellar constellations. Regardless of the stars' arrangements, people see patterns among them.

Psychologist Daniel Kahneman provides another great example (Kahneman, 2013, p. 115; Tversky & Kahneman, 1971) of how easily we're fooled by clustering illusions. Six babies are born in sequence. Are the following sequences of the sex of the babies equally likely?

Set 1: BBBGGG;   **Set 2:** GGGGGG;   **Set 3:** GGBGBG;   **Set 4:** BGBBGG.

The sequence in which boys and girls are born is obviously random, the previous birth having no influence over the next. Even though unlikely sequences contradict intuition, each sequence is equally as likely! Since each birth is independent, and the outcome of the sex is approximately 50 percent B/G, any random sequence is equally as likely. The same goes for coin flips, rolls of the roulette wheel, being struck by lightning, and any other occurrence. Even the so-called hot hands effect—when basketball players can't seem to miss a basket—is explained by our tendency to seek patterns (Gilovich, 1991, pp. 11–17; Gilovich, Vallone, & Tversky, 1985). Even knowing this doesn't help us believe it because once we see a pattern, our brains cannot unsee it. Research has even shown that skeptics and believers in ESP are affected differently by repetition avoidance, with believers finding greater meaning in random patterns than skeptics (Brugger, Landis, & Regard, 1990; Musch & Ehrenberg, 2002; Shermer, 2011a, p. 79).

Searching for Meaning

As we saw in chapter 8, our brains are constantly seeking patterns. This tendency likely exists because it has a number of survival advantages, such as assisting us with identifying rotten food, deadly predators, friends, foes, potential mates, and so on. Since identifying patterns is important for survival, our brains auto-matically imbue patterns with meaning so that we take them seriously. Skeptic Michael Shermer calls this "agenticity," or *the tendency to infuse patterns with meaning, intention, and agency, often invisible beings and from top down* (Shermer, 2010). Shermer agrees that believing patterns have meaning is an evolved sur-vival strategy. For instance, noticing that Mom got sick after eating moldy fruit is a pattern. The ability to identify that pattern prevents us from eating moldy fruit, thus increasing our chances of survival.

Detecting patterns and imbuing them with meaning is one reason beliefs such as "everything happens for a reason"[2] or "*they* control everything" are so powerful and convincing (see chapter 13). However, even if this inclination is innate, we certainly couldn't conclude that the patterns we see *actually* have meaning. The *meaning bias*, our need for the world to make sense or have some meaning, can lead us to false assumptions about the world, preventing us from noticing all the data our assumptions are wrapped in. By chance alone, random-ness produces apparently meaningful patterns. Science works out which patterns are meaningful—and which ones are not. "This is why we need science," as professor Steven Novella put it. "Science is partly the task of separating those patterns that are real from those that are accidents of random clumpiness. Sci-ence is formalized reality testing" (Novella, 2018, p. 130).

Say you're traveling in Europe and you run into your high school sweet-heart. A coincidence you're not likely to forget, but would it be any less shock-ing if you ran into your high school nemesis, your social studies teacher, or the school janitor? The chances of you running into any one of these people *in particular* are quite small, but the chances of you running into *any person* from your high school are much greater, though no less surprising. Once you think about the number of possible coincidental endpoints, opportunities for surprising coincidences become virtually endless.

What about thinking about a friend who calls seconds later or learning a new word that all of a sudden everyone is saying or talking about a movie only to find out that the movie is playing as soon as you turn the TV on? Such pairings—called *referential delusions* (Jaspers, 1913/1997, pp. 100–103; Kiran & Chaudhury, 2009)—seem odd but are fairly common. It's through the brain that we experience the world, including the thousands of things we hear, see, dream, and think every day. If we took stock of everything we experienced, we would see that some of those experiences inevitably line up with what happens in the world.

Psychologists David Marks and Richard Kammann explain innumeracy mathematically. If a person can recall 100 events in an ordinary day (we can recall many more), the total number of possible pairings for that person in one day is 4,950 (that is, $99 + 98 + 97 \ldots + 3 + 2 + 1 = 4,950$) (Schick & Vaughn, 2005, p. 78). Using this formula, in 10 years (roughly 3,650 days), just 1,000 people would generate more than 18 billion pairings ($4,950 \times 3,650 \times 1,000 = 18,067,500,000$). Accordingly, we should expect world events to align with our thoughts on a regular basis.

It's like when you're thinking about Sam and just then Sam calls. By itself it seems like a coincidence too great to have happened by chance alone, so naturally you think there's some sort of mind-to-mind communication or cosmic narrator pulling the strings. However, this coincidence is hidden in tons of unseen and unnoticed data (like 18 billion pairings!), which brings us back to the confirmation bias (see chapter 5). We don't usually think about the billions of phone calls made every day, nor do we remember all the times we thought about Sam and he didn't call. Even so, by sheer chance we should expect these two sets of data to line up on occasion, which is exactly why they do.

The belief that coincidences carry a deeper meaning is sometimes the result of asking the wrong question. When something highly improbable happens, we tend to think, "What are the chances that I would dream about Denise and see her obituary the next day?" when the question we should be asking is, "What are the chances that a coincidence like this would happen at all?" This thinking error is an example of the *lottery fallacy*, which goes something like this: if you won the lottery, you might be compelled to believe something that life-changing couldn't have happened to you simply by chance. The odds of it happening to *you* are literally hundreds of millions to one, which seems incredible. However, it's statistically certain that *someone* must eventually win. So the right question to ask is, "What are the chances that *anyone* would win the lottery?" and those odds are really good—so far, someone has won every lottery there's ever been.

American philosopher Daniel Dennett provides an excellent example of this all-too-common fallacy (Dennett, 1995, pp. 179–80). Let's say we conducted a ten-round coin-tossing tournament with 1,024 contestants. Each contestant has no idea that there are 1,023 other contestants. You tell each contestant that you're about to bestow great powers upon them, that with their new powers they'll win ten consecutive coin tosses without losing one. You then arrange for them to meet in pairs and compete until you have a final winner. Provided you don't allow any of them to talk to each other (and discover that you've told each one s/he has been bestowed great powers), whoever wins—and there *must be* a winner—will now have credible evidence that s/he has indeed been bestowed great powers. We could run this experiment one hundred times and get a new "winner" each time who would be equally convinced of their powers, never realizing the inevitability that someone would hit ten consecutive tosses.

Here's another example (Dobelli, 2014, pp. 58–59). We put a million monkeys on computers and allow them to trade on the stock market. In a week, about half of the monkeys turn a profit, and the other half don't. We keep the ones that made a profit and repeat the process for twenty weeks, at which point we'll have just one monkey remaining. This one monkey successfully turned a profit on the stock market for twenty consecutive weeks. Monkey genius? Obviously not. Statistically, if we repeat the process enough times, on average, we'd produce a monkey that by sheer chance made "good" decisions many times in succession.

Since their results are largely personal and rare events don't appear to have any causal explanation, statistically improbable coincidences affect us deeply. When we make connections between events and our personal lives, we completely ignore the billions of other people on the planet, which means a massive, active dataset is always in motion. Let's assume something with a chance of literally 1 in 10 billion happened to you. Our planet's population is approximately 8 billion people, so an event with the odds of 1 in 10 billion should occur roughly every 1.25 days. Something this improbable is bound to shock your senses.

> In reality, the most astonishingly incredible coincidence imaginable would be the complete absence of all coincidences.
> —John Allen Paulos

Additionally, people untrained in statistics, such as politicians and the media, are notoriously bad at conveying statistical information. For example, after a New Jersey woman won the lottery twice in four months, the *New York Times* reported (Samuels & McCabe, 1986) the odds of this happening were one in seventeen trillion! In fact, the odds of a specific person buying a single lottery ticket for each of two specific lotteries are indeed seventeen trillion to one. But here's an example of asking the wrong questions. The right question to ask is, "What are the odds that *anyone* would win a U.S. lottery twice in a four-month period?" Actually, the odds of someone winning the lottery twice in a seven-year period are better than 50-50, and winning twice in four months is 30 to 1 (Alcock, 2018, pp. 127–28).

To the untrained eye it seems like there must be some deep and hidden meaning to such unlikely coincidences, but the skeptic understands that the law of large numbers predicts that rare events occur frequently. Because we don't notice when likely events happen, our *selective memory bias* leads us to highlight unlikely events, so we notice rare events more than common events. Putting this into perspective, assume rare events like our 1 in 10 billion example never occurred—a highly suspicious situation that would defy the mathematical certainty of probabilities and require a really good explanation. As Aristotle once said, "It is likely that unlikely things should happen." Due to the huge numbers we're usually working with (nearly 8 billion people on the planet), rare events must regularly occur, so we shouldn't be too surprised when they do.

> Events may appear to us to be random, but this could be attributed to human ignorance about the details of the processes involved.
> —Brian S. Everitt

Critical Mass and the Hundredth Monkey

Another good example of a great coincidence that mystics have taken entirely out of context is the so-called hundredth monkey effect, which has become a bit of an urban legend. It was first introduced by the polymath and author Lyall Watson in his book *Lifetide* and is now championed by New Age believers in what skeptic James Randi famously calls woo-woo: paranormal, occult, and supernatural claims of all stripes.

In the 1950s, scientists were conducting studies of macaque monkeys on the Japanese island of Koshima. Despite being a highly qualified researcher, Watson claimed, without conducting any experiments, that after biologists introduced a new food to the macaques—raw sweet potatoes—a two-year-old macaque female, Imo, learned to wash the dirt from her potatoes before eating them. In the span of several years, Imo taught potato washing to other macaques, and once a "critical mass" of macaques began washing potatoes, that knowledge instantly spread to other monkeys on nearby islands through something like telepathy, the law of attraction, or quantum consciousness.

Investigators of Watson's claims couldn't confirm any of them. The Koshima macaques who began washing their potatoes never reached the number one hundred. After ten years, scientists carefully observing the behaviors of the monkeys knew that the number of potato-washing monkeys was exactly thirty-six (Shermer, 1997, p. 18). Once Imo learned to wash potatoes, potato washing spread throughout the population, although mostly among younger monkeys and females; the older males were too stubborn to adopt the new tradition.[3] When Imo was four, she learned another highly sophisticated food-gathering technique. Biologists would throw wheat grain in the sand for the monkeys to eat, and the monkeys had to manually (and literally) separate the wheat from the chaff. Imo discovered that by throwing handfuls of the mix into the water, the sand would sink and she could skim the wheat from the water's surface. This tactic also spread among the females and younger monkeys, though not as widely as potato washing, probably because it's a more sophisticated behavior (Shermer, 1997, pp. 17–18; Wilson, 1980, pp. 87–89).

Although some time later, macaques on the nearby island of Ohirayama were seen washing their sweet potatoes, no direct correlation between the Koshima monkeys and the Ohirayama residents was ever established. The local human population on Ohirayama relies heavily on sweet potatoes and washes them in rivers before preparing them to eat, so it's likely that the macaques picked up this behavior from watching humans, just as Imo's compatriots learned potato washing from her. If the hundredth monkey effect was real, one would expect the entire archipelago population of macaques to wash their potatoes, including the older males, and to practice the wheat trick many Koshima monkeys adopted. However, this isn't what happened.

This so-called critical mass idea of information transference is also the basis for James Redfield's book *The Celestine Prophecy* (discussed above), in which Redfield argues that, when a critical mass of people believe something, "the entire culture will begin to take these coincidental experiences seriously. We will wonder, in mass, what mysterious process underlies human life on this planet. And it will be this question, asked at the same time by enough people, that will allow the other insights also to come into consciousness" (Molé, 1999, p. 80). According to Redfield, if by chance some critical number of people simultaneously wonder the same thing, universal powers will hijack and replace all human thoughts.

No credible evidence has ever been found demonstrating that a collective belief can alter reality; reality exists independently of human thoughts (see chapter 2). And we already have really strong evidence that this critical mass effect doesn't work. For example, when everyone thought the Earth was flat, it didn't become flat. When everyone thought the sun orbited the Earth, the sun didn't orbit the Earth. And when everyone thought bloodletting was a legitimate medical practice, bloodletting still didn't heal people.

Believers in woo-woo cling to stories like the hundredth monkey and are swayed by similar stories like *The Celestine Prophecy* without investigating the details or applying basic skepticism to such bold claims. It's much easier—and more interesting—to propagate these stories, since the confirmation bias compels believers to cling to their beliefs while discarding any disconfirming evidence. The fact that other monkeys on another island picked up the same habit was coincidental and itself has a perfectly reasonable explanation that doesn't involve invoking paranormal non-explanations.

Research carried out by Susan Blackmore and Tom Troscianko led them to hypothesize that belief in ESP—extrasensory perception—originated because of our inability to judge the frequency of rare events and coincidences (Blackmore, 1990; Blackmore & Troscianko, 1985). Blackmore and Troscianko hypothesized that if their theory was true, we should expect that believers in ESP are worse than skeptics at judging probability. Using computer tests given to schoolchildren, university students, medical workers, and others, they found that skeptics were better at judging probability than believers, as predicted. Other research has lent credence to Blackmore's theory, and there appears to be a consistent pattern of evidence supporting it. Not a single paranormal mystery that's been properly investigated has ever ended by validating the existence of paranormal phenomena. If even the best cases cannot confirm any paranormality (Radford, 2010a), what's that say about the mysteries with weak evidence?

Counting the Hits and Ignoring the Misses

Another way we misjudge information is the *Jeane Dixon effect*, or counting the hits and ignoring the misses. According to Francis Bacon, "The general root of superstition is that [people] observe when things hit, and not when they miss; and commit to memory the one, and forget and pass over the other" (Law, 2011, p. 176). This bias is a specific type of System 1 error, as compulsive as it is common. For example, as an estimated ten million viewers watched a popular daytime talk show, a well-known psychic claimed that he could stop the watches of viewers with the power of his mind. Sure enough, in the next half hour, the show's phones boomed as the number of "hits" kept coming in. It's been estimated that the show heard from more than one hundred callers who were amazed that their wristwatch or grandfather clock or microwave oven timer stopped *just as predicted*! So how did the psychic know the viewers' watches were going to stop? Were these true "predictions," or should we be skeptical? Here, skepticism is our best bet.

Each "hit" is hidden in massive amounts of unseen data, such as how many watches people own, how often watches stop, how many viewers watch the show, and so on. In his book *Unweaving the Rainbow*, longtime skeptic Richard Dawkins explains:

If somebody's watch stopped three weeks after the spell was cast, even the most credulous would prefer to put it down to chance. We need to decide how large a delay would have been judged by the audience as sufficiently simultaneous with the psychic's announcement to impress. About five minutes is certainly safe, especially since he can keep talking to each caller for a few minutes before the next call ceases to seem roughly simultaneous. There are about 100,000 five-minute periods in a year. The probability that any given watch, say mine, will stop in a designated five-minute period is about 1 in 100,000. Low odds, but there are 10 million people watching the show. If only half of them are wearing watches, we could expect about 25 of those watches to stop in any given minute. If only a quarter of these ring into the studio, that is 6 calls, more than enough to dumbfound a naïve audience. Especially when you add in the calls from people whose watches stopped the day before, people whose watches didn't stop but whose grandfather clocks did, people who died of heart attacks and their bereaved relatives phoned in to say that their "ticker" gave out, and so on. (Dawkins, 2000, pp. 149–50)

A basic safeguard of Thinking Straight is "the requirement that the meaning of various outcomes be precisely specified (in advance if possible) and objectively determined [because] when we do not precisely specify the kind of evidence that will count as support for our position, we can end up 'detecting' too much evidence for our preconceptions" (Gilovich, 1991, pp. 57–58). The above example fails the test of objectivity because it lacks any specification as to what would count as "evidence" for the psychic's prediction.

Dawkins's estimations are extremely conservative, as he was exercising the *principle of charity* by giving his opponent the benefit of the best possible outcome. Even so, due to the huge number of variables, it's inevitable that many watches will stop working in any five-minute period. It may seem remarkable for those whose watches and clocks stopped, but vastly more watches and clocks didn't stop than ones that did. The miracle is even less remarkable when we consider that the same people who called in are not likely skeptics who reject the claims of psychics. Imagine how many stupefied callers the show would've had if the psychic predicted that in the next thirty minutes, viewers' watches *wouldn't* stop. Two main things are going on here that feed this cognitive illusion: (1) the tendency to count hits (the watches that stopped) and ignore misses (the millions of watches that didn't stop) and (2) a deceptive tactic known as **cold reading**.

Cold reading is a procedure by which "readers," such as tarot card readers, psychics, mediums, illusionists, and other mentalists, supposedly validate their paranormal abilities, like predicting the future or talking to dead people (Hyman, 1977). Because psychic abilities contradict what we know about the laws of nature, and because those laws are highly reliable, psychic powers automatically qualify as incredible claims. Especially given that after almost 140 years of research and investigation, not a single psychical phenomenon has ever been validated (Alcock, 2018, p. 449). Here's an example. You've all seen the psychic going on about "I'm getting an M; does her name begin with an M?" Think: why on earth would the spirit of a deceased loved one with an opportunity to communicate with the living whisper to a psychic, "My name begins with the letter M"? Why not just tell the psychic their name rather than play this weird guessing game? The answer is it's a trick, similar to what magicians do (Randi, 2007).

A seasoned psychic immediately picks up on information about a target from body language, age, clothing, fashion, hairstyle, gender, sexual orientation, religion, race or ethnicity, level of education, manner of speech, place of origin, and so forth, then feeds this information back to the target (Rowland, 2009). Then, making a series of vague and high-probability guesses—e.g., questions regarding leading causes of death, common names, items of sentiment—the reader tries eliciting a response to establish a platform to work from. Meanwhile—remember, most folks visit psychics because they already believe they're psychic—the target makes the "connections" on their own.[4] Biases lead them to focus only on the incidents in which the reader scores a hit, believing this is confirmation that the mediumship is working, completely ignoring those incidents where the medium "missed."[5]

Deceptive tactics of the cold reader:

- Make generalized assumptions
- Rely on high-probability guesses
- Ask vague questions
- Quickly size up their subjects to gain valuable information

In controlled conditions, psychic abilities have never been proven to be a real phenomenon.

ASTROLOGY AND THE BARNUM EFFECT

Although mediumship fails the test of science (O'Keeffe & Wiseman, 2005), anyone can test whether a supposed psychic is cold reading just by watching videos of "readings" and comparing the number of hits to misses. Psychics score vastly more misses than hits, yet the target completely ignores the misses and is genuinely mystified by the hits. Evidence shows that people who visit psychics are more prone to suggestion than skeptics (Hergovich, 2004; Wiseman, 2011; Wiseman, Greening, & Smith, 2003). Believers are seeking evidence that confirms their beliefs in much the same way as our brains actively seek patterns. This phenomenon is partially explained by what's known as the **Barnum effect**.

The Barnum effect is a widespread thinking error in which people rate descriptions of their personality that they think are tailored specifically to them as accurate, even when those descriptions are vague and general enough to apply to almost anyone (Johnson, Cain, Falke, Hayman, & Perillo, 1985). The Barnum effect is the main reason people believe in pseudosciences such as palm reading and astrology (Kelly, 1997; Kelly, 1998). Research shows that the Barnum effect works best if (1) you believe the statement applies to you personally, (2) you view the person making the statement as an authority, and (3) the statement lists many positive traits (Dickenson & Kelly, 1985). The Barnum effect works because people are excellent at interpreting information so that it aligns with how they see themselves. One surprising aspect of the Barnum effect is its ability to turn skeptics into believers! In a study that demonstrated the power of the Barnum effect (Alcock, 2018, p. 501; Glick, Gottesman, & Jolton, 1989; Vyse, 2014, pp. 163–64), psychologists placed high school students into two groups based on their level of belief in astrology: those who believed in the accuracy of horoscopes and skeptics who didn't.

Each group believed they'd be given a personalized horoscope reading by a "professional astrology service" based on information provided earlier. Of course, none of the horoscopes were "personal." Each student received one of two horoscopes: one was a positive personality assessment, and the other was a negative personality assessment. As expected, the believers rated the horoscopes as accurate regardless of whether the assessment was negative or positive. However, the skeptics who received the positive personality assessments rated them as more accurate than skeptics who received negative personality assessments. Even more surprisingly, after the horoscopes were given, the students were again asked to rate their level of belief in astrology. The believers, along with the skeptics who received negative assessments, were unchanged in their beliefs. However, skeptics who received positive assessments significantly increased their level of belief.

Another fascinating study (Hines, 2003, pp. 149–50) on astrology demonstrated the recurring theme that, on average, believers are less open-minded than skeptics. In this study, researchers had skeptics and believers in astrology test an astrological prediction that a particular individual was friendly, extroverted, and outgoing, according to his horoscope. The participants asked the individual questions about his personality and habits. Unbeknownst to them, the interviewee was a **confederate** who was instructed to answer questions in a specific way. If the confederate was asked a question that invited an extroverted response, he gave an extroverted answer. If he was asked a question that invited an introverted response, he gave an introverted answer. While there was some variation within each group, for the most part both skeptics and believers asked mostly questions that would confirm the hypothesis, thus overall, both groups were led to believe that astrology had accurately predicted the confederate's personality. The result was that skeptics tended to change their prior belief about astrology. When disconfirming evidence was looked at for the believer group, the data showed that believers maintained their prior beliefs.

The Barnum effect is also responsible for the widespread acceptance of other paranormal and occult beliefs such as fortune telling, graphology, and aura reading (Perez & Hines, 2011). Interestingly, the Barnum effect doesn't usually work unless participants believe the descriptions are really tailored to them. If they suspect the description is part of some test or gimmick, they're less likely to fall for it no matter how convincing the information sounds.

Following Bertram Forer's work (Forer, 1949), the conjurer, author, and skeptic James "The Amazing" Randi devised an experiment (Charlson & Apsell, 1993; Shermer, 2011b) where he told a group of twenty students that a world renowned astrologer would be coming in to give them readings. The astrologer only needed the students to send their birthdates and the place they were born. A week later, Randi handed out envelopes containing each student's reading and instructed them to rate the readings on a scale of 1 to 5, with 5 being highly accurate. The entire class rated the accuracy a 5—save one student who rated it a 4—believing the readings represented their individual personalities. The experiment, unbeknownst to the students, had a couple of twists.

After giving everyone a chance to review their readings, Randi instructed the class to pass their reading to the person behind them, at which point the students realized that everyone had been given *the exact same reading*. What's more, Randi ensured that 60 percent of everything in the reading contained statements *of things that were to happen in the future*. So the students scored the reading a 5 on accuracy, even though over half of it pertained to events that hadn't even happened! Other studies have produced similar results (Alcock, 2018, pp. 500–501; Dickenson & Kelly, 1985).

Astrology rests on the notion that celestial objects—which are sometimes tens of thousands of light-years away from us (and each other) and have never been shown to possess paranormal powers—somehow collaborate to reach across the cosmos and influence events on Earth, specifically events during the moment a human is born. Astrology has been around since the Babylonians first wrote about it over four thousand years ago (Smith, 2018, pp. 230–31). In the 1600s, a common cause of death was "planet" (Sagan, 1980, p. 51). Although it has precisely the predictive power of a fortune cookie, astrology remains a popular pastime and conversation starter that millions of people take seriously.

Remnants of cultures that held astrology in high esteem still exist in our modern lexicon. For example, the word "disaster" is Greek for "bad star." The Hebrew word "mazeltov" is Babylonian for "good constellation." The word "lunatic" is Latin for "moonstruck." Even the word "consider," a favorite of skeptics, has its roots in a Latin word meaning "with the planets" (Sagan, 1980, p. 49). We've come a long way since "planet" was an explanation for why someone died, so why does astrology continue to hold such an influence on the modern world? Are the claims astrology makes plausible or are they incredible? Does astrology work? If so, *exactly* how does it work? Does it make accurate predictions? What is the medium of communication between star (or planet) and human, and how do we detect it? How did ancient civilizations detect it? Can we run an experiment to confirm the existence of this medium (Harrison, 2012, p. 142)? These questions elude sensible answers, demonstrating that astrology is based on little more than ancient superstition and magical thinking.

Some have proposed that gravity, tidal/electromagnetic forces, magnetic fields, or the emission of some sort of particle is the mechanism by which astrology influences human behavior and destiny. Such incredible claims violate well-established scientific facts as well as the GASPs we use to investigate them (Dean, 2016). Even so, if astrology's claims didn't violate a single GASP, "all the proposed forces or fields are [still] far too weak to have any influence on human infants, let alone the massive influence that is required by astrology" (Hines, 2003, p. 212). The gravitational impact of the book in your hand is billions of times stronger than the gravity Mars exerts on you (Schick & Vaughn, 2005, p. 146). Likewise, the electromagnetic force of your smartphone or TV is millions of times stronger than that of Venus or Sirius (Smith, 2018, p. 232). So if stars and planets influence our lives by gravity or electromagnetism, does that mean books and TVs do as well? If so, how do they do it? If not, what's so special about orbiting balls of gas and rock? Gravity and electromagnetism are forces that can be scientifically measured, which means we can prove their existence and impact. Have astrologers demonstrated a specific force that can influence humans or shown exactly how that force exerts its influence? To date, astrology has failed every test it's ever been subjected to (Carlson, 1985; Hines, 2003, pp. 213–33). It makes zero novel predictions, and it doesn't explain anything (Wyman & Vyse, 2008). If astrology were true, it would mean cosmology, astrophysics, biology, and astronomy are all deeply flawed sciences. Given the huge success of these disciplines, as well as the criterion of conservation (chapter 10), it's much more likely that astrology is false, rather than several of our most successful scientific theories. Accordingly, we can safely classify astrology as an incredible claim.

Astrology failed to predict the collapse of the San Francisco Bay Bridge in 1989, it failed to predict the coronavirus pandemic, and the daily horoscopes of the 2,996 people who died in the 9/11 tragedy failed to reveal that these people would die horrifically in a violent attack that day. Can you think of a time you saw a warning of disaster written in someone's horoscope? Does astrology only predict positive events, such as lucky business or love encounters? Psychologists have shown that belief in astrology is strongly correlated with a desire for control (Landau, Kay, & Whitson, 2015; Lillquist & Lindeman, 1998), yet because of biases like the *self-serving bias*, the Barnum effect, and various forms of the confirmation bias, people continue believing in pseudosciences like astrology, mediumship, palm reading, and their counterparts (chapter 9).

MIRACLE WATER

Another example of counting the hits and ignoring the misses is miracle healing. We've all seen those infomercials requesting "donations" in exchange for "miracle" water. This scam works because when one's health is at stake, people go to extremes searching for a miraculous cure, whether from their favorite guru or a magical location (Nickell, 2019). Every year since 1858, after a fourteen-year old girl[6] claimed the Virgin Mary told her a spring had magical healing powers, some eighty thousand pilgrims travel to the Sanctuary of Our Lady of Lourdes, France, searching for miracles. In all reality, these pilgrims are "probably more likely to catch something from the thousands of other pilgrims who've wallowed in the same water" (Dawkins, 2006; Sagan, 1996, pp. 232–33). Some spend their entire life savings making the taxing trip to Lourdes in hopes that this magical destination will provide what they believe modern medicine cannot.

Out of more than two hundred million sick travelers, Lourdes officials have declared sixty-seven miraculous cures (Vyse, 2014, p. 7). That *only* sixty-seven cures have been officially declared is the real miracle here. Skeptics have observed that no miracle destination in the world, Lourdes included, has ever cured anyone of any permanent or irreversible injury like permanent blindness, missing teeth, or severed limbs:[7] people are only ever "cured" of maladies that bodies defeat naturally.

Toward a Better Understanding of What's Probable

On the popular daytime show *Let's Make a Deal*, host Wayne Brady offers an example of how understanding odds is counterinstinctual. Contestants are shown three doors, behind one of which is some sought-after prize. The contestant is asked to select a door. Brady knows which door the prize is behind, so regardless of which door is selected, he opens a door revealing a Zonk. Now the contestant is offered a chance to take home some cash or continue playing. If they continue, they're given a choice to stick with the door they originally chose or switch to the remaining door.

Originating from a 1975 letter posed by the American statistician Steve Selvin (Novella, 2018, p. 138), this example is now known as the *Monty Hall problem* (Mlowdinow, 2009, pp. 53–56). It reveals how even obvious problems of innumeracy are counterintuitive. Most of us in the position of the contestant would think it wouldn't matter if we switch doors or keep the original. Seeing that only two doors are left, we reason that the odds of getting a Zonk are 50-50. However, a little math shows the error of intuition. By switching to the new door, the chances of winning go from one-third to two-thirds (Novella, 2012, pp. 88–89; Scott, 2018). Here's how.

When initially choosing a door, there were three options, so the odds of choosing the correct door were 1 in 3. By revealing a Zonk, Brady revealed new information one can use to increase the chances of winning. Since the contestant now knows one door that contains a Zonk, by switching doors, the contestant is essentially selecting two doors—the new door plus the Zonk door Brady opened—increasing the chances of winning to two-thirds. If this still doesn't sound right, imagine there are one hundred doors rather than three, and after you select a door, Wayne Brady opens every door except one (Mlowdinow, 2009, p. 55). Either you were extremely lucky on your first pick, or Brady just showed you the door with the prize.

Tipping the Scales

Human brains evolved in what has been called "Middle World" (Dawkins, 2006, pp. 367–68) or "Middle Land" (Shermer, 2008a, p. 15)—that place between really fast and really slow, between really big and really small, and between brand new and really old. When things reach one extreme or another, contemplating them can be breathtaking and intuition goes out the window. Take cosmology for example. When we look up at the night sky, it's impossible to fathom just how many stars are "up" there. One estimate states that if a typical grain of sand represented every star in the universe, the sand would fill the Grand Canyon more than twice (Gray, 2018).

In his book *The Science of Good and Evil*, Michael Shermer points out the frequency of the number 150, suggesting that 150 represents an upper limit to human data processing; it can also shed some light on our innate innumeracy. Shermer, drawing from the work of the evolutionary psychologist Robin Dunbar, states that our brains are only capable of keeping so much data in working memory, and that this upper limit places restrictions on how much information we can keep stowed away. Certainly, *some* limit on how much information we can keep stored must exist, so the theory is plausible.

Why 150? According to Shermer, 150 "is roughly the number of living descendants (children, grandchildren, great-grandchildren) a Paleolithic couple would produce in four generations at the birthrate of hunter-gatherer peoples" (Shermer, 2004, pp. 40–43). It's also the average size of hunter-gatherer groups, the typical size of Near East agricultural communities, the average size of modern farming communities, the size of most small businesses (and departments within large corporations and efficiently run factories), and the number of people in the average person's address book. When research was done during WWII on what the best size of a military company is, the British Army decided on 130 and the United States decided on 223 (Shermer, 2004, p. 41). This number was supposed to represent a company where everyone knew one another and could develop social bonds, while still having some strength as a war force. Dunbar agrees, arguing that the number 150—now known as *Dunbar's number*—represents a sort of tipping point. In his book *Grooming, Gossip, and the Evolution of Language*, Dunbar explains that "at this size, orders can be implemented and unruly behavior controlled on the basis of personal loyalties and direct man-to-man contacts. With larger groups, this becomes impossible" (Dunbar, 1997, p. 76). If Shermer and Dunbar are correct, it would shed some light on why we have such a hard time with extreme variables.

Probability

Probability is the measure of the likelihood that an event will occur.[8] Over the last few centuries, scientists have investigated inevitabilities such as random chance, probability, human error, and other anomalies. Although the layperson's methods aren't as honed as scientists', everyone uses probability to draw inferences about the expected frequency of events.

For example, a coin has two sides. Thus, if we flip a fair coin, there are only two possible outcomes: heads or tails. Because our coin is fair, each outcome is equiprobable; that is, the probability of landing on heads is equal to that of it landing on tails. So far so good. But as we learned in chapter 8 when discussing the gambler's fallacy, we also tend to link previous events with future events when they're statistically independent, leading us to believe in things like hot or cold streaks. If we correctly guess heads five times in a row, we may think we're on a hot streak or due for tails, successfully predicting any outcome for the next toss. In reality, each flip is statistically independent, and previous flips cannot influence subsequent flips. Magical thinking works in a similar way. If you wish for something and you get what you wished for, you attribute it happening to the act of wishing. If you wish for something and you don't get what you wished for, you may just think the universe has a different plan. Either way, you've successfully predicted any possible outcome, and your belief system remains intact, yet another example of an unfalsifiable claim.

Although it illustrates a lot, this analogy is quite easy to understand; it's the confirmation bias at its best. However, thinking probabilistically is extremely unnatural, especially when the numbers get much bigger than those we're comfortable dealing with. Through a simple mathematical demonstration, mathematician John Allen Paulos showed that assuming Shakespeare accurately recounted Julius Caesar's last words, "You too, Brutus," and that after two thousand years Caesar's exhaled molecules are uniformly spread and float free throughout the world, there's a better than 99 percent chance that with each breath, we inhale at least one molecule Caesar exhaled with his dying words (Paulos, 2001, p. 32).

Another example is recalling someone you haven't thought of in many years—whether through a dream or just by popping into your head—and suddenly learning of that person's death. Coincidences like these stick out, especially to people who are already predisposed to connect things like dreams and death to a spiritual realm. As tempting as it is to declare that those who have this experience are capable of clairvoyance, a better explanation doesn't invoke any superpowers.

Motivated by the public's tendency to assign meaning to random events—as well as scientists' silence in offering satisfactory ways to explain such coincidences—Nobel laureate Luis Walter Alvarez had a similar experience and worked out the details. Alvarez was reading a newspaper when a familiar-sounding name appeared: Dr. Carleton S. Coon. He began thinking about a different Carleton Coon, who, thirty years ago, played in a band with one Joe Sanders. Not five minutes later, Alvarez read in the obituary that Joe Sanders had died. He then set out to answer the question, "What are the odds of this happening?" (Alvarez, 1965). After rigorous analyses, Alvarez's calculations show that the odds of this happening are minuscule: 0.00003. However, given the number of people on the planet and the number of names one would recognize from a thirty-year period, 0.00003 equals ten times every day. So coincidences like this may seem improbable, but it'd be even more improbable if they never occurred.

> A tendency to drastically underestimate the frequency of coincidences is a prime characteristic of innumerates.
> —John Allen Paulos

When dealing with statistics, probability, and coincidence, we'd do well to remember that the human brain didn't evolve to think in these terms, and while coincidences often have a certain degree of eeriness, oftentimes our intuition is wrong. As Barry Singer put it, "The mathematical probabilities of rare events, in particular, often run counter to intuition, but it is the mathematics, not our intuition, that is correct" (Abell & Singer, 1983, p. 13). So long as the probability of an unlikely event is greater than zero, it's not only possible; it's statistically demanded.

Wrapping Up

The psycholinguist and eminent skeptic Steven Pinker explains that "there is no such thing as fate, providence, karma, spells, curses, augury, divine retribution, or answered prayers—though the discrepancy between the laws of probability and the workings of cognition may explain why people are the way they are" (Pinker, 2018, p. 394). What Pinker is driving at is that it's reasonable to conclude that our beliefs in incredible claims are due to an inability to understand probability, combined with a reluctance to accept coincidence as just that. We may think a rare personal event couldn't be explained by coincidence alone, simply because we can't think of a more satisfying explanation. We may even data mine or manipulate facts to suit our pet idea or preferred explanation.

A lot of the time, understanding coincidences is about asking the right questions, such as "What are the chances that X would happen to *anyone?*" instead of asking "What are the chances that X would happen to *me?*" In a big rainstorm, what are the odds a particular raindrop would leave the sky, fall for minutes, only to cross paths with you at the perfect moment and land on the tip of your nose? Vanishingly low, indeed. But what are the odds that during a rainstorm *any* raindrop would hit you on the tip of your nose? Much, much better. If that raindrop was a cure for an untreatable ailment, a million-dollar lottery ticket, or an unexpected text from an ex-lover you'd been thinking about, you'd be tempted to assign the event more meaning than you do the random falling of a raindrop.

Along with all the biases that come with the territory of being human, we're inherently threatened by large-scale variables and have a difficult time understanding the mundane nature of coincidences. However, educating ourselves, raising our awareness, and using the tools of CT will decrease the chances we fall victim to probabilistic errors and keep us more firmly grounded in reality.

CHAPTER 10

Practically Thinking

We cannot have CT without science, or vice versa. The two are mutually rein-forcing, and in this chapter, I combine the powers of these two amazingly reli-able paths to discovering reality. We'll learn about the foundations and philoso-phy of science and what "the" scientific method is. We'll also cover two specific methods of scientific reasoning that overlap with CT, learning some of the limits of science and why it's helpful for skeptics to understand those limits. By the end of this chapter, readers should have a good idea of just how important science is in our everyday lives and be better prepared to more thoroughly explain the methods and limits of scientific methodology.

Building Vocabulary

Anecdote:	Hypothesis:
A story or personal account of something.	A proposed explanation that makes specific, testable predictions that are revealed by observation, investiga-tion, or experimentation.

Paradigm:	Provisional:
A body of knowledge in a speci-fied field that is generally accepted by the scientists who work in that specified field.	Temporary acceptance of a claim, with an understanding that the claim may be changed or amended.

The Scaffolds of Science

Visualizing scientists as nerds in lab coats, handling test tubes and Pyrex bea-kers, is fairly common. Some view science as a belief system, while others see it as a set of procedures. Some might simply picture the famed Albert Einstein

(1879–1955) or Charles Darwin (1809–1882), both of which are remembered by history as first-rate scientists. Science incorporates all the above, but it's an error to try and narrow science down or pigeonhole it into any one thing.

Indeed, some scientists wear white lab coats and work with beakers full of mysterious liquids (e.g., chemists, biologists). Others wear hiking boots and carry backpacks on field missions (e.g., geologists, paleontologists). And others listen intently with a pen and a clipboard (e.g., psychologists, psychiatrists). One method may work for some, and an entirely different method may be needed for others. For instance, the methods physicists use differ significantly from those used by sociologists, whose methods look nothing like the anthropologists'. Accordingly, most scientists agree that *the* scientific method is more myth than fact (Feyerabend, 1975/2010).[1] Instead, the term *scientific methodology* incorporates the methods used in pursuit of scientific knowledge.

Although the phrase "the scientific method" is often overused and even more often misunderstood, the term itself is quite instructive. The word "science" comes from the Latin word *scire*, which simply means "to know," while the word "method" is derived from the Greek word *hodós*, for "way, motion, journey." The British biologist Lewis Wolpert doubts there's such a thing as a scientific method except "in very broad and general terms" (quoted in Baggini, 2017). Accordingly, I've chosen the phrase "scientific methodology" when referring to scientific thinking and discovery.

In their book *UFOs, Chemtrails and Aliens*, authors Donald Prothero and Tim Callahan put it thus: "What makes someone a scientist is not a white lab coat or lab equipment but rather *how he or she asks questions about nature* and what thought processes he or she employs to solve problems. Science is about suggesting an explanation (a hypothesis) to understand some phenomenon, then *testing* that explanation by examining evidence that might show us whether the hypothesis is right or wrong" (Prothero & Callahan, 2017, p. 4). When we talk about scientific methodology, we're not talking about *the things we know*. Instead, we're talking about *a way to know* or *a path to knowledge*, understanding that scientists use any path that works.

The philosopher of science Paul Feyerabend once pointed out that scientists will use any reliable method until it's no longer useful. Over the centuries, scientists and researchers have used many methods in their pursuits—including aesthetic beauty of the theory (Gell-Mann, 2019; Weinberg, 1994, pp. 90–165)—and so science is better characterized as an approach than as a method. To clarify, science isn't a thing to believe in; it's a tool humans use to guide us toward knowledge. It also provides the foundation upon which CT takes place. Science embodies a collection of systematic methods that are designed to help illustrate and interpret the natural world. Put differently, science is best understood as a verb, rather than a noun.

Science is open to everyone, it's self-correcting, it's rigorous and quantitative, and it compensates for the failings of those undisciplined brains we've been learning about. The self-correcting nature of science serves to preserve the gains and eliminate the mistakes, so that what we're left with when we have a good scientific theory are the fruits of hard-won advances, not merely the opinions of an elite group of intellectuals.

Tools of Science

While there isn't any specific scientific method, scientists employ an array of GASPs— such as data collection, **hypothesis** forming, attempts at falsification, replication, open debate, and peer review in their pursuit to understand reality. When we want to fly from point A to point B, we take an airplane, not a magic carpet. When we want to send robots to distant planets, we use astronomy, not astrology. When we're sick or injured, we see a doctor, not a shaman. The reasons are simple: science works, while magic carpets, astrology, and shamans don't.

Any truth-seeking endeavor requires GASPs. For example, you could learn about prison by reading biographies written by (ex-)prisoners, biographers, or journalists; talking to prisoners; watching TV documentaries; or doing prison time yourself. Just as science doesn't declare a universal truth after one test or study (chapter 3), before you could learn reliable truths about prison, you'd need to use all reliable information at your disposal. Doing science is a matter of systematically gathering evidence that can be used for or against some hypothesis regardless of whether you are a "scientist." Adam Savage, from the popular show *Mythbusters*, is a prime example of a nonscientist who does science.

Sometimes scientists must work from theories that cannot be directly tested or observed, such as evolution or the Big Bang. We cannot reverse the tapes[2]

Generally Accepted Standards of Proof (GASPs):

- Observation
- Hypothesis testing
- Data collection
- Falsification
- Replication
- Blind/Double-blind experiments
- Open debate

- Random assignment
- Control groups
- Operational definitions
- Systematic empiricism
- Statistical analyses
- Peer review
- Expert consensus

These are some of the tools used by scientists, but scientists will use any method at their disposal that is a reliable path to the truth.

and reconstruct evolution or watch the Big Bang. What scientists do in these situations is ask questions about the nature of those theories, creating tests that inform their judgment. If "X" is true, what does "X" predict? If "X" happened, what should we expect to see? Do our findings align with our theory? For example, we can observe the expansion of the universe or the fossil record and test those observations against what we would expect if the Big Bang and evolutionary models are true.

These theories shouldn't be thought of as "just a theory." The concepts of ghosts, Bigfoot, space aliens, and the gamut of incredible claims are accepted by millions of people as universal facts, even though the evidence supporting such claims amounts to little more than vague dreams, blurry photos, ancient stories, or inner convictions. If our theories of, say, gravity, heliocentrism, evolution, Big Bang cosmology, plate tectonics, or the germ theory of disease were rested on blurry photos and voices in our heads, do you think anyone would've taken them seriously? More importantly, do you think we would've gotten as far as we have in medicine and technology if scientific theories didn't produce reliable results? Have ghosts ever given anyone any useful information? Has any Bigfoot hunter ever produced a corpse? Has any alien abductee ever scratched one of her abductors and brought back skin samples to investigate? Have any incredible claims contributed to the advancement of knowledge?

A Fly in the Ointment

Science is an extremely reliable source of discovery, though it isn't entirely without flaws or limitations. Science may be able to provide a neurological, chemical, or biological explanation for the experience of love, for example, but it cannot decide whether love is good or bad. Value judgments such as these are not currently accepted as legitimate scientific questions (although some seek to change that [Harris, 2010; Harris, 2011; Shermer, 2004; Shermer, 2013]). Additionally, science can only test what exists within nature, since anything outside of nature cannot be empirically examined. Thus, anything science can test, including bizarre phenomena such as quantum entanglement, would be consistent with naturalism (chapter 2).

Perhaps science's greatest weakness is its human origins; scientists are humans, and good and bad scientists exist in every field (Harrison, 2012, pp. 148–49). Nonscientists sometimes have trouble separating "sci*ence*" from "scien*tist.*" Science, as a pursuit of knowledge, is about the discover*ies*, not the discover*ers*. As the philosopher Julian Baggini explains, "Science as a pursuit is [colored] by the personalities of scientists but science as a set of results is [true for everyone]" (Baggini, 2017, pp. 78–79). Thus, science is a messy process that's riddled with errors and imperfections, including biases, preferences, mistakes, distortions, errors, lies, frauds, and all the other shortcomings we find in any human endeavor. Nevertheless, replacing science with something even *less* reliable would be as stupid as walking around blindfolded because vision doesn't work 100 percent of the time, or walking around barefoot because shoes don't last forever (for an approachable philosophical treatment of the issue, see Baggini, 2003, pp. 143–69).

Part of science is dedicated to discovering how human shortcomings can contaminate scientific experiments. As a result, over time, scientists have woven several self-correcting procedures into the fabric of scientific methodology, including replication and peer review (chapter 3). People often lack trust in scientific findings because the research was paid for by the company making the product (think of Moderna funding their own studies on vaccines for COVID 19). When companies pay for their own research, skepticism is warranted, which is why skeptics leave it to experts to hash out the details. For instance, in a study of forensic psychologists and psychiatrists (Murrie et al., 2013; Tavris & Aronson, 2019, pp. 62–63), researchers gave participants four identical case files of real sex offenders and asked them to assess the likelihood that the offenders would reoffend using standard risk assessment measures. Using these measures typically produces a high level of agreement among experts. However, in this study, some of the participants were told they were hired by the prosecution, while others were told they were hired by the defense. Expectedly, the findings were slanted in favor of the participant's presumed employer; those "working" for the defense scored lower on risk assessment, and those "working" for the prosecutor scored higher.

Throughout the centuries of scientific progress, we've discovered that we haven't been just learning about the things we were studying; we've also been learning a great deal about ourselves. As the above study demonstrates, we've learned that humans are prone to false beliefs and misjudgment and that we're biased beyond control. So skeptics are cautious, trying hard not to perform the all-too-easy task of fooling ourselves (Nuzzo, 2015). "The real purpose of scientific method," as Robert Pirsig explains, "is to make sure Nature hasn't misled you into thinking you know something that you actually don't" (Pirsig, 1974). Or, as the late theoretical physicist Richard Feynman famously said, "The first principle is that you must not fool yourself—and you are the easiest person to fool" (Feynman, 1974).

The Progressive Nature of Science

Scientific illiteracy can sow a great deal of distrust in science's abilities, and lead people to reason that since some scientific truth today might be overturned by another scientific truth tomorrow, science itself is wholly unreliable. By this reasoning, "believing in science" is every bit as arbitrary as believing in anything else. However, science isn't something to believe in. Indeed, viewing science as just one among many worldviews is a rather unfair and inaccurate accusation. While it's true that science is progressive, the tentative nature of scientific facts is one of science's strengths, not one of its weaknesses. Moreover, science and critical thinking are the only reliable truth-discovering methods we know of, and neither has any competitors.

In his 1962 book *The Structure of Scientific Revolutions*, the physicist and philosopher of science Thomas Kuhn explained the progressive nature of science by introducing the idea that all science takes place within a **paradigm**, an overarching view that embodies the current state of scientific thought (Kuhn, 1962/2012, pp. 10–11). Since Kuhn, others have described a paradigm in similar fashion, including the prominent skeptic Michael Shermer, who said a paradigm is "a model shared by most but not all members of a scientific community, designed to describe and interpret observed phenomena, past or present, and aimed at building a testable body of knowledge open to rejection or confirmation" (Shermer, 1997, p. 39). Not all scientists believe the same thing. In fact, getting scientists to agree on anything is a lot like trying to herd cats. Accordingly, we can look at a paradigm as the majority view within a particular field of study.

Let's return to our Street World metaphor. Think of a paradigm as a sort of map that details the roads of Street World. In this case, the roads represent the current state of scientific knowledge.[3] If we're navigating our way around, say, Memory City, we want to know if taking Repression Avenue can get us to Truth Boulevard. If our map was detailed in the late 1980s, it might show that Repression Avenue is under current construction. Today it'd show us that Repression Avenue is a dead-end street. Here, it's important to note that the theoretical models incorporated within a scientific paradigm are neither accepted nor rejected by popular vote. Science is not democratic, and truths are not voted into existence. In fact, most scientific models are rejected; it's not until theoretical models are empirically validated by replication that they're able to gain traction.

Repression Avenue couldn't get us to Truth Boulevard, not because commuters didn't like the scenery, but because there just wasn't a way to get from one to the other by following the rules of the road. Just as maps change based on transforming topography, sometimes an existing paradigm can't explain some aspect of how things work. At this point, a new paradigm is needed to explain the data. When this happens, science undergoes what Kuhn called a *paradigm shift*—a revolution in knowledge that changes how scientists approach their field of study. Kuhn argued that the history of science shows gradual progress from one paradigm to the next, each paradigm getting us closer to an accurate understanding of reality. In the beginning, scientific disciplines go through a period of confusion and uncertainty before a paradigm is created by a consensus among experts (Ioannidis, 2004). Perhaps the reason for this slow progression is because, as the famous physicist Max Planck put it, "a scientific truth does not triumph by convincing its opponents and making them see the light, but rather because its opponents eventually die and a new generation grows up that's familiar with it."[4]

In any event, when a new paradigm is introduced, it usually gains traction by showing that the evidence does a better job of explaining some phenomenon than the existing model. The new paradigm must explain everything the prior paradigm explains but more accurately. For example, Einstein's theory of relativity did a better job of explaining the orbit (or perihelion) of Mercury than did Newton's laws, and so physicists shifted from the classical pre-quantum mechanics paradigm to what they now call *the standard model* paradigm (Greene, 2003; Krauss, 2013; Krauss, 2017; Sagan, 1980; Susskind, 2006; Weinberg, 1994; Wolfson, 2000). Centuries earlier, Nicolaus Copernicus (1473–1543) challenged Ptolemy's (100–170) geocentric model of the universe. Galileo followed Copernicus with his observations of the moons of Jupiter and the phases of Venus, leading us to accept a heliocentric (sun-centered) paradigm. Prior to Darwin's discovery of evolution by natural selection, almost everyone endorsed a view that life was created. After Darwin, most people accept the fact of evolution by natural selection.

It isn't that reality actually changes once a new paradigm is accepted. Rather, new information and new ways of interpreting information lead progressively to better ways of explaining it. Einstein explained the introduction of a new paradigm this way: "Creating a new theory is not like destroying an old barn and erecting a skyscraper in its place. It is rather like climbing a mountain, gaining new and wider views, discovering unexpected connections between our starting point and its rich environment. But the point from which we started out still exists and can be seen, although it appears smaller and forms a tiny part of our broad view gained by the mastery of the obstacles on our adventurous way up" (Shermer, 1997, p. 41).

> Science is simply common sense at its best—that is, rigidly accurate in observation and merciless to fallacy in logic.
> —Thomas H. Huxley

In science, it doesn't matter how new ideas come about or who proposes them. Eventually all ideas are subjected to the rigors of observation, experimentation, and critical review. In 1869, the well-respected co-discoverer of evolution Alfred Russel Wallace proposed that the human brain couldn't be a product of evolution because he couldn't think of a natural reason for a brain as capable as ours to exist. Despite his insights on evolution and prominence as a first-rate scientist, this idea failed to gain traction. Ideas that are tested against reality boost our confidence that what we accept as factual actually is. Hypotheses that fail these tests, we forget (e.g., alchemy), while hypotheses that pass, we remember (e.g., radiometric dating).

Scientific theories are only useful to the extent that they accurately explain things, but just being able to explain something still falls a bit short. Theories about nature are only valid if they're capable of making testable predictions about the past or future. Compare the following examples (from Schick & Vaughn, 2005, pp. 65–67). The famous "prophet" Nostradamus is said to have predicted all manner of world events, from WWII, to the atomic bomb, to 9/11. Setting aside the fact that his writings have only ever been said to predict something after it happened (Randi, 1982–1983), even Nostradamus "experts" disagree as to what his predictions mean, such as Henry Roberts and Erika Cheetham, who disagree about the same Nostradamus verse:

> Roberts: "A Remarkable prophetic description of the role of Emperor Haile Selassie, in World War II."

> Cheetham: "Lines 1–2 . . . refer . . . to Henry IV. The man who troubles him from the East is the Duke of Parma. . . . Lines 3–4 most probably refer to the siege of Malta in 1565." (Schick & Vaughn, 2005, p. 66)

The reason there's no consistency between Nostradamus experts is because Nostradamus's writings are so vague that in hindsight they can be retrofitted and credited to an array of events as successful predictions. Nobody has ever used Nostradamus's predictions to predict a future event—his predictions are always interpreted with hindsight to retrofit some event after it occurs.

Herein lies a stark contrast between science and pseudoscience. The ability to develop, test, and confirm or falsify a prediction is one of the most fundamental components of scientific methodology. Scientific theories predict future events with great precision. In 1705, the English polymath Edmond Halley (1656–1742) successfully predicted that in 1758, at a precise location in the sky following a particular trajectory, what is now known as Halley's Comet (in his honor) would return. It did, exactly as predicted. In 1859, Darwin predicted the existence of a shared unit of biological inheritance that all earthbound life-forms must carry. This prediction was borne out in 1944 by three relatively unknown bacteriologists named Oswald Avery, Maclyn McCarty, and Colin MacLeod.[5] In 1916, Einstein's epoch-making theory of general relativity predicted the existence of gravitational waves—ripples in the curvature of space-time caused by colliding black holes—and one hundred years later, the technology was such that scientists were finally able to detect them.

By contrast, pseudoscientific predictions, such as those made by psychics or Nostradamus scholars, are typically vague, open to interpretation, retrofitted to fit the data, and unprovable or unfalsifiable. When predictions are specific—as when Edgar Casey (known as "the sleeping psychic" [Sugrue, 1973]) falsely predicted the destruction of LA, San Francisco, and New York; declared Piltdown Man was genuine; or his prophecy that in 1958, the United States would rediscover a death ray that had been used on Atlantis—they end up being false or unreliable (Lease, 1999).

The scientific enterprise is a precautionary tale, and while progress is necessary for success, scientific facts must be understood as **provisional**. Therefore, all scientific findings are only as good as our subjective interpretation of the data we currently have, along with the limited tools we use to interrogate nature. The provisional nature of knowledge can be explained by the following example. Let's say you have two children who are having a friend sleep over. You suspect that the kids are up late playing video games, so you quietly sneak up to the boys' bedroom door and listen in. You hear silence, but that could just mean that the kids are using headsets. You crack the door and find an apparent room full of sleeping children, but that could just mean they knew you were coming and are playing possum. So, you decide to sneak around and peep in through a window and again find a room full of sleeping children. The children could be sleeping, or they could've spotted you and turned the games off before you could catch them. You listen in on a monitor, but it could be hidden in the bathroom. You check the surveillance video of the boys' room, but that could've been hacked.

So how do you know what the boys are doing? Having exhausted all possibilities, is it reasonable to say that you *know* the boys are asleep? The answer is yes. All the best available evidence has pointed to the conclusion that the boys are sound asleep. However, as noted in the example, the evidence could've been manipulated, and there may be other ways of monitoring the boys' room that you haven't thought of, which is why the conclusion must be provisional. And so it is with science. We use what we have at our disposal to rule out variables and get closer and closer to what's true. We once believed that the Earth was a perfect sphere based on our interpretation of the best available data. We only later discovered that it was more accurately described as an "oblate spheroid"— in other words, the Earth is a bit flattened at the poles. It isn't that the idea that the Earth is a perfect sphere is completely false. It's just that with better technology, we discovered new information that enhanced our understanding of how the Earth is shaped. Or, as prolific author and scientist Isaac Asimov (1920–1992) put it, "When people thought the earth was flat, they were wrong. When people thought the earth was spherical, they were wrong. But if you think that thinking the earth is spherical is just as wrong as thinking the earth is flat, then your view is wronger than both of them put together" (Asimov, 1989b).

When science changes its position on something, that doesn't mean science was lying about the previous position; it means we learned more about it.

It takes a lot of research before something is considered a scientific fact, and even then, scientific facts are still provisional. Even so, as far as facts go, scientific facts are the crème de la crème. The research scientific facts are grounded on provides ample reason to accept them as true based on the best evidence available. The provisional nature of scientific facts can be compared to the iPhone. At any given time, the latest iPhone is the best thing smoking. Does the fact that there will soon be a new and improved iPhone give us a reason not to use the latest model? Refusing to enjoy the latest iPhone while holding out for the *best* possible iPhone would be as pointless as trying to find a leprechaun at the end of a rainbow. We have good grounds for enjoying the latest iPhone and even better grounds for accepting scientific facts. If we reject scientific facts, we might as well refuse to believe anything on the grounds that scientific facts could one day change.

Scientists are not concerned with what authorities have to say. To be sure, we should turn to experts when we have a question that's relevant to their field of expertise, but an expert's opinion is still an opinion. The only thing that counts in science is evidence, which is why any expert's finding can be overturned if the evidence demands as much. If new evidence is discovered, if we find new ways of interpreting information, or if we discover new tools that allow us to probe deeper than we could before, previous findings can be placed on the chopping block. Due to our limited abilities as humans, all data come with a margin for error, so a Straight Thinker should feel comfortable with uncertainty. It's perfectly OK to say, "I don't know." As Nobel laureate and renowned physicist Richard Feynman (1918–1988) put it, "I can live with doubt and uncertainty and not knowing. I think it is much more interesting to live not knowing than to have answers that might be wrong" (quoted in Shermer, 2004, p. 2). As soon as we believe we have all the answers, we stop looking for the truth.

As I've said before, certainty is the enemy of discovery. We don't discover something when things are as expected; we discover something when the unexpected occurs. It is ignorance, rather than certainty, that is the key to learning. Admitting uncertainty while honestly seeking the answer is a far more respectable pursuit than pretending to know something you don't know.

Where Science Meets Skepticism

Since the beginning of the scientific revolution nearly five hundred years ago, the methods of scientific discovery have gotten progressively better, and scientific facts have become more and more reliable. If we want to understand reality, we need methods we can trust, and this applies with equal force to the application of CT. We can investigate reality in a number of specific ways, two of which we'll explore with the remainder of this chapter: the SEARCH method and the Socratic method.

> The SEARCH Method:
>
> - State the claim
> - Examine the Evidence for the claim
> - Consider Alternative hypotheses
> - Rate, according to the Criteria of adequacy, each Hypothesis
>
> The SEARCH method is designed to separate bad claims from good claims and measure the reliability of a particular claim.

The SEARCH Method

The **SEARCH** method was developed by Theodore Schick Jr. and Lewis Vaughn in their seminal book *How to Think about Weird Things* (Schick & Vaughn, 2005, pp. 273–78). SEARCH is an acronym that describes an easy-to-follow four-step procedure:

1. **S**tate the claim.
2. Examine the **E**vidence for the claim.
3. Consider **A**lternative hypotheses.
4. **R**ate, according to the **C**riteria of adequacy, each **H**ypothesis.

Anytime you're presented with an incredible claim is an opportunity to use the SEARCH method to help evaluate the truth of the claim.[6] Once you've gone carefully through each step, you'll have good grounds for believing the claim or better reasons for rejecting it. With that said, we're ready to find out how to SEARCH for the truth.

STEP 1: STATE THE CLAIM

First, it's important to ensure that you understand exactly what the claim being made is *at the beginning of your examination*. Incredible claims are typically vague and confusing, so stating a claim requires clarity and specificity. For example, the claim "Our lives are controlled by the rich and powerful" is vague and nonspecific, so it wouldn't be a good starting point for examination. Controlled how, by whom, and in what ways? How powerful must one be to control our lives? A better way to state this claim would be "People of great wealth use their money to control what people do." You can go further by specifying any vague terms such as "people" or "control" or "great wealth" so that the claim is stated as specifically as possible. Not until a claim is sufficiently stated so that it is clear and everyone understands it can it be evaluated on its merits.

STEP 2: EXAMINE THE EVIDENCE FOR THE CLAIM

Before a claim can be considered on its merits, it must have merit; that is, there must be some evidence supporting it. What you've learned in the previous chapter are your guideposts to determine whether any evidence or logical arguments provide a reason to accept a claim. And don't forget that the available facts represent but one line of evidence, and evidence can be ad hoc, narrowly applicable, slanted, or even incomplete. For example, the multiverse theory says that the known universe might be but one universe that's nested within an infinite multitude of universes. While empirically testing the multiverse theory may be possible (Smolin, 1997), it could still be wrong. As we saw in chapter 4, evidence alone isn't enough to accept a claim as true. The word "evidence" isn't synonymous with the word "fact," which is why evaluating evidence is a specific skill set that's required to conduct a proper evaluation.

A. Determining the exact nature and limitations of the empirical evidence.

Just because evidence supports a claim doesn't mean that evidence is reliable; among others, dishonest prosecutors, SCAM practitioners, science deniers, and public officials all present "evidence" to support their cases. Confident eyewitness testimony frequently puts people in prison, even though eyewitness testimony is massively flawed by memory and perception. The situation is complicated when those same eyewitnesses are known for their honesty and credibility. When one personally attests to something, they are offering an **anecdote**. Anecdotes are generally seen as extremely weak forms of evidence for several reasons. For instance, anecdotes depend entirely on the accuracy of a person's explanation. Sometimes memories are contaminated; the event could've been misperceived; details could've been lost or reconstructed; people's recollections suffer from selective attention to details; and so on (Carroll, 2005). The synonym of anecdote is story, not data, thus they cannot rightfully be considered "evidence" any more than the book *Charlotte's Web* can be considered evidence for talking animals.

Let's say a friend who is otherwise honest, knowledgeable, and not prone to delusions told you she had a conversation with her houseplant. Do we have good reason to believe her? The answer is no and for several reasons. For one, talking plants would contradict massive amounts of established science. For another, if her claim can't be replicated, we shouldn't assume that this onetime occurrence is real. And as we've seen in previous chapters, everyone is susceptible to illusions and flawed thinking. Our friend making a mistake is a much better explanation than a talking plant. Nevertheless, we tend to take testimonials as much stronger evidence than they are. Even though the science says glucosamine doesn't treat joint pain and damage (Hespel, Maughan, & Greenhaff, 2006; Offit, 2014, p. 102, citing Clegg et al., 2006), a friend says he started taking glucosamine and his joints feel much better, so we believe it works. While the science shows that vaccines are safe and effective (Nirenberg, 2020; Offit, 2015), we hear an anecdote about a coworker's brother's daughter who was diagnosed with autism after receiving a vaccine and we believe the vaccine caused her autism. If a friend knows someone who knows someone who bought car X and had tons of trouble with it, we'll never buy a car X, despite the thousands of positive reviews about it (Stanovich, 2013, pp. 59–60).

> Anecdotal evidence leads us to conclusions that we wish to be true, not conclusions that actually are.—Barry Beyerstein

Readers of chapter 5 will recognize this sort of thinking as the representativeness heuristic, the tendency to ignore general information in favor of specific information. Moreover, in chapter 6, we saw that eyewitness testimony is unreliable for multiple reasons, including that it's based on anecdotal reports. As A. Leo Levin explains, "Eyewitness testimony is, at best, evidence of what the witness believes to have occurred. It may or may not tell what actually happened. The familiar problems of perception, of gauging time, speed, height, weight, of accurate identification of persons accused of crime all contribute to making honest testimony something less than completely credible" (Levin, 1968, p. 269). Millions of honest people can offer anecdotes of having witnessed space aliens and their crafts, just as millions of people in the past offered anecdotes of having witnessed witchcraft. Since we can be reasonably certain that no one has ever seen an alien or a witch, we can also be reasonably certain that evidence alone doesn't tell the whole story.

Examining evidence is an art form that requires evaluating whether the evidence is undermined by any of the deficiencies we've covered, such as human perception, beliefs, or memories. It also requires looking for other shortcomings, such as cognitive biases, pseudologic, vested interests, or proneness to belief in this sort of phenomenon. Ask questions such as "Did the claimant gather her evidence in a systematic way?" "Do experts agree with the claim?" or "Have experts already evaluated the evidence, and if so, what are their findings?" Be as critical of the evidence as the situation demands, and remember Hume's maxim: wise people proportion their beliefs to the evidence. Despite its appeal, a claim is only as credible as the evidence that supports it.

B. Discovering if any of these reasons deserve to be disqualified.

Humans are guided by their brains, and, as we've seen, brains can be extraordinarily unreliable. For this reason, people may offer considerations to support a hypothesis that we know should be disregarded, such as the claim that "mediums" can mediate communication between the living and the dead. The truth of mediumship would spell disaster for most of our theories of physics, which is a good reason to disqualify mediumship as an actual phenomenon. In making objective assessments of evidence, it's important to detect unsupported notions such as wishful thinking, faith, intuition, or certainty. Claims that are unsupported by evidence or reason are just that: unsupported.

C. Deciding whether the hypothesis in question actually explains the evidence.

Even after you've gone over the evidence and objectively weighed it against surrounding facts, competing hypotheses, and the susceptibility to human error, the work has just begun. Next you must determine whether the evidence offered supports the claim. If not, you have no reason to accept it and can move on without further consideration. Evidence must be relevant to the claim it is said to support; otherwise there's no reason to pay it any attention.

STEP 3: CONSIDER ALTERNATIVE HYPOTHESES

Step 3 is crucial but unfortunately is often skipped. So-called ghost hunters are notorious for failing to consider alternative hypotheses. While on a ghost hunt, any noise or shadow becomes evidence of ghosts (Michno, 2014; Radford, 2010a). Hypotheses are *proposed* explanations. As we saw in chapter 1, skeptics are relentlessly inquisitive, and so part of the process of discovery is coming up with alternative hypotheses that might better explain the data. We tend to test our own ideas but not others': the *congruence bias*, as it's called. For example, it would be an error to conclude that you can project your "astral body" because you've had an experience that fit other descriptions of astral projection. This experience may be explained some other way that has nothing to do with astral projection, such as hypnogogic or hypnopompic hallucinations. Step 3 is an important step toward avoiding the congruence bias. By thinking of and testing the different ways in which a hypothesis *could be* explained, you greatly increase the chances of finding what *does* explain the data.

Consider the following hypothesis: the Easter Bunny, the rabbit that leaves behind colorful eggs and baskets full of chocolates on Easter Day, is real and exists in the world. To support this hypothesis, we could draw on the following evidence: millions of people believe the Easter Bunny is real; the Easter Bunny leaves evidence behind; given the number of rabbits in the world, the age of the rabbit species, and the rate of genetic mutation, it is possible that one line of rabbits evolved to cooperate with humans and now we celebrate a holiday together; some people claim to have seen the Easter Bunny; the story of the Easter Bunny had to have begun somewhere, and it's reasonable to believe it's based on true events. Evidence for Santa can be just as compelling (Cuno, 2019).

If we ended our inquiry there without looking into any possible counterevidence, even with this incredible claim, we could build a reasonably convincing case. It isn't until we stack it against alternative explanations—that the Easter Bunny is a folkloric figure whose legacy is kept alive in the spirit of fun and celebration—that it loses its promise. The folkloric hypothesis is well attested to and supported by multiple lines of independent evidence; it doesn't conflict with well-established theories of biology; and it doesn't require the existence of any new entities—namely an overgrown rabbit species capable of planning, interspecies communication, supersonic travel, and a seemingly endless supply of delicious chocolate replicas of itself!

Step 3 requires Examining the **E**vidence (Step 2) and approaching the claim with a creative and open mind. Step 3 involves brainstorming alternative ways to explain the claim and examining whether there are reasons to favor an alternative explanation. We should also remember that when confronted with an incredible claim, it's natural to want to commit the argument from ignorance fallacy, and jump directly from an inability to immediately explain something to drawing a specific conclusion. It cannot be overstated: unexplained does not mean unexplainable. Just because we cannot instantly offer a natural explanation does not mean one doesn't exist. Like juries that make up their minds during opening statements and from there on retain only evidence that supports their preference, our inclination is to latch onto the explanation we prefer.

Criteria of Adequacy:

- Testability
- Fruitfulness
- Scope
- Simplicity
- Conservatism

Using the criteria of adequacy as a tool can help determine whether the reasons offered to explain some phenomena are acceptable.

STEP 4: RATE, ACCORDING TO THE CRITERIA OF ADEQUACY, EACH HYPOTHESIS

With alternative explanations in tow, we now turn to the criteria of adequacy. Step 4 is about sifting through the alternatives offered in Step 3 to decide which explanation to accept. To discover a truth, we cannot just list all the evidence, put it on a scale, and form a conclusion based on which way the scale tips. We need objective standards to help put all the evidence—or lack thereof—into perspective. Our maxim *incredible claims require credible evidence* doesn't mean incredible claims require *a lot* of evidence, for a lot of bad evidence is still just bad evidence. This maxim means that stronger claims demand reliable evidence. We evaluate the strength of evidence using the criteria of adequacy.

And what are the criteria of adequacy? The criteria of adequacy are standards that guide us toward whatever explanation the data support. These criteria are testability, fruitfulness, scope, simplicity, and conservatism. Hypotheses that make novel predictions are testable. Hypotheses that make accurate predictions are fruitful. Hypotheses that explain a range of data have scope. Hypotheses rested on the least number of assumptions are simple. And finally, hypotheses that cohere with well-known facts about the universe are conservative. Let's see how it works.

Testability

The first thing we need to find out is whether the hypothesis can be tested. An untestable hypothesis is one in which no procedure does or can exist for checking its truth. Thus a testable hypothesis is one in which some process can actually or theoretically determine its truth. Moreover, a hypothesis is testable if it predicts more than what it intended to explain (Vaughn, 2019, p. 332). Darwin's theory of evolution was introduced to explain the diversity of life, yet it predicted DNA, an approximate age of Earth, and the order of the geologic columns, among other things. Einstein's theory of general relativity was introduced to explain how objects influence other objects by manipulating the fabric of space, yet it predicted black holes, gravity waves, stellar parallax, the aberration of starlight, and more. Unlike testable hypotheses, untestable hypotheses explain nothing since they cannot be confirmed or refuted.

Testability has two parts: replication—the ability to reproduce findings—and falsification—the process of disproving a hypothesis. Replication ensures that the results of an experiment weren't due to confounding variables (chapter 3). If independent researchers investigate a phenomenon and get the same or similar results, this is good evidence in favor of the hypothesis. However, if independent researchers investigate a phenomenon and their results vary from those of other researchers, then something is either wrong with the methodology or something is wrong with the hypothesis.

Falsification is equally as important. Researchers need to know whether they can test the truth value of a claim. Incredible claims typically cannot be tested and thus cannot be falsified. If it turns out that a claim is unfalsifiable, we're justified in disregarding the claim altogether for the simple reason that if we cannot test it, there's no reason to assume it's true. Unfalsifiable means immune to evidence. In addition, a claim isn't considered useful unless it makes specific predictions about what *will* and *will not* happen (Stanovich, 2013, p. 22). Let's say I claim that criminals can't follow rules and that prison is the antidote to misbehavior. Any observed misbehavior of prisoners will seem to offer supporting evidence for my claim, but so would positive behavior, as I could attribute positive behavior to the success of the penal system. My claim successfully predicts both good and bad behavior, rendering it unfalsifiable and therefore useless.

Fruitfulness

The next question to ask is, "Does the hypothesis make observable predictions that explain reality?" The importance of the question is simple: any claim that makes accurate predictions has a greater chance of being true than one that doesn't. Dowsing is a good example of a claim that makes predictions. *Dowsing* is the practice of hovering a dowsing rod—anything from an antenna-looking contraption to a Y- or L-shaped stick—over the ground to locate things of value, such as ground water, metals, minerals, precious stones, or oil. (Some dowsers claim they can find valuables by hovering dowsing rods or a pendulum over a map rather than physically dowsing a particular location. Frighteningly, hovering a pendulum over an accused person has also been used to determine guilt or innocence.) When the stick makes some sort of movement, the dowser is said to have located the sought valuable (Hines, 2003, pp. 418–22). Dowsing therefore predicts that hovering dowsing rods over the ground will cause valuables to somehow communicate their location. Scientific evidence that dowsing works is lacking (Alcock, 2018, pp. 273–74), and evidence suggesting that dowsing is explained by the *ideomotor effect* abounds. If dowsing ever does work, it works by accident: it doesn't take a dowser suspending sticks over the only green bush in the middle of a desert to "divine" that water is near.

By contrast, Darwin's theory of evolution by natural selection is a theory that bears fruit. Before there was much in the way of a fossil record, and long before any studies in molecular genetics were under way, Darwin's theory predicted gradual complexity throughout the geologic column,[7] as well as a shared unit of biological inheritance. Alas, Darwin's predictions bear fruit: no anachronistic—i.e., misplaced in history—fossils have ever been found, and so far, DNA shows that all life on Earth is biologically related. A hypothesis that doesn't make any predictions is not necessarily wrong; it's just not as credible as one that does. And if two such hypotheses are in competition, we should accept only the one that bears fruit.

Scope

Next, we determine the breadth, depth, or reach of the hypothesis. That is, we need to know how many different phenomena the hypothesis explains and predicts. The more it explains or predicts, the more it broadens our understanding and provides better reason for believing it. Furthermore, the more a hypothesis explains, the more likely it is to be true since the more it explains, the more evidence it has in its favor. In chapters 6 through 8, we developed the hypothesis that human beliefs, perceptions, and memories are constructed by error-prone brains that distort reality in numerous ways. Our hypothesis explains many phenomena, including pareidolia, hallucinations, confabulation, perceptual constancies, constrictive perceptions, conspiracy thinking, false memories, and more. If our hypothesis explained only one or two phenomena, and not the others, we'd have fewer reasons to believe it over an alternative.

Additionally, as Lewis Vaughn observes, juries often decide guilt based on which theory of the case has the most scope. If the prosecution's theory explains the skid marks leaving the scene, the blood on the defendant's clothing, and the knife found in his pocket, while the defendant's theory raises more questions than it answers, the verdict is guilty.

Simplicity

Simplicity asks which hypothesis explains the data in the simplest manner. Simplicity doesn't make a hypothesis correct. However, since simpler hypotheses have fewer opportunities to go wrong, and less evidence is needed to support them, they have a much better chance of being true. Here, simple means *makes the fewest assumptions*. Suppose you moved into a new house and found it unbearably cold the first night, and, in trying to warm up, you discovered that you couldn't get the furnace pilot lit. Perhaps the gas had not yet been turned on.

Perhaps ghosts from past tenants are blocking the gas lines, cooling down the temperature of the house as a deterrent to would-be tenants.

The gas hypothesis is the simplest because, in addition to being testable, capable of yielding predictions, and able to explain several phenomena, it doesn't require that we add another hard-to-explain phenomenon. The ghost hypothesis, on the other hand, cannot be tested, it yields no novel predictions, it doesn't explain how the house got cold, and it proposes the existence of sentient ghosts that have wants, desires, and the ability to manipulate the physical world. The criterion of simplicity shows that the gas hypothesis is the more likely explanation, while not ruling out the incredible ghost hypothesis.

Conservatism

Lastly, we must ask whether a claim conflicts with well-established facts. If a hypothesis conflicts with well-attested-to scientific theories or natural laws, we have fewer reasons to believe it than if it's consistent with these laws. Our job, then, is to assign weight to the claim given all the available evidence. In the end, a claim that's consistent with the best evidence is the one we should believe. This reasoning makes good sense. If a hypothesis flies in the face of exceedingly well-established evidence, its likelihood of being true would be much smaller than the likelihood of one that didn't. Thus, when a proposed hypothesis conflicts with empirically confirmed theories, it must be regarded as improbable unless additional evidence demonstrates that those empirically confirmed theories are wrong.

What is considered a "well-established" fact can also be objectively determined (Vaughn, 2019, p. 340). We can start with beliefs that we have little reason to doubt, such as, "It's sunny outside," or "My car won't start." We can also add facts derived from reliable sources, such as "The Egyptian pyramids were built by humans," or "The Earth orbits the Sun." Finally, we could also add facts that have been demonstrated by empirical sciences, such as "Vaccines are overwhelmingly safe and effective," or "Human activity is negatively impacting the environment."

The Socratic Method

The SEARCH method is useful for investigating any incredible claim. Sometimes, however, we're confronted with claims during casual conversation, and the party making the claim may be unwilling to sit down and go through the steps of conducting a proper examination. Indeed, that same unwillingness may explain why people come to believe incredible claims in the first place! By contrast, the Socratic Method doesn't require such a rigorous or time-consuming approach. It is fun, it can be deployed at any time, and it takes minimal effort.

The Socratic Method is all about asking questions. While the SEARCH method investigates by systematic examination, the Socratic Method investigates strictly by Q&A. Socrates used this method to help people evaluate their beliefs and subject them to rigorous examination. The Socratic Method is usually aimed at questioning one particular claim, belief, or point of view using a cooperative, argumentative dialogue between individuals. Its focus is to stimulate CT and draw out ideas and underlying presumptions. With this **dialectic** and unthreatening approach, it's possible to both challenge unreasonable beliefs and evaluate claims advanced by anyone. With this understanding, I'll go over the five stages of the Socratic Method while applying it to a specific claim (adapted from Boghossian, 2013, pp. 105–27).

STAGE 1: WONDER

The Socratic Method has just five stages from start to finish. The first stage begins with wonder, where thoughts are put into words in the form of a question. The wonder stage begins with questions like "What is X?" Examples of wonder might be, "What is beauty?" "Are people responsible for who they become?" or "What's worth dying for?" Wonder leads to a hypothesis, which is the second stage of the Socratic Method. In this example, we're going to wonder, "Why is Earth teeming with life?"

STAGE 2: HYPOTHESIS

Earlier we defined a hypothesis as a proposed explanation for some phenomenon, so we'll apply our definition to the question posed in Stage 1. An emerging hypothesis to the question "Why is Earth so teeming with life?" might be *Evolution by natural selection led to the diversity of life on our planet.* Another hypothesis could be *Life is an illusory aspect of the computer simulation we're in.*

STAGE 3: ELENCHUS (Q&A)

Elenchus (plural elenchi) is the meat and potatoes of the Socratic Method. The elenchus is a Q&A approach to logical refutation. This approach uses the method of counterexamples to question a claim with the aim of showing that the claim may be false or otherwise unreasonable. In response to our earlier example:

> Person 1: Why is Earth so teeming with life? [Wonder stage]
>
> Person 2: Darwin showed that life evolved from a single ancestor via evolution by natural selection. [Hypothesis stage]

Socratic Interlocutor: Could life be a technically advanced computer simulation that came from an all-powerful computer programmer? [Counterexample]

The Socratic interlocutor's intention is to come up with one or more ways that show the hypothesis *could* be false, not necessarily that it is. I should stress here the important difference between showing the hypothesis *is* false and showing that it *could* be false, since showing something *is* wrong doesn't explain how it *could* be wrong. If the interlocutor forces the other party to defend their position by claiming their hypothesis *is* wrong, nothing is gained. There must be open dialogue between the parties. By showing that the hypothesis could be wrong, the Socratic interlocutor encourages the other party to question their hypothesis rather than defend it.

In our example, one condition that could render the hypothesis false would be that a technologically advanced computer programmer created all life as part of a simulation and made it appear as though it evolved through natural selection (Kassan, 2016). Here we have a good counterexample because it places doubt on the hypothesis by providing an example of how all life might have come about. Other counterexamples could be that life was seeded by intelligent extraterrestrials or that an all-powerful deity created life from nothing. The point is to create legitimate counterexamples that call into question the hypothesis being offered so that the one making the claim can no longer hold it with certainty.

> Just as the body is exposed to toxins so is the mind.
> —Peter Boghossian

Hypotheses are never proven true. If a hypothesis survives repeated iterations in Stage 3, we can still only say that so far it has withstood the process of falsification, not that it's true. To use the standard example, let's say you hypothesized that all swans are white. In formulating this hypothesis, you searched the American continents, Asia, Africa, and Europe, and on every continent you found white swans. Despite your efforts, you haven't proven that all swans *are* white. In fact, a surviving species of black swans (see figure 10.1) lives in Australia.[8] People hold many irrational beliefs for the same reason; they've never looked in Australia!

Even though millions of confirming instances cannot confirm a hypothesis as true (such as all the supporting evidence for evolution by natural selection), it only takes one counterexample to prove a hypothesis false, such as the documented observation of a single black swan (or, in the case of evolution, a single anachronistic fossil). Stage 3 is simply a way to undermine a hypothesis by bringing to light internal contradictions or logical inconsistencies.

Figure 10.1. An Australian black swan (*Cygnus atratus*) with her chick.
Wikimedia.org.

STAGE 4: ACCEPT OR REVISE HYPOTHESIS

Stage 4 has two options: accept the hypothesis as temporarily true, or reject it outright. If it's accepted as true, we move directly to Stage 5. If it's rejected, we go back to Stage 2 to form another hypothesis and then on to Stage 3 to conduct the elenchus again. If the person advancing the hypothesis is unable to overcome the Socratic interlocutor's argument made during the elenchus, the hypothesis must be revised. In our current example, if she couldn't rebut any of the counterexamples, she might have to amend her hypothesis to something like, "Several possibilities exist for how life could have begun."

On the other hand, if the Socratic interlocutor is unable to refute the hypothesis, it should be accepted provisionally because the fact that it's unrefuted doesn't mean it's true, only that it couldn't be falsified. It could mean the interlocutor wasn't creative enough, our current knowledge is incomplete, or any number of other factors. Remember, unexplained does not mean unexplainable.

STAGE 5: ACT ACCORDINGLY

Since the Socratic Method is intended to challenge beliefs, if a successful intervention has occurred, ideally the claimant should act accordingly. What does that mean? Acting accordingly could range from changing one's beliefs and adopting a new outlook to taking some specific action like using the Socratic Method on others who hold the belief you have now abandoned. Stage 5 is more about the consequences of one's examination than the application of the method.

Wrapping Up

We've seen that science is a limited, flawed, imperfect process that cannot explain every aspect of reality. We've also seen that the self-correcting and progressive nature of science has proven both useful and reliable and that no competing methods to discovering facts about nature exist. Science has uncovered the structure of DNA, the composition of stars, the age of distant galaxies, the causes of disease, and how to develop cures and vaccines. For all its flaws, science is the best method we have for obtaining knowledge about reality.

We've also learned that the use of science is not strictly limited to scientists alone. The two methods of examination introduced in this chapter are available to and can be carried out by almost anyone. These methods will help you, the skeptic, investigate all manner of claims. However, don't be surprised if convincing believers requires other steps, such as prior agreement on what to accept as evidence or who should be counted as reliable sources.

CHAPTER 11

The Fringes of Science

An often-overlooked aspect of science is understanding what sort of claims are within its domain and what sort of claims are beyond its ability to explore. A common misconception that science should be able to explain everything prevails, which often leads to the argument from ignorance fallacy, where people believe that the unexplained is unexplainable and thus supernatural.

In this chapter, I'll cover the differences between those things that are within the arena of science and those that are forever beyond its reach. I'll cover both how and why something nonphysical couldn't interact with something physical and discuss why some phenomena are strictly off limits to science. I've chosen three topics for their breadth and popularity and because they offer a chance to put our CT skills to the test.

Building Vocabulary

Burden of proof:	Cause and effect:
A disputer's duty to provide sufficient grounds for establishing their position.	Influence by which one action, process, or event in some way affects another action, process, or event.

Supernatural:	Paranormal:
Any force, process, entity, or agent that does not exist in nature, usually in reference to a spiritual or magical being; beyond the physical, material realm.	Alleged phenomena, often strange or bizarre, that cannot be explained by scientific means, for example, ghosts, telepathy, or extrasensory perception.

Supernatural

Supernatural entities are commonly thought of as strictly religious, though not all believers in the supernatural are religious, and not all religious people believe in the supernatural. When people talk about the supernatural, they're usually talking about something like deities, curses, angels, demons, miracles, spirits, afterlife, and, one of my personal favorites, karma trolls. The existence of supernatural entities, however, is beyond the reach of science (chapter 2), since they exist outside of nature and are therefore not constrained by natural laws. Here we can draw a parallel between the laws of nature and the laws of the land. Wherever you find yourself, you're legally obligated to follow the law. Similarly, any supernatural entity entering the natural world would be bound by the laws of nature. Since these constraints would deprive the entity of its supernatural status, anything found within nature by definition cannot be supernatural. However, an important difference between human-made laws and the laws of nature is that the human laws *can* be violated. By contrast, the laws of nature cannot be broken, even if we don't know what those laws are or we don't understand them. The laws of nature establish strict limitations to what can occur, which means that a violation of natural laws is a contradiction in terms. That is, if something is allowed to happen in nature, that thing will always be governed by physical limits, and these physical limits are as fundamental to our universe as numbers are to mathematics.

Paranormal

Even though most **paranormal** phenomena contradict massive amounts of science, people all over the world believe them anyway. As the psychologist Bruce Hood says, "Just because something is nonsense doesn't stop people believing in it" (Hood, 2009, p. 159). No solved mystery has ever concluded by affirming the supernatural or paranormal, yet belief persists. And even if some paranormal phenomena exist, by definition those phenomena are beyond the scope of scientific inquiry.

> People in virtually every society exist in a state of contradiction, valuing and promoting reason while at the same time harboring strong transcendental beliefs that defy it.
> —James E. Alcock

More importantly, if any paranormal claims were true—such as ESP, telekinesis, or astral projection—many well-grounded theories in physics would melt away (Reber & Alcock, 2019; Reber & Alcock, 2020). As it turns out, all our firmly grounded scientific theories remain on firm ground. Recall the criterion of conservatism, which states that if a claim conflicts with well-established laws of nature, we have good cause for skepticism. The physicist Steven Weinberg undertook an exhaustive search looking for any experiments over the past one hundred years that refuted the consensus of the physics community, and he found exactly zero (Weinberg, 1994, pp. 129–30). Popper made a similar observation almost fifty years earlier (Popper, 1959/2014, p. 252). Not only are Weinberg's and Popper's findings a testament to the strength of a well-established theory, they're a testament to the extreme unlikelihood of incredible claims.

So why does belief in the supernatural and paranormal persist? When believers are asked why they believe, the number one reason given is personal experience (Alcock & Otis, 1980; Clarke, 1995). As table 11.1 reveals (Alcock, 2018, pp. 455–56), personal experience can be marred by our beliefs, biases, flawed perceptions, and mental health. Moreover, research reveals that belief in the paranormal is correlated with intuitive, rather than analytical, thinking (Lindman & Aarnio, 2007). Thus, while personal experience can be extremely compelling, it isn't the type of reliable evidence that's used to determine what's true. Nevertheless, some paranormal ideas are extremely popular, ESP probably being the most well-known. A 1987 survey of more than fourteen thousand Americans found that fully 67 percent believed they had experienced ESP (Greeley, 1987). A year later, the National Research Council (NRC) looked at the 130-year record of ESP research and found that despite over a century of scientific inquiry, there is "no scientific justification for the existence of phenomena such as extrasensory perception, mental telepathy, or 'mind over matter' exercises. . . . Evaluation of a large body of the best available evidence simply does not support the contention that these phenomena exist" (Gilovich, 1991, p. 160). The reasons people believe in ESP range from uncritical treatment in the popular media to personal experience, such as having a dream or a hunch that seemingly came true (Hines, 2003, p. 147). No new or compelling evidence has surfaced since the NRC conducted their study, and ESP, telepathy, and other "psi phenomena" still remain both scientifically invalid and logically incoherent (Reber & Alcock, 2019; Reber & Alcock, 2020). Even the best evidence for the existence of ESP has been invalidated beyond any reasonable doubt (Gilovich, 1991, pp. 159–67).

The same could be said for every paranormal phenomenon that's ever been investigated (Alcock, 2018; Gilovich, 1991; Hines, 2003; Wiseman, 2011). Nonetheless, in the third decade of the twenty-first century, roughly 75 percent of Americans endorse some sort of paranormal belief, including ESP (41 percent), astrology (25 percent), precognition (26 percent), and the idea that human spirits can return to earth (32 percent) (Alcock, 2018, p. 452; Cooper, 2010). People are certainly entitled to believe what they want, but they do so at the risk of missing

Table 11.1. Poor Thinking = Poor Beliefs

Fantasy proneness	Those who believe in paranormal phenomena tend to be higher in measures of fantasy proneness,[1] and research has reported an association between fantasy proneness and reports of out-of-body experiences.[2] People who present themselves as mediums or psychics are often fantasy-prone individuals.[3]
Magical thinking	People who report having had paranormal experiences tend to score higher in measures of magical thinking and are more likely to automatically attribute causality when two events occur one after the other.[4]
The search for meaning	People vary in terms of how important it is to find meaning in life. Whether based in reality or not, paranormal experiences and beliefs can provide meaningfulness and a sense of purpose in life.[5]
Weakness in understanding probabilities	When faced with ambiguity and uncertainty, we often fall back on cognitive heuristics, or "rules of thumb," in assessing the probability of events. Most people are demonstrably poor at estimating probability, but believers in the paranormal have been found to be particularly prone to errors and biases in probabilistic reasoning.[6] Other research suggests that those who believe in the paranormal tend to have a specific weakness with regard to the perception of randomness, making it difficult for them to appreciate the likelihood of chance events.[7]
Memory distortion	Our memories are all vulnerable to distortion, but some research suggests that believers in the paranormal may be more susceptible than nonbelievers to forming false memories surrounding events that they have interpreted as being paranormal.[8]
Brain damage	While there is no reason to think that believers and nonbelievers in the paranormal differ in terms of brain health, there is some evidence that brain dysfunction may play a role in the perception that one personally has psychic powers. For example, a study compared seventeen "sensitives" from the College of Psychic Studies in England with a matched set of seventeen controls.[9] It was found that the sensitives had more frequent histories of head injuries and serious illness, and two-thirds of them showed evidence of right temporal lobe dysfunction.

1. Irwin, H. J. (1990). Fantasy proneness and paranormal beliefs. *Psychological Reports, 66*(2), 655–58.
2. Gow, K., Lang, T., & Chant, D. (2004). Fantasy proneness, paranormal beliefs, and personality features in out-of-body experiences. *Contemporary Hypnosis, 21*(3), 107–25.
3. Spanos, N. P., Cross, A., Dixon, K., & DuBreuil, S. C. (1993). Close encounters: An examination of UFO experiences. *Journal of Abnormal Psychology, 102*, 624–32.
4. Thalbourne, M. A., & French, C. C. (1995). Paranormal belief, manic depressants, and magical radiation: A replication. *Personality and Individual Differences, 18*(2), 291–92.
5. Kennedy, J. E. (2005). Personality and motivations to believe, misbelieve, and disbelieve in paranormal phenomena. *Journal of Parapsychology, 69*, 263–92.
6. Wiseman, R., & Watt, C. (2006). Belief in psychic ability and the misattribution hypothesis: A qualitative review. *British Journal of Psychology, 97*, 323–38.
7. Dagnall, N., Parker, A., & Munley, G. (2007). Paranormal belief and reasoning. *Personality and Individual Differences, 43*(6), 1406–15.
8. Wilson, K., & French, C. C. (2006). The relationship between susceptibility to false memories, dissociativity, and paranormal belief and experience. *Personality and Individual Differences, 41*(8), 1493–1502.
9. Fenwick, P., et al. (1985). Psychic sensitivity, mystical experiences, head injury, and brain pathology. *British Journal of Medical Psychology, 58*(1), 35–44.

out on an already wonderous reality. When someone is wrongly convicted of a crime, the guilty party runs free and people stop looking. The same goes for paranormal beliefs. If we were to explain some strange event as being "paranormal," we wouldn't have explained anything, as the true explanation would be missing and people would no longer be looking (Alcock, 2018, p. 457).

The Boundaries of Science

It may seem obvious that science can only deal with scientific claims. The problem is what makes a claim scientific isn't always clear. We've seen that scientific claims must be testable. Here I'll go a step further and state that scientific claims must be at least "theoretically and practically testable and falsifiable" (Novella, 2012, p. 111). That is, even if we couldn't empirically test a claim because of some limit in our abilities, it could still be scientific if it *could* be tested—and thus falsified—in theory.

Which brings us to the paranormal and supernatural. Everything within the natural world is considered natural, thus nothing natural can also be *su*-*per*natural. We faced this problem in chapter 2 when discussing relativism and concluded that something cannot be both X and not X. Likewise, paranormal claims contradict the laws of nature and so couldn't exist within a naturalistic framework. Therefore, since supernatural and paranormal phenomena aren't testable, scientifically speaking, they're indistinguishable from one another.

Incredible claims generally fall into two categories: those outside the arena of science and those within. Supernatural and paranormal phenomena are forever outside of that arena, since they cannot be tested or verified using scientific instruments or methods. However, it's important to stress that this doesn't mean such phenomena don't exist, only that they aren't scientific. If ghosts existed, their presence should be obvious to all observers, not "discovered" under questionable circumstances and "proven" by unreliable evidence. Ghosts, it is claimed, are active participants in our world with abilities such as physically moving objects, making noises, or tampering with electronic devices. If ghosts really could interfere with the physical world, they would be investigable using scientific methodology (Radford, 2017).

Moreover, to prove anything, the thing to be proven must be defined, yet so-called ghost experts never define what they mean by "ghost"; they even disagree on what sort of ghosts exist (Radford, 2018). Ghost hunters attempt using scientific instruments like infrared cameras or Geiger counters to bolster their credibility, while committing blatant logical errors, such as searching for ghosts in dark rooms with the lights out.[1] The problem with looking for ghosts with the lights out is obvious: *looking* requires light, and less light means a lesser chance that anyone will *see* anything. As far as using scientific instruments goes, "until someone can reliably demonstrate that ghosts have certain measurable characteristics, devices that measure those characteristics are irrelevant" (Radford, 2010a).

Skeptics, in attempting to rectify the disconnect between themselves and proponents of the paranormal, conducted a joint study with believers in EVP (electronic voice phenomena) to see if there was any merit to the claim. EVP proponents claim they can record voices of the dead electronically by going to a graveyard or haunted destination with recording devices, turning the volume all the way up, and hitting record (Hines, 2003, pp. 110–11). When the recording is replayed, EVP proponents claim they can make out voices and even specific messages from otherworldly realms. Unfortunately, the evidence doesn't support their incredible claims. The joint study mentioned above showed that believers using the same data failed to reach consistent agreement on the meaning of research data taken from the field (Buckner & Buckner, 2012). In this study, four groups independently recorded a total of 153 alleged EVPs; in other words, noises allegedly made by ghosts. Each group analyzed its own recordings and those of two other groups, with none of the groups having knowledge of any other group's results. Although none of the groups had any difficulty finding instances of EVP, of the 153 alleged EVPs identified, no three groups had matching analyses of the data. That is, much like people interpret Rorschach inkblot tests differently, there wasn't even one instance where all three groups identified the same EVPs. If EVPs were caused by human telepathy, deceased persons, extraterrestrials, or people trapped in other times or dimensions, as EVP proponents believe, we should expect different EVP experts to get the same results from the same data. However, if the interpretations of supposed EVPs are due to physical explanations such as outside interference and human error, or there exist psychological explanations, such as audio pareidolia, apophenia, or patternicity (recall the studies from chapter 8, where nonsense noise was played to participants and, despite there being no human words in the recordings, many of them still heard language), we should expect vastly different results. For better or worse, the latter is exactly what the study found. EVP is no longer taken seriously other than by a few dedicated believers. Even sincere attempts to replicate pro-EVP findings have found no evidence that it's a real phenomenon (Alcock, 2018, p. 493; Baruš, 2001; Nees & Phillips, 2015).

Unless science can disprove EVP—unlikely, since science cannot prove negatives—all that can be said is the best attempts to discover evidence of EVPs have failed to yield positive evidence. If we really lived in a universe that was teeming with paranormal activity, we'd still be unable to produce evidence of the paranormal because any phenomenon we could identify and explain would be considered part of our natural world and thus would be normal, not *para*normal.

Although scientists cannot currently explain every phenomenon, to conclude that something was paranormal *because science couldn't explain it* would be committing the argument from ignorance fallacy (chapter 5) by drawing a specific conclusion from an absence of evidence. Drawing such conclusions would also commit the either/or fallacy (chapter 4) by assuming only two options exist: either science can explain it, or it's paranormal.

The question has been posed, "If some unexplained phenomenon persisted and continued to evade scientific explanation, would that constitute evidence that we lived in a universe where the supernatural and paranormal were real?" (Novella, 2012, pp. 111–12). No doubt, many scientists would be intrigued by the challenge since, after all, investigating and explaining mysteries is precisely what scientists do. It should be remembered, however, that the history of science is full of explained phenomena that were once unexplained.

For decades scientists couldn't explain the orbit of Mercury: Newtonian mechanics couldn't account for an irregularity in Mercury's orbit. The irregularity was only later explained by Einstein's theory of general relativity (Weinberg, 1994, pp. 91–94). Similarly, the theory of what was then called "continental drift" was first proposed in 1912, though it wasn't fully explained until the theory of *plate tectonics* was validated in the 1950s and 1960s. During the time scientists couldn't explain Mercury's orbit or the movement of the continents, they didn't resort to suggesting magical gnomes move planets and continents; that would've been unreasonable and inconclusive. Rather than resorting to magical gnomes, scientists suspended judgment and worked hard until a proper explanation came along. Physicists know that unknown forces drive the expansion of the universe and control how stars and galaxies move about. Instead of declaring that these unknown forces are evidence of something specific, they give them names like "dark matter" or "dark energy," suspending judgment until they develop scientific explanations for what they are.

In some ways, science progresses from one unexplained anomaly to the next, much like Kuhn's idea of science progressing from one paradigm to another. Shortly after the Copernican hypothesis of heliocentrism was introduced, there was a problem with what's called *stellar parallax*. If the Earth orbits the sun, as the theory proposed, during its orbit, the farthest stars should appear to shift positions with respect to the closer stars, as when an object appears to move when you alternately look at it from one eye to another. Observers couldn't detect any shift, and the puzzle persisted for nearly three hundred years until powerful telescopes were invented that allowed scientists to finally detect the parallax. Accordingly, if some unexplained phenomenon remained unexplained, it wouldn't be positive evidence of anything, much like an unsolved murder isn't evidence that there was no murderer. Remember, unexplained doesn't mean unexplainable, and we're ultimately limited by our own imaginations and technological abilities to conduct tests and produce explanations.

The scientific way is to continue searching for answers, even when we think we already have them. Vaccine safety and efficacy continue to be tested for decades after they've gone to market. Einstein's theory of general relativity is still being tested more than one hundred years after he proposed it. Appealing to the paranormal or supernatural for explanations would be to give up on discovering an actual explanation because invoking the paranormal and supernatural explains exactly nothing. If spirits did cause noises in the woods or magic gnomes do perturb a planet's orbit, the question of *how* they do it still remains. Moreover, invoking the paranormal or supernatural as an explanation is a combination of wishful thinking and a failure to explain, not a conclusion one arrives at through an exhaustive, objective analysis of facts and evidence.

Proof and Probability

As we've seen, some claims are unfalsifiable, rendering them unscientific. Does this mean we should only take something as true if it has scientific support? If so, how do we know what has scientific support and what doesn't? If not, how do we choose what to believe? For starters, we need to keep in mind that the person making the claim bears the **burden of proof**.

Figure 11.1. Welsh polymath Bertrand Russell. *Wikimedia.org.*

Bertrand Russell argued that if he suggested a celestial teapot were orbiting the sun between Earth and Mars, nobody could disprove his claim, provided the teapot was too small for scientific instruments to detect it. Now, does the fact that one couldn't *dis*prove Russell's claim provide sufficient grounds for believing it? Should we believe a teapot is orbiting the sun simply because we cannot prove it isn't? Certainly not, and the reason is an inability to prove or disprove a claim isn't a good reason to believe it.

Comedian John Oliver provided an excellent example of why we shouldn't believe something just because it hasn't been disproven. Former congressman

Dan Burton made the argument that even though there are no studies that prove the mercury in vaccines cause autism, "there are no studies that disprove it either." In responding, Oliver explained that "proving a negative is an impossible standard." Why? Oliver's explanation is far more colorful than I care to be here, but I can make the point equally as clear. Let's say I told Dan Burton that every day in the shower he sings "I'm a Little Teapot." Burton could argue that there's absolutely no evidence that he sings "I'm a Little Teapot" every day in the shower. Yes, Mr. Burton, but there's no evidence proving that you don't either (Oliver, 2017).

Take the "brain in a vat" philosophical thought experiment, for example. According to this experiment, there's no physical, logical, or theoretical proof demonstrating that we aren't really brains in vats and that our reality is perfectly simulated. So do we say that, since it cannot be disproven that we are brains in vats, we should suspend judgment, conceding that it's possible? The answer is clearly no because the brain in a vat hypothesis doesn't offer the best explanation. If proving the nonexistence of something is impossible, the best stance is showing that its existence is impossible or unlikely given our current understanding of reality. If we had to suspend judgment every time we couldn't prove a negative, we'd have to suspend judgment on many incredible claims. As the American philosopher Guy Elgat put it, "We would thus have to admit that we can't really say whether unicorns are real or not, whether there is or there is not a troupe of invisible leprechauns dancing the hora behind our backs, or whether or not we are professional assassins whose incriminating memories are erased by our employers, the undetectable aliens from planet Xanadu. This would be utter epistemological bankruptcy" (Elgat, 2019).

Just because imaginary friends' existence cannot be disproven doesn't mean it's equally likely that they do exist. Dawkins explains that "the fact that we can neither prove nor disprove the existence of something does not put existence and non-existence on an even footing" (Dawkins, 2006, p. 72). Claims worth taking seriously must be either physically or logically possible. Anything that violates a logical principle is logically impossible, while anything that violates a law of nature is physically impossible (Vaughn, 2019, p. 391). Accordingly, anything that's physically or logically impossible cannot exist. Einstein's theory of special relativity says it's physically impossible for macroscopic objects to exceed (or even reach) the speed of light. Heisenberg's *uncertainty principle* says it's physically impossible to simultaneously know a subatomic particle's location and speed (Wolfson, 2000). Saying two words at once is also physically impossible, while making 2 + 2 = 5 is logically impossible. Moreover, a four-sided triangle is physically *and* logically impossible. Importantly, just because something is logically possible doesn't mean it's physically possible. Astral projection (chapter

12)—the belief that people can leave their physical body and travel in their astral body—presents no logical contradictions, yet it cannot exist because leaving your physical body is physically impossible (Loxton, 2018c; Reber & Alcock, 2019; Reber & Alcock, 2020). Therefore, although the saying "anything is possible" is extremely popular, it is also extremely wrong.

Believing Is Not Enough

A popular belief persists that if one channels their spiritual powers inward, meditates, thinks positive thoughts, and connects with a "universal energy," it's possible to defy the laws of nature and walk barefoot across scorching hot coals without burning the skin. This belief is similar to the concept of "manifesting" or the "law of attraction" that was popularized by the film (and later book) *The Secret* (Byrne & Heriot, 2006). By believing something strongly enough, by repeatedly thinking about it, it's possible to "manifest" it into existence, so the argument goes. No innocent prisoner sitting on death row ever manifested his or her pardon. No desperate, starving mother ever manifested food to feed her children. If manifesting were possible, there'd be no shortage of evidence.

Returning to firewalking, those who believe that supernatural forces are at play believe that firewalkers can stroll across burning coals and "magically" avoid burns. Unfortunately for believers, no paranormal magic is required to explain firewalking: physics works just fine (Alcock, 2018, pp. 504–6; Coe, 1975). Several reasons exist for how people can walk across lit coals without injury, the first being the *Leidenfrost effect*. Anyone who has ever watched droplets of water jump around a hot frying pan without evaporating instantly has witnessed the Leidenfrost effect. When the water meets a surface that's significantly hotter than the water's boiling point, an insulating vapor is created between the water and the pan, protecting the water from the intense heat. On the popular TV show *Mythbusters*, the team showed that the Leidenfrost effect allowed people to wet their hand and dip it into molten lead momentarily without sustaining any injuries.[2]

> If we mistakenly assume something to be impossible, then its occurrence mistakenly appears miraculous.
> —James E. Alcock

Firewalking is further explained by what's called heat conductivity. While burning wood coals can reach upward of 1,000° Fahrenheit, wood coals aren't very good at transferring—or *conducting*—that heat, which is why your hand won't burn if you stick it into a 450° oven to remove a cake. Although the air in the oven is also 450°, your hand doesn't burn because air isn't very conductive. You can grab the 450° cake as well without being burned, but touch the metal rack and suffer the consequences because, unlike cake or air, metal is highly conductive (Leikind & McCarthy, 1985–1986; Shermer, 1997, pp. 52–53).

What's more, some people *do* get burned. As it turns out, those who avoid burns either have thicker calluses on their feet or higher arches, so the softer skin doesn't come into contact with the coals (Shermer, 1999). Professional magician and skeptic James "The Amazing" Randi challenged firewalkers to walk not across the traditional fifteen-foot coal bed but instead a sixty-foot coal bed. As predicted, several firewalkers showed up, prepared with the usual pre-walk meditating and chanting, and made it about eighteen to twenty feet before receiving second-degree burns. Whether people receive burns depends on moisture, toughness of skin, and the duration of contact with the embers, not on magical universal energy tapped into by a chosen few.

Superstition

Many would agree that when beliefs are based on the personal experience of individuals, they aren't as reliable as those that are based on evidence that's available to everyone. The main reason to trust scientific facts is because they're both public (objective observers will all witness the same thing) and replicable (under the same or similar conditions, the same results are found). Accordingly, people over the age of ten typically don't avoid stepping on cracks for fear they might break their mama's back because that superstition is neither public nor replicable. The same holds true for other superstitions. Nonetheless, superstitions have become deeply imbedded in our culture and persist on the fringes of science.

Superstitions are found everywhere from sports and test-taking to religion and gambling and are taken quite seriously by millions of people throughout the world. Superstitious behavior can be something as simple as avoiding thoughts of desirable events for fear of "jinxing" or crossing one's fingers to prevent some ill fate. The baseball player who never washes his hat and the test taker who always answers odd-numbered questions first are examples of people who engage in superstitious behavior. For our purposes, I define superstition as *beliefs or behaviors that haven't been empirically validated though are intended to influence one's fortune* (Vyse, 2014, p. 24). Superstitions can generally be attributed to a lack of specific information, scientific illiteracy, magical/wishful thinking, lack of control, or fear of the unknown.

Some superstitions are relatively harmless, such as believing that crossing one's fingers brings good luck. Crossing fingers is relatively benign, since it doesn't cause the believer to act in destructive ways. Other superstitions, such as belief in hexes, spirits, demons, witches, curses, or voodoo, cause unspeakable misery and strife (Alcock, 2018, pp. 314–16). In the twenty-first century, human beings are regularly kidnapped, raped, killed, and dismembered because superstitions lead others to believe that body parts of albinos can be used in witch doctors' healing potions or that having sex with an albinistic woman cures HIV. Superstitions drive animal capture, exploitation, torture, and extinctions (Ladendorf & Ladendorf, 2018); they create untold amounts of personal anxiety; and they consume precious resources that could otherwise be spent on more fruitful endeavors.

Superstitions include belief in omens, karma, or fate; that walking under a ladder or a black cat crossing one's path portends doom; or that someone can "jinx" you by speaking bad luck into existence. Superstitions also include various rituals and behaviors—like wearing the same T-shirt during sports games or carrying around a severed rabbit's foot—that people think have the power to influence future events. Others are based on tradition and folklore—such as the idea that snorting crushed rhino horns boosts virility—while others still are based on ignorance and scientific illiteracy—such as the Madagascan military prohibition against eating the knee of an ox for fear the soldier will acquire weak knees (Vyse, 2014, p. 8). Looking at superstition from a scientific perspective illuminates why we should strive to rid our culture of such beliefs. One way to do that is by better understanding **cause and effect**.

> A superstition is a premature explanation that overstays its time.
> —George Iles

Cause and Effect

One of the main contributors to superstition is fundamentally misunderstanding cause and effect. When an outcome quickly and regularly follows an action, the cause is easy to identify. If every time I tap the Google icon on my phone, Google appears on my screen, I know what caused its appearance. But when outcomes lag behind their cause—as when we see an alluring Coke advertisement and two days later we've got a hankering for a Coke, or we take a medication and two weeks later we feel better—it's harder for us to pin down what caused what (Jenkins & Ward, 1965). Additionally, sometimes events occur without an identifiable cause—as when we want a Coke and we didn't see an ad, or when we feel better but we didn't take any medicine—and sometimes the cause doesn't consistently produce the intended effect—as when the billboard fails to entice us, or the medication doesn't work. Moreover, sometimes effects are the result of multiple causes—billboard + opportunity; medication + healthy diet. And finally, as the bumper sticker has it, sometimes shit just happens with no cause, purpose, or meaning involved.

Science is largely about figuring out what causes what and *how*. Do vaccines cause autism, and if so, how? Is human activity warming the planet, and if so, how? If knocking on wood prevents bad luck, how does it work? Do supernatural gremlins listen for wishful statements, hoping their plans to ruin those wishes aren't foiled by someone knocking on wood? If so, how do they ruin people's wishes? The human brain can be easily fooled by the complicated nature of cause and effect, which is why figuring it out takes so much effort.

Cause and effect—or *causality*—refers to the process of explaining the influence one event (a cause) has on another (the effect). In cause–effect relationships, it's always the case that the effect is dependent on the cause; that is, the observed outcome would be impossible without the preceding event. Researchers conduct correlational studies (chapter 3) to try explaining causal relationships by isolating and manipulating one variable (the independent variable) to see what happens to another (the dependent variable). If a correlational experiment is properly controlled and has been repeated enough times, researchers confidently infer causality. That is, if the dependent variable is repeatedly affected in the same way by introducing the independent variable, an inference that the latter caused the former is justifiable. Another way of putting this is to say that if it were not for the independent variable, the dependent variable wouldn't have changed; its effect *depends* on introducing the independent variable.

Throughout this book, I've repeatedly said that correlation does not mean causation. To understand causality, it helps to understand its three principles: priority, contiguity, and contingency (Blanco & Matute, 2019, pp. 50–51). The following example demonstrates these principles. Say I smack my head on a counter (the cause) while bending over to tie my shoe, and then I feel a sharp pain in the middle of my forehead (the effect). The principle of priority says the cause will come before the effect. If I experienced pain before I smacked my head rather than after, we wouldn't say the smack caused the pain. The principle of contiguity says the cause and effect should occur close to each other in space and time. If a month after smacking my head, I got a sharp pain in the middle of my forehead, I wouldn't even consider that smacking my head a month ago caused the pain. Lastly, the principle of contingency simply states that effects depend on their cause. Here, the pain wouldn't have occurred without the smack. If one of these three principles is violated or missing, then there is no cause–effect relationship between variables.

Sometimes events *appear* correlated even though they're not. Thus, detecting patterns (i.e., patternicity) is a healthy cognitive function. A consequence of pattern detection, then, is *mis*identifying cause and effect, which can easily lead to false conclusions, such as belief in pseudoscience (Blanco, Barberia, & Matute, 2015) and the effectiveness of pseudomedicine (Blanco, Barberia, & Matute, 2014). Consuming more ice cream is strongly correlated with increases in drowning. Is there something in ice cream that causes people to drown? Probably not. More likely, people buy ice cream when it's hot outside, which is also when more people are near water, thus more opportunities for people to drown. Therefore, one way to minimize false beliefs that are based on misunderstanding reality is by understanding cause and effect (Matute, Yarritu, & Vadillo, 2011).

Causality cannot usually be deduced from events alone. If police ask, "Why did the house burn down?" they're looking for a more satisfactory answer than "Because it was flammable." They want to know the cause, and the fact that the house was flammable, or that it burned because there was plenty of available oxygen, are not the sort of causal explanations anyone interested in determining a causal relationship cares about. Even so, determining causation is tricky, and causality isn't always clear-cut. For instance, if I was climbing a tree and a branch broke, which caused me to fall and break my leg, was the cause of my broken leg the height from which I fell? Or the branch itself breaking? Or the strength of my leg bone? Perhaps the branch broke because the tree had been infested with invasive parasites, so would parasites be to blame? The only way to pin down a cause–effect relationship is to repeatedly test the hypothesis until the chances that X does not cause Y are negligible (Chabris & Simons, 2010, p. 161); in other words, collecting the best evidence available and subjecting it to rigorous scrutiny.

Karma—the notion that behaving kindly magically brings good fortune, while behaving badly magically brings misfortune—presents a good example of a persistent and widespread superstition that lacks any objective evidence. Does the moral character of our deeds have a causal effect on future events? If so, how is that information carried through space and time? When we look around, we see bad people who live wonderful lives, and great people who live miserable lives. While this in no way proves karma doesn't exist, it is evidence against the existence of karma that shouldn't be ignored. The burden to prove a causal connection between behavior and consequences is on those making specific claims, and assuming that because "Y" follows "X" that "X" caused "Y" is a logical error. If three earthquakes followed the release of Taylor Swift's last three albums, we don't conclude that Taylor Swift songs cause earthquakes.[3]

Let's say we conducted a *double-blind* (chapter 3), placebo-controlled study on a pill that was supposed to make people sad. We might measure sadness by giving participants the pill (the independent variable) and later asking questions about how they feel (the dependent variable). If participants consistently reported being depressed or suicidal after taking the pill but not prior to, we'd be justified in inferring that taking the pill caused those people to feel sad. On the other hand, if someone was in a car wreck after walking under a ladder, we wouldn't conclude that walking under the ladder caused the wreck, for we would lack a justifiable reason to infer causality.

Believing in Everything

We saw in chapter 11 that believers in psychics and mediums are more suggestible than skeptics. We also saw in chapter 6 that those who report being abducted by aliens were especially prone to develop false memories in the lab. Research also shows that belief in one paranormal claim accompanies belief in others, such that those who believe in, say, ghosts and aliens are likely to believe other incredible claims like the existence of Bigfoot or Atlantis (Sagan, 1996, p. 154; Spanos, Cross, Dickenson, & DuBreuil, 1993). Susan Krause Whitbourne, reporting for *Psychology Today*, noted that "the strongest believers in alien visitations and cover-ups were in fact more likely to have paranormal and superstitions beliefs, particularly beliefs in ESP, the afterlife, and to have had unusual perceptual experiences" (Whitbourne, 2012). Further research shows that paranormal beliefs are also associated with belief in conspiracies and difficulty understanding science (Lobato, Mendoza, Sims, & Chin, 2014). That such beliefs tend to run together suggests that what skeptics would chalk up to hallucinations, dreams, or the foibles of the human brain, less critical minds give credit to the supernatural or paranormal. As one woman put it, "Over the years I have seen and talked to 'ghosts,' been visited (though not yet abducted) by aliens, seen 3-dimensional heads floating by my bed, heard knocks on my door. . . . These experiences seemed as real as life. I have never thought of these experiences as anything more than what they certainly are: my mind playing tricks on itself" (Sagan, 1996, p. 198).

Magical thinking can even outweigh direct evidence, as psychologists Barry Singer and Victor Benassi demonstrated in their classrooms. In a series of experiments (Singer & Benassi, 1980), their students were repeatedly told, in very clear language, that a magician would be *posing* as a psychic and what they were about to witness was nothing more than conjuring tricks. Despite these clear instructions, in trial after trial, the students concluded that the magician really had psychic powers. Some students went so far as to conclude that the magician was an agent of Satan. It is, therefore, less surprising that facts, evidence, and reason alone are insufficient tools to remove superstitious beliefs.

In ancient times, people offered sacrifices to the gods in hopes of receiving some favor, such as rain for crops, healthy soil, or victory in battle. When it rained after a sacrificial offering, people took the rain as confirmation that the sacrifice caused the rain. If it didn't rain, they took this to mean they had failed to please their gods, and oftentimes they continued to offer greater and greater sacrifices until the gods were appeased and provided rain (another example of counting the hits and ignoring the misses). These examples are all cases of misunderstanding cause and effect, and although nobody is immune from uncritical thinking, there is hope in education. Studies have found that analytical (i.e., critical) thinking reduces the type of thinking that leads to belief in conspiracies or the paranormal (Gervais, 2015; Swami, Voracek, Stieger, Tran, & Furnham, 2014).

So are superstitions scientific? Carl Sagan once said that "superstition is marked not by its pretension to a body of knowledge but by its method of seeking truth" (Sagan, 2007, p. 1). We've seen that superstition, like paranormal and supernatural beliefs, is defined by its predisposition to belief without evidence. And throughout this book, we've seen that science isn't a body of facts; it's a collection of methods used to interrogate nature until it reveals its secrets. Superstitious beliefs require no systematic approach to confirmation.

REGRESSION TO THE MEAN

Another way we misconstrue cause and effect is by failing to understand *regression to the mean*. Regression to the mean (regression for short) describes a phenomenon where if one variable is measured at some extreme, the second measurement tends to come closer to the average. Think about visiting a chiropractor. You go see the chiropractor because you have pain in your back. After your visit, your pain subsides. The chiropractor worked. Perhaps, but regression is a much more likely explanation. Everyone is susceptible to making regression errors, and nobody is immune. Even professional researchers make mistakes if they draw conclusions without taking regression into consideration (Barnett, van der Pols, & Dobson, 2005).

We see regression every day, such as when short people have tall children or when we perform better (or worse) than we did on a previous test. Regression provides statistical rather than situational explanations of events. "The key to not being deceived by regression," says Gary Smith, "is to look behind the luck and to recognize that when we see something remarkable, luck was most likely involved, and so the underlying phenomenon is probably not as remarkable as it seems" (Smith, 2016).

A popular example of regression is the so-called *Madden Curse* or *Sports Illustrated Curse*. The Madden Curse is based on the belief that athletes experience a decline in performance the year after appearing on the cover of the video game *Madden NFL*. To the superstitious, the "curse" is attributed to magic or witches. To skeptics, it's explained by regression. Here's how. Athletes selected to appear on the cover of *Madden* are those whose performance exceeded expectations that year. Such high-level performance cannot possibly be maintained; otherwise athletes would just get continually better and better and there wouldn't be any such thing as a performer's average. Thus, an athlete's performance must regress toward more average performances, which is exactly what happens. Athletes' careers wax and wane, from exceptional years to unexceptional years, great games to not-so-great games, always maintaining a rough average. We don't have to believe in curses to understand that following a year of outstanding performance, the next year should be closer to the average.

While giving a lecture to the Israeli Air Force on the evidence that rewards for improved performance are more effective than punishments for mistakes, psychologist Daniel Kahneman had an opportunity to explain regression (Kahneman, 2013, pp. 175–76; Tversky & Kahneman, 1974). One of his listeners (a flight instructor) objected to Kahneman's explanation on the grounds that after giving flight cadets praise for exceptionally smooth landings, their landings were usually much worse the next try. By contrast, screaming at his cadets for poor landings usually meant better performance on the next flight. Indeed, the instructor was seeing a pattern but not the right one. He was correct to recognize that after praising a performance, the cadet's performance would usually decrease, and when he scolded poor performance, the cadet's performance usually increased. However, his observations led to inaccurate conclusions.

Obviously, the instructor praised a cadet when his performance was praiseworthy; the performance was above average. Regardless of whether the cadet received praise, regression predicts that on the next attempt, his performance would've lapsed anyway, which is exactly what the instructor observed. When a cadet's performance was particularly bad (below average), the instructor would chastise him by screaming over his earpiece. Again, regardless of the instructor's behavior,

> When man seized the loadstone of science, the loadstar of superstition vanished in the clouds.
> —William R. Alger

regression predicts that the cadet's performance would improve after a bad run and deteriorate after a good run. Failing to understand regression, the instructor wrongly assumed that the cadets' performance (the effect) was a result of his instructions (the cause).

Oftentimes, cause and effect defy our intuition and even our own experience, which is why what the cadet instructor experienced for himself and saw with his own eyes was simply misconstruing reality. Without training, we shouldn't naïvely assume we understand how it works. In life, as well as in science, post hoc must be separated from propter hoc before conclusions are drawn. "We encounter regression almost daily," says Smith. "Yet, we seldom recognize it, are often surprised when it occurs, and frequently draw erroneous conclusions. We can do better" (Smith, 2016).

Wrapping Up

This chapter covered three belief systems that persist on the fringes of science—the supernatural, the paranormal, and superstitions—exploring why they fail as explanatory models of reality and why they cannot be investigated scientifically. We saw that when beliefs are grounded in evidence anyone can access, we have more reason to accept them than we do beliefs that are based on the personal experiences of individuals. The key to understanding what makes something scientific isn't what the claim is or what its consequences are but, rather, whether those claims can be practically or theoretically tested.

It's worth noting that I haven't tried showing that supernatural or paranormal phenomena don't exist. What I've tried to do is demonstrate that if any such phenomena exist, those phenomena are beyond the limits of scientific investigation. Lastly, even harmless superstitious or paranormal beliefs should be discouraged because failing to do so promotes a culture of non-thinking and the acceptance of baseless claims without evidence.

CHAPTER 12

Standing in Science's Way

Readers who have made it this far understand the strength and power of scientific discovery. The mysteries of science are far more inspiring than any ghost tale, afterlife fiction, or conversation with the dead, and they teach us about reality in a way that pseudoscience never will. In this chapter, I pit science against pseudoscience, evaluating how they compare to one another. We're almost at the end of our journey, and we'd be remiss without covering pseudoscience in detail, along with several key features of the pseudoscientific worldview and why science is the superior method for evaluating reality.

Building Vocabulary

Null hypothesis:	Credulity:
The hypothesis that no relationship between two variables exists.	The willingness to believe, even without reasonable proof.

Pseudoscience:	Quackery:
A body of information that purports to be scientific or supported by science but that doesn't adhere to scientific methodology.	The practice of fraudulent medicine, usually to make money or for ego gratification and power; health fraud.

Meet Pseudoscience

Although it has the word "science" in its name, **pseudoscience** is a far cry from being science's counterpart. The word, however, comes with a warning, for pseudoscience "achieves meaning only through a comparison (always unfavorable) with the thing that it purports to be" (Hecht, 2019). It's more accurate to describe the two as opposite ends of a continuing spectrum (Walton, 1996). The word "pseudo" derives from the Greek word for "fake, or lying," so a simple definition of pseudoscience is fake science. But this vague definition wouldn't give us any real understanding of what pseudoscience is, so we must understand those characteristics of pseudoscience that distinguish it as an independent, albeit massively flawed, epistemology (Hines, 2003, pp. 13–19).

Pseudoscience can be identified by a suite of themes, including virtual unfalsifiability, the failure to progress, avoidance of criticism, the pursuit of confirmation over truth, burden shifting, the lack of explanatory paradigms, low (or nonexistent) evidentiary standards, and misuse of scientific terms (Lilienfeld, Lynn, & Lohr, 2014). To demonstrate pseudoscience's failure to progress, I'll compare Darwin's theory of evolution with Samuel Hahnemann's beliefs about homeopathy. In 1859, Darwin proposed his theory of evolution by natural selection, basing it on field observations, comparative anatomy, and the similarities of species found on different continents, among other things. If the evidence presented in his book *On the Origin of Species* were all that supported his theory, we'd have plenty of evidence for natural selection. However, if any evolutionary theorist from Darwin's era were to take a biology exam today, she would surely flunk. While Darwin's insights are as timeless as they are brilliant, his theory has moved far beyond the body of evidence he was able to gather. Nowadays, evolutionary theory explains much of biology, geography, paleontology, and even cosmology and has made numerous accurate predictions that can only have been successful if Darwin's theory is correct.

Popular Examples of Pseudoscience:

- Chiropractic
- Facilitated communication
- Healing crystals
- Telepathy
- Polygraphy
- Ghost hunting
- Alternative medicine
- Quantum consciousness
- Iridology
- Numerology
- Dowsing
- Acupuncture
- Ancient aliens
- Free energy
- Flat Earth
- Ufology
- Homeopathy
- Mediumship
- Crop circles
- Reiki
- Astrology
- Flood geology
- Bermuda Triangle
- Feng shui
- Aura
- Tarot
- Astral projection
- Earthing

This list is eerily similar to our list of paranormal claims. But while all paranormal claims are pseudoscientific, all pseudosciences are not paranormal.

By contrast, Samuel Hahnemann created homeopathy in 1796 based on the ideas that substances that cause certain symptoms can be diluted and used to treat those symptoms and that the more diluted a substance is, the more potent it becomes. As far as progress is concerned, homeopaths today believe Hahnemann's ideas are ironclad and unchangeable, despite that they contradict much of physics, chemistry, biology, and other successful sciences. Once a pseudoscience takes root, what typically follows is strong advocacy of the claim despite any disconfirming evidence. If some pesky fact destroys a pseudoscience, its advocates don't generally admit the error and abandon the claim. They usually find some rationalization they believe adequately dismisses the criticism. "Homeopathy conflicts with chemistry. Then we must not fully understand chemistry because homeopathy works."

In chapter 3, we learned the importance of falsification. Unfalsifiable claims are incredible because no evidence could ever show the claim is wrong. Unfalsifiability is one of the most common characteristics of pseudoscience. If I said I can move objects with my mind and I agreed to test my claim, if I failed to mentally move any objects, rather than admit I don't have that ability, I can claim, "My powers come and go, and this time they went" or "My powers don't work around skeptics" or "Telekinesis doesn't work in controlled studies." I could always come up with some explanation (move the goalpost) for why my claim cannot be proven.

If one could call pseudoscience a discipline at all, it's a discipline that proceeds similarly to how lawyers argue cases. Lawyers portray the facts in a specific way regardless of countervailing evidence or original context, seeking only confirmatory evidence and discounting the rest. Whereas science is a process of obtaining knowledge, pseudoscience lacks anything resembling systematic methodology. Pseudoscience touts incredible claims as scientific, while simultaneously lacking the constraints that allowed science to flourish (Shermer, 1997, p. 33). In other words, pseudoscience wants to have its cake and eat it too.

From the way pseudoscience is sometimes portrayed, it can be easy to assume that pseudoscience is just bad science. This assumption is a mistake. Scientific conclusions are drawn provisionally, and when errors are generated, those errors are systematically identified and removed by the self-correcting mechanisms that fuel the credibility of science and make it such a reliable pursuit. Scientists frame hypotheses that are subjected to disproof, and no developing science banks on just one hypothesis. Many hypotheses are offered and, within certain ethical, financial, and other constraints, are tested through experimentation and observation. This process has been improving our understanding of how the world works since science began. Bad science may follow scientific methodology while omitting crucial steps or succumbing to human errors such as biases or greed, but such errors are eventually discovered through the self-correcting processes of peer review and replication. Pseudoscience skips the entire process of *doing* science, boldly making claims based on bogus ideas and bad (or no) evidence.

Objections to pseudoscience range from lacking logical support for its far-reaching conclusions, to the poor methods used in seeking evidence, to the absence of supporting evidence. For example, when a scientific hypothesis is introduced, its every instantiation is attacked by qualified professionals in relevant fields (chapter 3), including by the proponents of the hypothesis being proposed. Their best efforts are spent trying to *dis*prove the claim (aka trying to prove the **null hypothesis**). By contrast, when a new pseudoscience claim is introduced, the only evidence of interest is that which supposedly confirms the phenomenon. Any evidence that people believe suffices. If claims aren't systematically challenged, there's no real way to know if they're correct, which is why pseudoscience is riddled with false and incredible claims.

The number of pseudoscientific claims is astounding. New pseudosciences pop up while old ones are continuously resurrected. For every new pseudoscience, thousands of eager believers are ready to embrace it. As the famous motto credited to P. T. Barnum goes,[1] "There's a sucker born every minute" (Gilovich, 1991, p. 60; Meehl, 1956).[2] Nevertheless, despite the lack of credible evidence supporting pseudoscientific claims and the extraordinary success of science, pseudoscience is thriving. A small sample of pseudoscience includes: Bigfoot, alchemy, the lunar effect, flood geology, the Bermuda Triangle, perpetual motion machines, Ouija boards, near-death and out-of-body experiences (NDEs and OBEs), hollow Earth, flat Earth, water memory, faith healing, chemtrails, therapeutic touch, graphology, the law of attraction, phrenology, and electronic voice phenomena (EVP). As this list demonstrates, claims that are pseudoscientific aren't necessarily paranormal, and claims that are paranormal aren't necessarily pseudoscientific.

Because science and pseudoscience are part of a continuing spectrum, identifying exactly where science ends and pseudoscience begins can be challenging. This challenge, known as the *demarcation problem*, is what the philosopher of science Karl Popper tried to solve by introducing his concept of falsification (Popper, 1959/2014). The demarcation problem results from the difficulty in drawing the blurry lines separating science from pseudoscience, but it isn't limited to science. Turning bald or becoming middle aged doesn't happen at a precise moment. The fallacy here, known as the *decision-point fallacy*, rests on the assumption that things must be clearly separated in black-and-white terms. However, everyday experience tells us that not everything *is* clearly separated. The difference between cold and hot, night and day, young and old, tall and short all demonstrate the difficulty in drawing hard lines when things are not binary, but rather exist on a spectrum.

To be sure, some areas of investigation are vague or unclear. However, insisting that since some things are vague we cannot make clear distinctions is untrue and irrational. Accordingly, even without a bright-line rule separating science from pseudoscience, we can still tell the difference between them, just as we can tell the difference between microscopic and macroscopic or a beard and a five o'clock shadow (Walton, 1996). Failing to grasp this point invites the *false continuum fallacy*, which basically says that vagueness alone doesn't imply that something is invalid. We have no difficulty seeing the colors of a rainbow, but identifying exactly where red ends and orange begins is impossible. Of course, this doesn't mean red or orange don't exist or that calling some things orange and others red isn't useful. Our job as informed skeptics is to learn how to label something as properly scientific or strictly pseudoscience, without making the mistake of thinking that because the area in the middle is blurred, there's no difference between what is clearly scientific and what clearly is not.

Fake Science in Action

Once a pseudoscientific claim is introduced, it lingers around well after its expiration date, oftentimes for no better reason than people's desire to believe it. For example, James Randi exposed Uri Geller on national television as a fraudulent spoon bender; weeks later, Geller was selling out theaters (Measom, Weinstein, & O'Toole, 2014).

A dead giveaway that a claim is pseudoscientific is its inability to progress (Smith, 2018, p. 94). When Isaac Newton first defined gravity mathematically, this was striking progress from our previous ignorance. When Albert Einstein published his theory of general relativity, our understanding of gravity moved ever forward. When Fritz Zwicky discovered the existence of what he called "dark matter," a new chapter was opened in our understanding of gravity. And once physicists develop a quantum theory of gravity, they'll continue to seek important discoveries that explain gravitational effects. Pseudoscience makes no such progress. The ghost stories of the past are the same ghost stories of the present (and, despite monumental advances in technological abilities, the evidence hasn't gotten any better). Mediums never produce any useful information (Randi, 2007). Cryptozoologists—hunters of creatures such as Bigfoot, chupacabras, aliens, or the Loch Ness Monster—never recover physical evidence. Pseudosciences remain stagnant, ceaselessly chasing rainbows and the leprechauns beneath them.

Proponents of pseudoscience impress upon people that what they're doing is scientific. They do so because scientific discovery enjoys a high level of intellectual respect. When "science" has proven something, people listen. Why? Because "we live in the age of science [where] pseudoscientists know that their ideas must at least *appear* scientific because science is the touchstone of truth in our culture"

(Shermer, 1997, pp. 6–7, emphasis in original). Thus, pseudoscientists attempt to present themselves and their claims as scientific to piggyback on science's success, while completely ignoring its bedrock principles.

A good example is the *polygraph*, or "lie detector." The polygraph, a device invented by the same guy who invented Wonder Woman, records different physiological—i.e., autonomic—reactions during questioning, supposedly to detect dishonesty. The polygraph measures several autonomic reactions including sweat, heart rate, and respiration. The concept of gauging deception by measuring physiology traces back to ancient Chinese and Hindu civilizations. Suspects were told to chew on a grain of rice wherein a dry grain indicated the dry mouth of a liar (Sahu, Naidu, & Sankar, 2014). Polygraphs certainly measure autonomic reactions. The problem, however, is that some people are completely calm when telling a lie, and some people get nervous when telling the truth. Autonomic reactions are the result of multiple factors, some of which have nothing to do with one's present circumstances. Accordingly, no direct correlation between a particular autonomic response and what someone is thinking or saying has been empirically verified. Nonetheless, these facts haven't stopped polygraphy proponents—who mainly point to personal experience as their evidence that they work—from believing in the efficacy of the device (Hines, 2003, p. 431). Like the gambler whose occasional win keeps him pulling the lever, polygraph examiners solicit enough confessions to keep them believing. However, in science, all the data must be considered before drawing conclusions. Thus, the "misses" as well as the "hits" must be tallied and analyzed statistically to determine whether the polygraph "worked" or whether something else elicited those confessions. That some people confess under torture doesn't mean everything people say under torture is true.

The device itself creates an appearance of scientific rigor, though it's far from scientific. Polygraphs are incapable of detecting lies and are rejected by a vast majority of experts (Brett, Phillips, & Beary, 1986; Iacono & Lykken, 1997; Saxe, Dougherty, & Cross, 1985) and skeptics (Dunning, 2014; Hines, 2003, pp. 428–32; Randi, 2017a) alike. Polygraphs are about as successful as Tarot card readers, Ouija boards, astrologers, or tea leaf readings. Philosophers debate whether it's even *possible* to know what goes on inside someone's head (Dennett, 1996, pp. 153–68), and the polygraph has made zero useful contributions to that debate. Using a polygraph to expose lies is like using the *Malleus Maleficurum* to expose witches, rendering the statement "I passed a polygraph" as meaningful as the statement "I saw the color four" or "I'm slightly pregnant."

In 2001, former CIA-operative-turned-KGB-spy Aldrich Ames called polygraphy "a superstition" and "junk science" (Ames, 2001). In a 2001 study, the psychologist and neuroscientist William Iacono stated that "members of scientific organizations who have the requisite background to evaluate [polygraph testing techniques] are overwhelmingly skeptical of the claims made by polygraph proponents" (Iacono, 2001, p. 84). Iacono, an expert on the science and methodology of polygraphy, cites multiple studies demonstrating that polygraphy doesn't pass muster as a scientifically credible method because it's based on naïve and implausible assumptions indicating that (a) it's biased against innocent individuals and (b) it can be beaten simply by artificially augmenting responses to control questions (Iacono, 2001; Iacono & Lykken, 1997), such as by controlling breathing, tensing and untensing various muscle groups, or keeping a tack in your shoe and pushing on it to stimulate an autonomic response (Biddle, 1986).

In 2003, the National Academy of Sciences stated that polygraphy is "unreliable, unscientific, and biased" and that for law enforcement agencies, "its accuracy in distinguishing actual or potential security violators from innocent test takers is insufficient to justify reliance on its use" (National Research Council, 2003). Like most pseudosciences, the polygraph has far outlasted the disproof of its efficacy, and unless serious objections are raised, millions of tax dollars and government resources will continue being wasted on this pseudoscientific device. Worse than wasting tax dollars, polygraphy creates a false sense of security that leads to people like Ames evading detection and others being falsely convicted of serious crimes, such as murder (Dunning, 2014).

If it can't detect lies, why is the polygraph still around? Several reasons exist, but the one that concerns us is the confirmation bias. We don't have an objective methodology for interpreting the results of a polygraph, so the "findings" are simply the subjective opinion of the examiner (To, 2002). Since examiners are humans, and thus are plagued by biases and other human errors, their subjective opinions inevitably lead to inaccurate conclusions. If an examiner believes a suspect is guilty, he or she will likely interpret their findings in a way that supports that belief. If the examiner believes a suspect is innocent—a highly unlikely scenario, given their status as a "suspect"—he or she will find ways to interpret their findings accordingly (Tavris & Aronson, 2019, p. 346, n. 14). Compounding the problem, research shows that the more an interrogator believes his suspect is guilty, the better the chances of obtaining a false confession (Kassin, 2005).

Like polygraph examiners favorably interpreting the results of a polygraph test, interpreting events in light most favorable to our own beliefs and preferences is human nature, and many biases contribute to this sort of *motivated reasoning* (Kunda, 1990). Since we're excellent at retrofitting evidence to suit our preferences, it should be no surprise that people untrained in CT ignore these biases or are completely oblivious to them.

In chapter 4, we learned that reasoning works from the premise(s) of an argument and seeks to support a conclusion, whereas rationalizing works from a conclusion and seeks to justify the premise(s). Pseudoscience works more like rationalizing, whereas science works like reasoning. Pseudoscientific investigators tend to work backward from some idea or conclusion, then "discover" evidence that supports their preconceived conclusion. One example is New Age advocates who claim OBEs (out-of-body experiences) constitute proof of astral projection—the idea that humans have an "astral body" that can leave the physical body and travel about. Such "evidence" includes many people independently having similar OBE or astral experiences or being able to recall events that happened at or around one's physical body while asleep or under anesthetics. However, this reasoning is ad hoc (to a particular end) and amounts to nothing more than weak, anecdotal evidence. Despite a long history of extensive research on OBEs, including research showing that OBEs can be induced in the laboratory, nothing paranormal has ever been established (Wiseman, 2011, pp. 57–93). In fact, researchers have induced OBEs with a little suggestion and virtual-reality goggles (Ehrsson, 2007; Lenggenhager, Tadi, Metzinger, & Blanke, 2007). OBEs prove that the brain is extremely complicated, not that paranormal phenomena are real.

"Science . . . requires the proof to come before the conclusion" (Harrison, 2013, p. 110). To test astral projection, a scientific approach might be to hide specific information inside a vault and tell an astral projector where the information is hidden, then, controlling for access to the vault, seeing if the astral projector can relay the information. If the astral projector relays the information, we would have at least some grounds for supposing astral projection might be real, and we could then devise more precise testing methods to investigate the phenomenon. However, researchers have conducted similar experiments (Hines, 2003, p. 104; Sagan, 1996, pp. 223–24), and so far, astral projectors cannot pass these simple tests. One can pose a similar question to mediums like, "Why don't you ever return any useful information?" such as "Where did grandma hide the key for the safety deposit box that stores the family inheritance?"

> We also know how cruel the truth often is, and we wonder whether delusion is not more consoling.
> —Henri Poincaré

Among other violations of the known laws of physics, astral projectors also have the problem of light to contend with. We saw in chapter 8 that a visual image is created when photons disrupt electrons in the photoreceptors of our eyes. Our brains then translate this information into visual data that is represented in the visual cortices of our brains. If astral bodies aren't made of matter and lack physical eyes, how can they see (Loxton, 2018c)? What absorbs the photons? And without brains, where is visual information processed? Since the act of observing is a physical process, the idea of astral projection—a nonphysical means of travel and observation—violates basic laws of physics as well as simple logic. If astral bodies *are* made of matter, scientists would be able to detect them by, say, measuring the refraction of photons or disruption of electrons that astral bodies would cause. With no possibility of detecting astral bodies, how can anyone know they exist? The fact that scientists cannot detect astral bodies casts serious doubt on their existence. If astral bodies are nonphysical, they couldn't have physical traits, such as shape or a position in space and time, thus they couldn't secure information about the physical world.

Pseudojargon

Another ploy found throughout pseudoscience is using scientific terminology to feign scientific credibility. James Clerk Maxwell—the discoverer of electromagnetic waves—realized this in 1871 when he said it "is the respect paid to science that the most absurd opinions may become current, provided they are expressed in language, the sound of which recalls some well-known scientific phrase" (Law, 2011, p. 165). We saw this in chapter 5 with pseudoprofundity. In chapter 1, we also saw a similar instance of intellectual dishonesty with the deliberate misuse of the word "theory." The effectiveness of pseudojargon is demonstrated when, for example, marketers make millions selling patches or plastic bands that are "infused" with magnets or holograms specially designed to tap into your "life force" using "resonance" to match your "natural vibrational frequency" (see, e.g., Hall, 2017; Hall, 2021). Such magic bracelets are claimed to do everything from alleviate headaches and menstrual pain to increase balance and athletic ability. The evidence supporting the benefits of hologram-infused plastic bands is, unsurprisingly, entirely lacking (Hall, 2010).

When scientists use specific language, they do so because making a clear, precise argument is important when establishing scientific facts and because everyday language can cause confusion when precision is needed. Accordingly, scientists use *operational definitions* when pursuing a line of argument. Operational definitions specify how a concept is measured. For example, when physicists use the term "wavelength," they mean something very precise, namely the distance between two points of a wave, such as the distance between two crests. Pseudoscience literature is littered with hijacked terms including "vibration," "energy," "magnetism," "frequency," "wavelength," "dimension," "negative ions," "quantum," "force," "fields," and don't forget "force fields." These terms are often mixed with others like "essential," "spiritual," or "natural." Spiritualists and marketers are known for using nonsense language such as "This essential oil corrects your natural wavelength frequency by balancing your essential energy vibrations." Such language appeals to those predisposed to believe that some natural connection between spiritualism and the findings of science exists, when really, it's just marketing nonsense that's specifically intended to make people buy worthless products. What's an "essential energy vibration" and how does one correct a "natural wavelength frequency"?

Scientific terms have precise meanings when real scientists use them to describe natural phenomena—and are utterly meaningless when taken out of context. Hearing these terms from spiritual gurus should ring your skeptical alarm bells. Are these terms being used correctly? Does the person using them have any credentials in physics? Is this person an expert in the field being discussed? How does the person's argument relate to the study of the physical world? Are there supporting mathematics? Do the terms change meaning from usage to usage? Do experts in relevant fields agree?

The fraudulent use of scientific language is becoming increasingly popular, especially with the commercial success of shows like *Ghost Hunters* and *Ancient Aliens*. The point here is so important it's worth belaboring it with a lengthy quote:

> Dressing up a belief system in the trappings of science by using scientistic language and jargon, as in "creation-science," means nothing without evidence, experimental testing, and corroboration because science has such a powerful mystique in our society, those who wish to gain respectability but do not have evidence try to do an end run around the missing evidence by looking and sounding "scientific." Here is a classic example from a New Age column in the Santa Monica News: "This planet has been slumbering for eons and with the inception of higher energy frequencies is about to awaken in terms of consciousness and spirituality. Masters of limitation and masters of divination use the same creative force to manifest their realities, however, one moves in a downward spiral and the latter moves in an upward spiral, each increasing the resonant vibration inherent in them." How's that again? I have no idea what this means, but it has the language components of a physics experiment: "higher energy frequencies," "down-ward and upward spirals," and "resonant vibration." Yet these phrases mean nothing because they have no precise and operational definitions. How do you measure a planet's higher energy frequencies or the resonant vibration of masters of divination? For that matter, what is a master of divination? (Shermer, 1997, p. 49)

Evidence is the currency of scientific facts. Incredible claims without evidence are just that: incredible. It isn't until those claims accompany supporting evidence that we call them science or even consider believing them. Moreover, pseudoscientists commit the *fallacy of inconsistency* (or equivocation) when they apply scientific criteria to one thing but not another, like using the term "frequency" in its proper context, only to use it out of context later in the same argument. "In order to have a coherent discussion or formulate a valid scientific hypothesis," says Professor Steven Novella, "all terms must be unambiguously defined" (Novella, 2018, p. 77). If terms are not operationally defined and consistently applied, what you have is neither science nor scientific regardless of who is saying it or how convincing they make it sound.

David Icke, one of the planet's most extreme conspiracy theorists, is one of the worst culprits. To convince his audience that civilization is now and has always been controlled by extraterrestrial lizard-people who masquerade as politicians and businesspeople, Icke insists that

> Isn't it enough to see that a garden is beautiful without having to believe that there are fairies at the bottom of it too?—Douglas Adams

> the Anunnaki created DNA streams or bloodlines to suit their agenda and they have continued to infuse their DNA into human blood streams [sic]. They rewire the DNA to close down humanity's interdimensional communication and telepathic powers. This puts us in a vibrational prison in which we can perceive only the very narrow frequency range accessed by our physical senses. (Icke, 2001, p. 74)

Readers will recall from chapter 4 that statements, assertions, opinions, or explanations aren't arguments, which seek to prove something. What Icke accomplishes is writing a string of assertions and opinions without attempting to prove his claims. In more than four hundred pages, Icke fails to state exactly *how* the Anunnaki create DNA streams, *how* doing so serves any purpose, *how* extraterrestrial DNA is infused into human bloodstreams, *how* DNA is rewritten, *how* interdimensional communication or telepathy work, or *what* a "vibrational prison" is. Having failed to follow a single GASP, Icke's claims cannot possibly be taken seriously.

When you combine spiritual gurus with alternative medicine, you get something like Therapeutic Touch (TT for short)—a mysterious healing practice that uses so-called energy therapy wherein the therapist doesn't physically touch the patient. Proponents of TT claim there is something called the "human energy field" that affects both spiritual and physical health, which, of course, they alone have the power to adjust and restore—for a modest hourly rate. Having seen multiple instances of New Age and paranormal beliefs running in pairs, it should come as no surprise that research shows that believers in a human energy field are especially prone to belief in the paranormal (Wilson, 2013).

Scientists can (and do) measure energies down to the atomic and subatomic levels (Stenger, 2000). For example, an electron emits a magnetic field that can be measured to an accuracy of one ten-billionth. The idea that a human energy field is so strong that it can be detected by wavy hands, yet so weak that even our most sensitive energy-measuring devices cannot detect it, is nothing short of delusional. As the psychologist and skeptic Terence Hines observes,

> If the claim made by Therapeutic Touch supporters is true, the discovery of a strong but previously unknown type of energy would be a major scientific breakthrough, worthy of a Nobel Prize. Given the potential importance of such a discovery, both theoretically and practically, one would expect that the developers of Therapeutic Touch would have spared no expense in doing elegant experiments clearly demonstrating that this new energy really exists. . . . [However,] Therapeutic Touch proponents have never, [sic] done a single experiment to try to conclusively show that this energy field exists, or what its characteristics are—nor have they ever done a single study aimed at determining whether practitioners could really detect the alleged energy field. (Hines, 2003, p. 364)

If human energy fields exist, scientists would be able to detect, measure, and study them and would do so as enthusiastically as they do with every other energy they study. Moreover, since Einstein, we've understood energy as the equivalent of mass times the speed of light squared; hence Einstein's famous equation $E = mc^2$. Therefore, by definition, energy cannot be immaterial. Evidence supporting the existence of a human energy field (Rosa, Rosa, Sarner, & Barrett, 1998) has never been found, and scientists studying the supposed manipulation of human energy fields haven't detected any effects on the body (Hall, 2012). The incredible claims behind TT have survived up until present day, even though the evidence supporting them is as weak now as it ever was (Bleske-Rechek, Paulich, & Jorgensen, 2019).

Biases in Pseudoscience

The practice of science is the process of skeptically examining all claims regardless of whom or where they come from. When a scientific hypothesis is proposed, it volunteers itself to the gauntlet of experimentation, refutation, replication, and peer review. In this process, scientists are looking for, among other things, any inconsistencies with known laws of nature, logical errors, mistakes in methodology, confounding variables, conflicts with well-grounded theories, or biases that may have contributed to the findings. Disconfirmation is the name of the game. If experts cannot disprove a claim, the claim is temporarily accepted as true, with an understanding that someone might later falsify it. Pseudoscience adamantly avoids this entire process and, intentionally or not, often commits the *congruence bias*, which we first encountered in chapter 10.

No pseudoscience willingly submits to the rigors of scientific scrutiny, not without rejecting its conclusions, of course. Claims that do are scientific, not pseudoscientific, because pseudoscience lacks trained professionals who investigate phenomena using GASPs. For example, no ghost experts that publish the findings of their investigations in peer-reviewed ghost journals exist. Rather, pseudoscience is comprised of untrained and undisciplined individuals and groups that willingly ignore basic rules of logic and accept bottom-of-the-barrel evidence to support their incredible claims. Pseudoscience amounts to an enterprise dedicated solely to convincing people of the supposed truth of incredible claims. When scientists do take the time to investigate pseudoscientific claims, the result is usually a confirmation of the null hypothesis through a demonstration of how the claim being tested fails to comport with reality. The way scientists and pseudoscientists pursue investigations is markedly different. Pseudoscientists make no attempts to disprove their claims and often try shifting the burden of proof or engage in special pleading.

Markers of Pseudoscience:

- Failure to progress
- Lack of systematic methodology
- Immunity to disproof
- Claims exceed the evidence
- Contradiction of known laws of nature
- Misuse and abuse of scientific terms
- Resistant to counterevidence
- Pretentions to science
- Rests on logical fallacies

While it can be hard to identify a particular pseudoscience, if it contains some or all of these characteristics, it is probably pseudo-scientific.

Another bias that plagues pseudoscience is the confirmation bias, which, as readers will recall, happens when one seeks out and recognizes evidence that confirms previously held or favored beliefs and suppresses, rejects, and ignores all disconfirming evidence. For example, proponents of the Bermuda Triangle will point to plane or ship disappearances in a particular area of the Atlantic Ocean as evidence, while ignoring all the disappearances in other geographic locations (Hines, 2003, pp. 314–23).

An especially weak and dubious form of evidence pseudoscientists love are anecdotes. In fact, some pseudosciences are based entirely on anecdotes, such as alien abduction or astral projection. Anecdotes are extremely emotionally compelling, especially given our predisposition to being captivated by stories. Anecdotes are especially compelling when the person relaying an event doesn't have a reputation for lying or making things up. As we've seen, people can just be plain mistaken, and when we have predilections toward certain beliefs, we're much more likely to genuinely think we experienced something we didn't. Aliens, for example, never abduct skeptics; they abduct people predisposed to belief in alien abduction. Demons never possess people who don't believe in demon possession (strange, since these are the people we'd expect demons would *want* to possess); they possess devoted believers. If you were an alien and wanted people to know you existed—as some aliens surely would—would you waste your time convincing those who already believe in alien visitation or seek out people who don't?

For a variety of reasons, science cannot use anecdotes as evidence. One, it's impossible to control the variables—e.g., you cannot make an alien abduction or demon possession happen—and two, the observations are subject to all the flaws and failings of the human brain. It is commonly said that "the plural of anecdote is not data." In this regard, anecdotes are like the popular AA mantra regarding alcoholic drinks: "One is too many and a thousand is never enough." If you can call them evidence at all, anecdotes are of the weakest variety and should never be used to confirm a hypothesis. Rather than seek reliable evidence, pseudoscientists prefer any evidence regardless of how substandard it is. They then use poor evidence to weave an argument that seemingly supports their belief, instead of accepting disconfirming evidence and challenging the incredible claim.

Folk Knowledge

Folk knowledge—cultural or intuitive beliefs—also drives belief in pseudoscience. For instance, eyewitnesses can *see* the Earth is flat, *see* the sun orbits the Earth, and *feel* the Earth is motionless, the same way research participants *know* a sugar pill cured their ailment and alien advocates *know* aliens abducted the neighbor's cat. Many folk beliefs seem self-evident, which leads to unjustified conviction in their truth. Folk beliefs include the belief that Toronto is farther north than Portland, Montreal is farther north than Seattle, and San Diego is farther west than Reno; that matter is solid; or that joint pain is caused by cold weather, even though none of this is true (Redelmeier & Tversky, 1996; Smedslund & Hagen, 2011). While finding examples of false beliefs that *seem* true is easy, the point is to demand evidence before forming beliefs, since even the most self-evident truth can be completely wrong.

Evidence in Pseudoscience

Contrary to the empirical sciences, pseudosciences don't stand on logical arguments or quality evidence; they survive by preying on **credulity**—that part of us that wants to believe. Making nonsense sound plausible is an art pseudoscientists are well versed in. Carl Sagan elucidated this tactic with his "dragon in my garage" analogy in which he showed that nothing could disprove the existence of a fire-breathing dragon, provided one continually moves the goalpost (Sagan, 1996, p. 171).

Let's say I argue that fluorescent lightbulbs are filled with thousands of magical dwarves who wear special boots that, when rubbed against glass, create millions of sparks that make up the light that's emitted from the bulb (adapted from Schick & Vaughn, 2005, pp. 187–88). Naturally, you're skeptical, and you'll probably demand evidence. So we seek out the nearest fluorescent light to investigate. I turn on the light and say, "See. The dwarves are in there running now." You insist they aren't there because you can't see them. "Oh," I respond. "That's because they're microscopic." Well, you insist, then let's get a microscope, to which I retort, "They're so small that even electron microscopes aren't powerful enough to detect them." You then suggest we fill the tube with powder so we can see their footprints. "Magical dwarves don't leave footprints," I say. Well, maybe we could get an infrared camera and detect their body heat. "Sorry," I counter. "Magic dwarves don't emit body heat." Perhaps we could empty the tube and spray-paint the dwarves, you propose. "Not possible," I explain. "Magic dwarves are incorporeal, so paint doesn't stick."

And on and on this goes. Any attempt you make to verify the existence of magic dwarves, I refuse on some ground that cannot be challenged. Readers may have noticed that I've committed another error by shifting the burden of proof or that my "magic dwarves" claim is unfalsifiable. Pseudoscientists enjoy these tactics of burden shifting, goalpost moving, and unfalsifiability because they leave their incredible claims intact and unrefuted. However, these tactics are no match for skepticism. When skeptics challenge the claims by contesting the strategies used to advance them, pseudoscientists claim science cannot solve all mysteries or make the even more far-reaching assertion that their incredible claim is correct and it's science that's wrong. Pseudoscientists generally refuse to admit defeat, even in the presence of strong disconfirming evidence or bulletproof logic.

As far as we know, no pseudoscientists are currently vouching for the existence of fluorescent light–dwelling dwarves, though equally as outlandish claims can be found almost anywhere people are found. For example, some mythologies and folklores maintain that the Earth is hollow (Schick & Vaughn, 2005,

p. 185). These stories are given credence by such ideas as the Greek underworld or the Christian hell. Others claim omnipotent governments are hiding entire cities inside of Earth. Indeed, testimonials and ancient lore are evidence for a hollow Earth, albeit of a very weak variety. Does this evidence justify belief in the hollow Earth hypothesis? If we investigate what science has to say about a hollow Earth, we would learn that the speed of seismic waves traveling through and around the planet, the density of the planet, and the strength of Earth's gravity all contradict the hollow Earth hypothesis.[3] Weighing whether our theories of seismic wave propagation and gravity are wrong against ancient folklore and conspiracy theories, the hollow Earth hypothesis fails to hold water.

If we proportion our beliefs to the evidence, we should conclude that a hollow Earth is unlikely, simply because the evidence is weak. Pseudoscientists traffic in making a claim so broad that it becomes irrefutable. This tactic is known as "torturing" data (Smith, 2017)—e.g., data mining; selecting *some* data rather than *all* data; using questionable statistical analyses; lumping, ignoring, and/or cherry-picking data; limiting the time, date(s), place, or length of the area of study—until they confess. As the British economist and Nobel laureate Ronald Coase put it, "Torture the data, and it will confess to anything" (Novella, 2018, p. 129). It might even confess that the Earth is hollow.

Science, done properly, strictly adheres to Hume's maxim that it is wise to proportion one's beliefs to the evidence. Scientific facts are held in close pro-

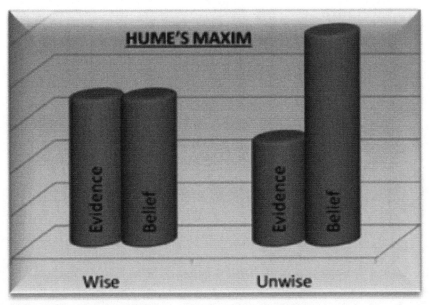

Figure 12.1.   Hume's Maxim. *Courtesy of the author.*

portion to the evidence supporting them. When scientists make unsupported claims, they suffer consequences ranging from being ridiculed by their colleagues and losing grant money to public embarrassment and being consigned to the fringes of science. Although wacky articles do make their way into respected science journals from time to time (Gorski, 2021b), a science paper must survive scrutiny before it's even considered for publication (chapter 3). As Professor Novella explains, "In the process of peer review, when experts in a field review a paper submitted by one of their colleagues in the hopes of being published, one of the specific things they have to decide is whether the conclusions of the researcher extend from the evidence. If the authors are making conclusions that go beyond the evidence, they will often be required to fix that before the paper can be accepted for publication" (Novella, 2012, p. 136). No such standards exist in any pseudoscientific discipline: in pseudoscience, there are no standards at all.

Questioning Evidence

Science thrives on self-examination, on staying open to new evidence and new arguments, and on the willingness to consider the possibility that a theory is wrong. By contrast, pseudoscientists tend to respond to scientific criticism with outright hostility. Harsh criticism is part and parcel to the scientific process. When new scientific ideas are proposed, if they make it into a peer-reviewed journal at all, they're rigorously scrutinized by qualified scientists in relevant field(s). Any tiny problem with a hypothesis becomes a target, and critics love showing off their intelligence by exposing where another intellectual went wrong. When Einstein proposed his theory of special relativity, hundreds of scientists attacked both him and his theory (Cuntz, 2020). When Darwin first proposed his theory of evolution by natural selection,[4] he was attacked by people in nearly every intellectual field, including biology, physics, geology, and (especially) theology. He was accused of being a second-rate scientist and an atheist, motivated by his disbelief to upset the theory of divine creation.[5] It took somewhere between forty and seventy-five years (Gregory, 2008; Larson, 2002)[6] after the publication of *On the Origin of Species* before Darwin's theory gained acceptance by the scientific community!

This brutal process is how hard-won gains are made and how scientific facts are separated from pseudoscientific fictions. Through the process of scientific debate, natural selection is now considered a theory on par with the theory of gravity. Reaching the status of theory is a lot like meeting a tribal elder or an OG (original/old-school gangster): we give them our utmost respect by virtue of them having earned their status. Elders and OGs achieve their status with know-how, street smarts, surviving attacks by rivals, and "keeping it gangster" in the process. Darwin's theory of natural selection is the ultimate OG of scientific theories in that it was wise beyond its years; it's remained unfalsified (yet falsifiable) for nearly two centuries; it's withstood attacks from every one of its rivals; and it stayed true to the scientific ideals of testability, falsification, and defeasibility—the hallmarks of scientific methodology.

Scientists thus prepare themselves for the rebuke that often follows publication of a new idea. Pseudoscientific ideas generally fly in the face of natural laws that were established by this very process, and so any pseudoscientific claim is wont to experience extremely harsh criticism. Rather than develop a thick skin, weigh the points of the critics, and return to the drawing board—as scientists do—pseudoscientists usually reject mainstream criticism and cling to their beliefs regardless of the evidence. Fringe groups go on insisting they've been made the victim of a conspiracy to suppress their ideas—a tactic that was most notably undertaken by the intelligent design movement in Ben Stein's pseudodocumentary *Expelled: No Intelligence Allowed* (Craft, Ruloff, & Sullivan, 2008).

Health and Pseudoscience

Scientists in any given field have their own pseudosciences tugging at their sleeves. Chemists have alchemists. Astronomers have astrologers. Geologists have flat- and hollow-Earthers. Biologists have creationists. Medicine has SCAM. Botanists have biocommunication (the idea that plants are sentient and experience emotions like pain and fear). Cosmologists have intelligent designers. Anthropologists have cryptozoologists. Archeologists have ancient astronaut and Atlantian "theorists." Psychologists have Freudian psychoanalysis and parapsychology. Optometry has iridology. Even CT has a fringe counterpart: denialism.

Most fringe sciences are touted by the media and sold to the public with complete disregard for the truth, which is one reason they're so prevalent. Moreover, the simple explanations offered by fringe science are more readily accepted by the public. Unfortunately, simplistic explanations are hardly ever useful tools for gauging truth, which is especially dangerous when it comes to our health. Literally thousands of SCAMs (Ernst, 2018), although completely ineffective, manage to compete with science-based medicine. What's even scarier, some mainstream medical institutions, such as the Cancer Treatment Center of America, the Mayo Clinic, and the WHO, promote SCAMs while being fully aware of their ineffectiveness and dangers (Atwood, 2004; Hagen, 2016; Renckens & Dorlo, 2019; Stefanek & Jordan, 2020). Naturopathy, a cult-like branch of pseudomedicine, claims to have discovered the ability to cure nearly any human ailment by manipulating a magical energy using "natural" remedies (Hermes, 2019; Hermes, 2020). As we'll see, if SCAM practitioners were as capable as they claim, we'd have no need for MDs, hospitals, medical schools, treatment facilities, research laboratories, or anything resembling science-based medicine.

Pseudomedicine

Belief in the efficacy of SCAMs correlates with intuitive, rather than analytic, thinking styles (Lindeman, 2011); a frightening prospect, considering that with modern medicine, the average person lives to about eighty, while the average life expectancy for prescientific people relying on folk remedies was about thirty (Pinker, 2019). Clearly, we've come a long way since the days that quacks and snake-oil salespeople ruled the market for medical treatment. Nevertheless, a vocal minority still prefers folk remedies over science-based medicine. Let's discuss some of these people and their incredible claims.

Once upon a time, medicine that didn't work was called **quackery**. Quackery is risky, untested, unproven, or disproven medicine (Angell & Kassirer, 1998, Fontanarosa & Lundberg, 1998); the polar opposite of mainstream, science-based medicine. The word *quack* derives from a Dutch word meaning "hawker of salve." Quacks spoke rapidly in a high-pitched, nasally voice, like a duck, claiming to possess medical skills, and dispensed potions and ointments that supposedly had curative powers (Offit, 2014, pp. 199–201). Quacks were known frauds and snake-oil salespeople that appealed to the desperate and credulous. It should be no surprise, then, that although there are many reasons for believing in quackery (Saher & Lindeman, 2005), such beliefs are correlated with paranormal beliefs (Ernst, 2018, pp. 111–12).

Quackery never went away and likely never will. However, rather than stick with the label "quack," quacks have rechristened themselves and now prefer legitimate-sounding names such as naturopath, holistic healer, chiropractor, acupuncturist, homeopath, or functional medicine practitioner. Like the phrase "alternative fact," the very phrase "alternative medicine" is self-contradictory. An alternative fact is something that is not factual, and an alternative medicine is treatment that is not effectual. Specifically, SCAMs are characterized by meaningless pseudojargon, anecdotes, and testimonials (Hines, 2003, pp. 351, 356), as well as "treatments that have never been tested or haven't been tested adequately; treatments that have been tested and shown *not* to work; treatments based on nonexistent phenomena such as human energy fields and acupoints; treatments such as homeopathy that if true would violate established scientific knowledge; and treatments that have been proven to work but that mainstream doctors have good reasons not to recommend" (Hall, 2019b). Thus, pseudomedicines have aptly been dubbed SCAMs: so-called alternative medicine. The world's first professor of complementary medicine calls them SCAMs because if a treatment doesn't work, it cannot be an alternative to medicine, and if a treatment does work, it belongs to medicine, not alternative medicine (Ernst, 2018, p. 1).

Literal hawkers of salve are still around—such as proponents of "black salve," also known as "compound-X"—promising everything from exfoliation to cancer cures (Novella, 2019). Regardless of what name it goes by, quackery is a fraud. Some quackery has a relatively low risk factor, such as taking multi- or megavitamins for brain health (Spector, 2019; but see Offit, 2014), and some is extremely dangerous. SCAM can (and has) cause symptoms ranging from allergic reactions and viral infections to paralysis (Harrison, 2015, p. 195; Offit, 2014, pp. 47–62), disfigurement, and even death (Harrison, 2013, pp. 97–99).

The world has no shortage of desperate and gullible people who drive the demand for quackery. A quick Google search for "alternative medicine" turns up hundreds of different SCAMs including Reiki, homeopathy, naturopathy, therapeutic touch, faith healing, ear candling, acupuncture, chiropractic, and electromagnetic therapy. In the United States alone, more than $34 billion is spent every year on quackery (Harrison, 2015, p. 196), with the world spending over $60 billion per year (Harrison, 2015, p. 211). That's right; the United States is responsible for purchasing over half of all quack medicines, meaning that in the United States, "more money is spent on quack medicine than on all medical research" (Sagan, 1996, pp. 399–400) combined, and the costs of quackery far exceed the amount of money spent on legitimate medical research (Eisenberg, Kessler, Foster, Norlock, Calkins, & Delbanco, 1993; Mielczarek & Engler, 2012).

Why are people attracted to SCAMs? The placebo response (chapter 3) is a real phenomenon that people taking SCAM experience, but the infamous placebo isn't the only player in town. Some people may have heard from a friend that Compound X removed a cancerous mole or a chiropractor cured a child's autism. Some people may think that since homeopathic products are sold in mainstream drug stores on the shelves next to science-based medicines, they're equally as effective (Little, 2019). And, although completely ineffective, sometimes SCAMs *seem* effective. Psychologist Barry Beyerstein lists ten reasons people believe SCAMs work, including the natural course of the disease, spontaneous remission, misdiagnosis, combining SCAMs with conventional treatments, and psychological influences, such as one's perception of sickness and health (Beyerstein, 1997). Moreover, during our sicknesses, we'll experience what's known as *cyclicity*, or "the ups and downs" of our condition (Alcock, 2018, pp. 339–40). Since we usually seek medicine at low points, even bogus

> Not until the empirical resources are exhausted need we pass on to the dreamy realm of speculation.—Edwin Hubble

treatments have numerous opportunities to *seem* effective. Additionally, regression (chapter 11) accounts for false correlations between taking SCAMs and improvement. To the untrained consumer, any one of these reasons can provide compelling "evidence" for the effectiveness of SCAMs.

To the extent that SCAMs have any effect at all, no research demonstrates that they work better than science-based treatments. As Ernst put it, "I have been in this business for a very long time now, I have published more papers on alternative medicine than any other researcher on this planet—and yet, I have never come across an alternative therapy that clearly and demonstrably outperforms conventional medicine" (quoted in Hall, 2019b, p. 21). Although many organizations out there are taking legal action to protect the general consumer from the deception of quack peddlers (Little, 2019), educating ourselves about SCAM and its potential effects remains a worthy endeavor.

> There's no such thing as complementary . . . medicine. There's only medicine that works and medicine that doesn't.
> Paul A. Offit

HOMEOPATHY

Despite that its truth would overturn all of chemistry and much of physics (Ernst, 2018, p. 41), homeopathy is perhaps the most popular and widely endorsed form of quackery. Homeopathy appeals to those who prefer believing that when it comes to medicine, "natural" means better and others who've had bad experiences with science-based medicine and find solace in its supposed safety (Lindeman, 2011). That homeopathic products usually come in fancy packaging making bold claims well beyond the evidence certainly can't hurt their sales, especially when we consider everything we're learning about the field of neuromarketing (Johnson & Ghuman, 2020). Homeopathy is made from everything from rhino horns and pangolin scales to witch-hazel and dog milk, and claims to cure every ailment imaginable, from loneliness and fear to addiction and cancer! In short, homeopathy is today's bloodletting and purging, just as conspiracy theories and alien visitation are today's witches and werewolves.

The thing is, homeopathy has zero ability to heal (Ernst, 2002; National Health and Medical Research Council, 2015) and performs no better than placebos (Hines, 2003, pp. 358–62). If homeopathy performed as claimed, we might have eradicated illness and disease by now. Unlike mainstream medicine where every therapy, medicine, or treatment must pass clinical trials before being sold, the homeopathy industry is completely unregulated. Consequently, homeopathic manufacturers can (and do) put anything they want into their products

(Offit, 2014, p. 92), make bogus claims of efficacy on the package, and sell it on the shelf next to mainstream medicine (Little, 2019).

Nicorette is a good example (Hood, 2009, p. 173). Before its manufacturers could sell it, Nicorette had to undergo clinical trials demonstrating that it helped people quit smoking. Nicorette's homeopathic rival, CigArrest, is a knock-off that didn't have to undergo any testing before its manufacture or sale. Homeopathic distributors opt out of clinical trials for several reasons. For one, they're expensive and time-consuming. For instance, during Phase III in clinical trials on a rotavirus vaccine, researchers spent over $350 million testing the vaccine over four years on more than seventy thousand children from eleven countries, just to get FDA approval for claiming the vaccine is safe and effective for one age group (Offit, 2014, pp. 95–96). A better reason homeopaths don't test their products is because doing so would prove they didn't work, and this multibillion-dollar mega-industry isn't about to commit suicide.

Homeopathy, from a German word meaning "similar suffering," traces back to the German physician Samuel Hahnemann (1755–1843), who thought that if someone was ill, giving them a diluted version of something that causes similar symptoms triggers the body's natural defenses and heals the patient—his so-called Law of Similarities. For example, if someone has an itchy rash, the cure is diluted poison oak (since poison oak makes the skin itch), or if someone has pink eye, the cure is diluted onions (since onions irritate eyes). The idea that small doses of something can activate the body's repair mechanisms is not entirely unscientific. In fact, scientists have dubbed just such a process as *hormesis*. Hormesis is a process where low exposure to radiation can have positive biological responses, though greater exposures to the same radiation can have the opposite effect (Hayes, 2007). However, the science on whether the theory of hormesis explains health benefits remains unsettled (Hall, 2019a). Even if hormesis ends up proving that tiny amounts of radiation have health benefits, homeopathy isn't vindicated since hormesis still requires *some* amount of radiation while homeopathy doesn't.

Which brings us to Hahnemann's "Law of Infinitesimals," his idea that the smaller the dosage, the greater its powers to heal. These days, homeopaths use a system of dilution wherein "X potency" means the substance has been diluted to one part substance, nine parts water and "C potency" is dilution to one in a hundred (see figure 12.2, adapted from Grams, 2019). The result is referred to as 1X or 1C, respectively. Many homeopathic products are diluted to as much as 200C. To get to 200C, one part of the original 99:1 dilution is then mixed with ninety-nine parts water, and this process is repeated 200 times. With a 200C dilution, the chances of ingesting a single molecule of the original substance are an astounding 1 in 10^{400}. When you consider that there are an estimated 10^{100} atoms in the known universe, homeopathy's claims seem extraordinarily unlikely

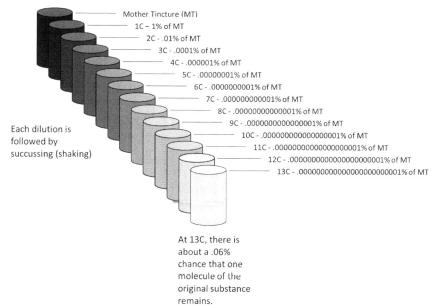

Mother Tincture (MT)
1C – 1% of MT
2C - .01% of MT
3C - .0001% of MT
4C - .000001% of MT
5C - .00000001% of MT
6C - .0000000001% of MT
7C - .000000000001% of MT
8C - .00000000000001% of MT
9C - .0000000000000001% of MT
10C - .000000000000000001% of MT
11C - .00000000000000000001% of MT
12C - .0000000000000000000001% of MT
13C - .000000000000000000000001% of MT

Each dilution is
followed by
succussing (shaking)

At 13C, there is
about a .06%
chance that one
molecule of the
original substance
remains.

Figure 12.2.   The making of homeopathy. *Adapted from Grams, N. (2019). The un-diluted truth. The Skeptic, 39(1), 28–32.*

(Alcock, 2018, pp. 359–72). This is no exaggeration. Homeopathic concoctions are diluted to the point that one would have to drink the equivalent of Lake Michigan to have a chance of consuming one molecule of the original ingredient (Harrison, 2013, p. 99; Randi, 2007).

Hahnemann thought—and modern-day homeopaths think—that homeopathy worked by the original substance leaving an untraceable "essence" of itself that tapped into the body's "vital force" (Schick & Vaughn, 2005, pp. 278–79). In other words, the water "remembers" the ingredient, and transmits the remembered information to the body spiritually. Interesting, but even if true, how exactly would it work? Couldn't scientists observe the transference of information and measure it? Perhaps even synthesize it? Wouldn't there be proof of this water memory? The fact is water has no ability to "remember" anything (Gardner, 1988–1989; Maddox, Randi, & Stewart, 1998), despite incredible claims to the contrary, and homeopaths have never shown *how* water "remembers," let alone how it remembers what it's supposed to and how it "forgets" everything else it's come into contact with, such as bacteria, trace contaminants, and so on.

Skeptics will recognize homeopathy's claims as incredible, for they contradict well-established GASPs. For instance, if a medicine got more potent the more it was diluted, why would such a product require overdose warnings on the package? The longtime (and late) skeptic and bulwark against homeopathy

James "The Amazing" Randi once quipped, "We heard about a guy in Florida, the poor man. He was on homeopathic medicine, he died of an overdose. He forgot to take his pill!" (Randi, 2007).

Appeal to Nature Fallacy

The claim that "homeopathy is safe because it's natural" is such a terrible argument, it's not even wrong. A 2012 systematic review that evaluated the evidence for possible adverse effects of homeopathy found that "homeopathy has the potential to harm patients and consumers in both direct and indirect ways" (Posadzki, Alotaibi, & Ernst, 2012). To be sure, in one sense homeopathy is just as safe as drinking water, since it *is* water. However, the combination of corporate greed and lack of oversight is a recipe that almost guarantees exploitation, and instances of adverse events abound. Too many people have died because they relied on advice from homeopaths (Bell, 2016; Hall, 2014; Offit, 2014) or chose homeopathy over conventional medicine (Steve Jobs, cofounder of Apple, was one such unfortunate person). The following is a truly heart-crushing example recounted by Guy P. Harrison:

> In Australia, a homeopathic doctor and his wife allowed their infant daughter to die a slow and painful death because they trusted in alternative medicine over evidence-based medicine. Thomas Sam, a "practicing homeopath," was advised to seek proper care for the baby's severe skin infection but refused. The parents' faith in this alternative medicine was so great that they refused to waver, even as they watched their child's hair turn white, body shrink, skin bleed, and corneas melt. The baby was in constant agony and often screaming. . . . After suffering for months, she finally died. (Harrison, 2013, p. 100; and see Offit, 2014, pp. 7–22)

What we've been discussing here is known as the *appeal to nature* fallacy. The appeal to nature fallacy is the erroneous idea that what's natural is good for us and what's unnatural is bad for us (Baggini, 2003, p. 182). Cosmic radiation is natural, and even minute amounts of it can cause serious nerve and brain damage—probably the biggest barrier that's currently stalling long-term space exploration. Moreover, snake venom, hungry lions, black holes, hurricanes, the sun, lightning, and viruses are all natural and are all positively detrimental to human well-being. Chemotherapy, genetically engineered crops (GEs), vaccines, hypodermic needles, heart transplants, and the internet are all unnatural, yet they save millions of lives every day all over the planet. Whether something is natural or unnatural tells us nothing about whether it's good for human health or well-being.

Homeopaths attempt to gain credibility by claiming that homeopathic treatments are infused with this-or-that herb, plant, extract, or whatever. However, multiple investigations show that homeopathic products taken from the shelves of major retail stores contained zero DNA from whatever was advertised on the package, had potentially harmful ingredients (such as black walnut, mercury, lead, and arsenic) (Harrison, 2015, pp. 208–11; Schneiderman, 2015), or contained prescription drugs (Offit, 2014, p. 92).

Figure 12.3.   The "Amazing Randi" warns a crowd. *Wikimedia.org.*

Consumers don't seem to notice because usually there's nothing to notice. Taking a pill or drinking water that has supposedly been infused with ginkgo biloba when it was really infused with houseplant will go largely unnoticed because neither have medicinal properties in the first place. Modern medicine understands that any treatments that have effects also have side effects. Homeopathic products lack side effects because they have no effects at all. Thus, homeopathy only comes under scrutiny when someone is hurt or killed by ingesting toxins or allergens.

Without the ingredients being tested or proven to treat or cure medical conditions, claiming a product is infused with some herb or natural extract is simply a marketing distraction. Millions of people believe ginkgo biloba cures brain diseases and improves intelligence, yet research shows that ginkgo biloba cannot make one smarter (Laws, Sweetnam, & Kondel, 2012) or treat cognitive impairment like dementia or Alzheimer's (Cooper, Li, Lyketsos, & Livingston, 2013; Offit, 2014, p. 99; but see Kanowski, Hermann, Stephan, Wierich, & Horr, 1996; Schacter, 2002, p. 37). What it can do is inhibit the blood's ability to clot (Cupp, 1999), a potentially dangerous outcome for anyone at risk of bleeding or undergoing surgery. Taking homeopathic treatments makes about as much sense as putting dog food into your gas tank and expecting your car to run. Unless trained experts have tested and confirmed that dog food can make cars run, we shouldn't trust that it can. The same logic applies to health. Until experts have tested and confirmed that poison ivy cures joint pain and fever or snorting rhino horns cures demonic possession, we shouldn't just assume they do on the say-so of quacks or snake-oil salespeople.

When it comes to health, thinking critically is a must, but an unfortunate number of us simply fail in this regard. Seeing homeopathic medicines with attractive packaging shelved next to science-based medicines is a recipe for unquestioning, System 1, veridical thinking. Don't fall for the trap! When it comes to taking medicine, any medicine, it's never smart to put your guard down. As

Harrison put it, "Alternative medicine encourages people to trust without evidence and to accept ideas without thinking. It encourages and feeds a culture of gullibility" (Harrison, 2013, p. 101). Therefore, since medical interventions touch our lives in matters ranging from our health to our wallets, we cannot be cognitive misers when it comes to our well-being.

A great question to ask anyone touting the benefits of SCAM is "How do you know?" When it comes to homeopathy, it's easy to fall for the *post hoc ergo propter hoc* ("after this, therefore because of this") fallacy, which I discussed in chapter 4. If after taking homeopathy, one's condition improves, the homeopathy is given credit. On the other hand, if after taking homeopathy, one's condition doesn't improve, it is always possible to claim that the dosage was incorrect or treatment was delayed too long or the illness was just too powerful. Thus, homeopathy is unfalsifiable. Nobel Prize–winning biologist and immunologist Peter Medawar—best known for his work on how the immune system responds to organ transplants—once stated that "If a person a) is poorly, b) receives treatment intended to make him better, and c) gets better, no power of reasoning known to medical science can convince him that it may not have been the treatment that restored his health" (Gilovich, 1991, p. 128). Medawar's reasoning, in conjunction with people's lack of understanding of cyclicity and regression (chapter 11), helps us understand how reasonable people reach false conclusions about SCAMs.

For the last several hundred million years, our ancestors have fought off infection and disease from the microbial world. As a result, about 50 to 80 percent of all human ailments go away on their own (Blaskiewicz & Jarsulic, 2018; Gilovich, 1991, pp. 127–28). With statistics like these, people taking homeopathic drugs should be "cured" 50 to 80 percent of the time. So how do you *know* a medicine or treatment cured an ailment? Without controlled, double-blind experiments, you simply cannot.

Natural as harmful:

- Bee stings
- Black holes
- Hurricanes
- Snakebites
- Wild lions
- Cancer
- Lead
- *E. coli*
- Poison oak
- Cosmic radiation
- Cyanide

Nature is full of things that are harmful to humans, and there's no reason to equate natural with better.

To be sure, problems with mainstream medicine persist. Some doctors are corrupt, some lie, some are incompetent, and everyone makes mistakes. A study conducted in the 1980s revealed that the average hospital erred in administering patient medication 12 percent of the time (Cialdini, 2007, p. 219). In the field of medicine, the stakes are high, which is exactly why we see so much oversight and huge lawsuit settlements when something goes awry.

Drab furniture and plain walls can make hospitals feel like jails. Additionally, many people believe mainstream doctors are in cahoots with "Big Pharma" and, as an alternative, offer bogus treatments that have serious consequences. By contrast, the offices of SCAM providers are warm and welcoming with burning incense, friendly receptionists, and uplifting music playing throughout the building. Is it any surprise that so many people are attracted to SCAMs?

In an ideal world, the time and efforts of scientists and skeptics would be used improving the problems of our medical systems. Unfortunately, much of their efforts are spent combating dis- and misinformation about medicine that powerful and influential quacks feed to the public by the barrel. One example is a recent study showing that during the coronavirus pandemic, 65 percent of vaccine misinformation came from just twelve people who were labeled the "Disinformation Dozen" (Crist, 2021; Gavura, 2021). As such, informed consumerism, promoting science-based medicine, opposition to quackery, and supporting progress in the medical field are desperately needed to combat misinformation. Calling attention to the problems with mainstream medicine is an important task that helps drive progress, but doing so has no bearing on the efficacy of homeopathy and doesn't make its problems disappear. To argue as much is to commit the *tu quoque* ("you too") fallacy discussed in chapter 5. Homeopaths can be just as corrupt, incompetent, and mistaken as anyone else. Problems within mainstream medicine—where treatments are rigorously tested, and ethical standards are higher than in any other field—persist. The homeopathy industry doesn't test their treatments, lacks oversight, and is still susceptible to all the problems of human error that plague mainstream medicine. Thus, the only possible thing that system can be better at is making mistakes.

Wrapping Up

As Straight Thinkers who care about the truth, we must learn how to strike a balance between being open to new and even extraordinary claims and skeptically evaluating all claims regardless of whom or where they come from. Scientific findings are stated in percentages and probabilities; for example, "What's the probability that our germ theory of disease is wrong?" or "What percent of experts currently working in the field accept human-caused global warming as factual?" When an incredible claim is touted as the absolute truth, it cannot be scientific because there are no absolute truths in science.

Whether talking about ghosts or space aliens, health or medicine, it's always appropriate to bring a healthy dose of skepticism and an open mind—just not so open that your brain falls out (Sagan, 1996, p. 219; Shermer, 2004, p. 263; Shermer, 2011a, pp. 336–37). As Professor Novella reminds us, "Skepticism isn't close-minded, and the opposite of skepticism is not open-mindedness" (Novella, 2018, p. 72). What Novella means is that skepticism is about being open to evaluating any claim based on its merits and that thinking without doubting is gullibility, not open-mindedness.

CHAPTER 13

Conspiracies—What "They" Want You to Think . . . Maybe

Nothing is particularly controversial about the fact that conspiracies really occur. Arguably, we're surrounded by an uncountable number of genuine conspiracies every day. However, the ways in which conspiracies are portrayed often place them into the realm of incredible claims, especially as they grow in size and reach. People throughout the world are willing to kill and be killed on account of conspiratorial beliefs, the majority of which are almost certainly false.

A variety of conspiracy theories and theorists exist. With the help of the internet, new ones pop up so quickly that it'd be impossible for anyone to keep up with them all.[1] This final chapter primarily focuses on *belief* in conspiracies, although the psychology behind conspiracy thinking, who perpetuates conspiracy theories, as well as who believes them and who doesn't will also be discussed.[2]

Building Vocabulary

Schema:	Conspiracy theory:
A mental model of preconceived ideas; a framework representing some aspect of the world; a system of organizing and perceiving new information.	An idea, suspicion, or conjecture that a real conspiracy has happened or is happening, especially one that contradicts the laws of nature or generally accepted standards of proof.

False flag:	Crisis actor:
An operation carried out by a group that seeks to hide its intention and involvement in the operation.	An actor pretending to be the victim of an event (e.g., a mass shooting) to push an agenda (e.g., more gun control).

Minding Conspiracy Theories

Conspiracy thinking has some of the same intellectual trappings of skepticism, such as doubt and a desire to find the truth. However, what separates the two are the methods used in seeking answers. Skepticism uses the methods of science and the rules of logic and evidence. By contrast, conspiracy theorists tend to rationalize backward from a foregone conclusion. Rather than rejecting an idea until it's proven beyond a reasonable doubt, those who believe in unlikely conspiracies veridically accept them until it's *dis*proven beyond any doubt. In this chapter, I show that in doing so, avid conspiracy theorists ignore evidence that doesn't validate their preconceived beliefs, they commit logical fallacies when furthering their arguments, and they're immune to the idea that they may be wrong.

Conspiracy theories differ significantly from genuine conspiracies in important ways.[3] Take the Watergate debacle, for example. This "million-dollar plan," executed by top officials of the most powerful government in the world, was foiled by a security guard because, like anything else, conspiracies are constrained by the vicissitudes of reality. Mistakes happen, and even the best plans hardly ever go, well, according to plans. Moreover, mistakes are just one of many ways conspiracies fail. Accidents, informants, serendipity, spies, or defectors can ruin even the best-planned conspiracy. Perhaps Michael Shermer captured this best when he said that conspiracies "are messy events that unfold according to real-time contingencies and often turn on the minutia of chance and the reality of human error" (Shermer, 2013, p. 100).

Furthermore, do we have any reason to believe that government conspiracies are special? Are government agents more competent or trustworthy than everyone else? Think about common criminal conspiracies. The mores of street culture shame anyone labeled a "rat," and those who rat on their fellow criminal do so at the risk of banishment, serious harm, and even death. Even these strict norms (the Mafia's *omerta* code of honor comes to mind) don't stop people from ratting; people rat for reasons that range from attention and revenge to cutting deals and getting out of jail. If gangsters are willing to rat out their best friends and crime lords, lesser criminals are certainly willing to rat out the government.

G. Gordon Liddy, one of the conspirators in the Watergate scandal, said, "The problem with government conspiracies is that bureaucrats are incompetent and people can't keep their mouths shut" (Shermer, 2011a, p. 225). It follows, then, that the more complex a conspiracy, the more people involved, the more susceptible it is to being exposed or discovered. Strangely, conspiracy theorists would have us believe that the bigger the conspiracy, the better "they" are at keeping it secret. Such thinking is backward logic at its worst.

A Web of Beliefs

From Watergate to Iran-Contra, MKUltra to the Tuskegee syphilis experiment, and Edward Snowden to the Pentagon Papers, powerful elites have engaged in numerous conspiracies (Olmsted, 2019; Shermer, 2020). It's these exposed conspiracies that lend credence to conspiracy theories that are invented and exist solely in minds (Cooper, 1991). The existence of conspiracies has created an air of mistrust in the power elite, and indeed, history has provided no shortage of reasons to mistrust powerful elites. But we must be careful in our thinking and avoid the Nirvana fallacy, which is committed when we make the logical leap that if one part of something isn't perfect, the entire edifice is bad. Because governments and powerful corporations have engaged in misdeeds in the past doesn't mean everything those governments and corporations do is sinister. This logical leap can also be described as the *fallacy of division* (what is true of the whole is true of its parts) or the *fallacy of composition* (inferring that something is true of the whole because something is true for one of its parts). That prisons are full of immoral people doesn't mean every person in prison is immoral. Just because airplanes are heavy doesn't mean every part they're built from is. Likewise, we must think of conspirators as individual actors because no one person is "the government" or "Big Pharma" or "Big Oil."

Nevertheless, although conspirators have done untold damage, conspiracy theorists' rhetoric is far from harmless. Conspiracy theorists have contributed, and continue to contribute, to the declining health and well-being of society by promoting racism, denying the benefits of water fluoridation, and refusing to vote or to reduce their carbon footprint (Douglas, Sutton, Jolley, & Wood, 2015; Jolley, 2013). The number of deaths caused by the promotion of baseless conspiracy theories is likely uncountable, but it is a certainty that, say, believing the coronavirus pandemic is a hoax or that vaccination is a dangerous ploy has led to the needless deaths of far too many (Vilches, Moghadas, Sah et al., 2022). Suspicion is not always bad, but the ability to tell the difference between misbehaving leaders and faulty pattern detection (chapter 8) is the difference between critical thinking and anti-thinking.

Conspiracy theories are an epistemology in that a conspiracy theory is a claim that one knows something specific occurred or is occurring. Anti-vaxxers claim to *know* vaccines are harmful (Halford, 2020), whereas researchers claim to *know* they're not (Offit, 2015; Offit et al., 2002). Climate change deniers claim to *know* humans aren't causing the planet to warm, whereas climate experts claim to *know*, albeit cautiously, they are (Mann & Toles, 2018; Oreskes & Conway, 2010b). Edgar Welch just knew Hillary Clinton was running a child sex-trafficking ring through a pizza shop in New Jersey. In each scenario, someone must be wrong. The ability to Think Straight imbues a responsibility to decide which side is right by carefully evaluating the available evidence. Therefore, conspiracy theories cannot be considered valid epistemologies until they adhere to GASPs. It's worth noting that the conspiracies everyone agrees happened—e.g., the Bay of Pigs Invasion or tobacco companies conspiring to keep the dangers of tobacco secret—were exposed by the same methods—e.g., whistle-blowing, mistakes, investigative journalists, insiders, and leaked documents—that conspiracy theorists mistrust (Prothero & Callahan, 2017, pp. 30–31). Anti-vaxxers claim the government cannot be trusted to report accurate information about vaccines, unless the government is reporting adverse reactions; then they point to those reports as true and simply claim the government is hiding other information.

As Hugo Drochon once observed, mistrust can be a healthy component of living in a thriving society. However, mistrust can be taken too far. Mistrusting the intentions or honesty of those in power without CT isn't skepticism—it's fear without direction; it's suspicion run amok. Some people take conspiracy theories so far that they even come to believe in contradictory conspiracy theories (Prothero & Callahan, 2017, p. 28; Uscinski, 2019c, p. 4), such as believing Osama bin Laden was dead before Seal Team Six got to him and that he's still alive or that Princess Diana was killed by the British government and that she faked her own death (Wood, Douglas, & Sutton, 2012).[4] Researchers

> Yes, conspiracy theorists tend to get it wrong, but the skeptical mindset they participate in, when used correctly, is essential to the proper functioning of modern democracies.
> —Hugo Drochon

have suggested (Goertzel, 1994a) and argued convincingly (Swami et al., 2011) that maintaining contradictory beliefs is a product of what's called a "monological belief system," a belief system in which the conspiracy theory worldview is maintained through a "network of mutually supportive beliefs" (Wood, Douglas, & Sutton, 2012). Of course, it's possible to believe just one conspiracy theory, and not everyone who believes conspiracy theories will develop a monological mindset (Franks, Bangerter, Bauer, Hall, & Noort, 2017). However, once begun, the confirmation bias (chapter 5), cognitive dissonance (chapter 7), and a general lack of CT can lead one down the primrose path of monologicality.

Monological belief systems are accountable only to themselves. Later, we'll discuss *Barkun's Script*, a monological belief system that insulates conspiratorial beliefs from logic and evidence. For now, our Street World metaphor helps explain how monological belief systems work. Street World represents an entire complex of interconnected roadways, like the network of interconnected neurons inside our brains. In far-off recesses of this complex, we find an isolated network that's connected to the main complex by a few loan interstates. Let's call it Conspiracy City. Information flows between the main complex (i.e., the rest of the world) and Conspiracy City by way of those lone interstates. But since Conspiracy City is monological, those interstates are regulated by tolls, roadblocks, security checks, border patrol, and customs agents. We might say Conspiracy City is run by communists who insist on monopolizing the flow of information, so to maintain stability, information is thoroughly vetted and screened before entering or exiting.

In Conspiracy City, roadways (i.e., our beliefs) are connected to other roadways (i.e., other beliefs), so that tearing down one (changing a belief) affects any others it's connected to (other beliefs are compromised). The bigger the roadway being demolished, the greater the numbers of other roadways affected by its destruction. For example, if you believe that the U.S. government was behind 9/11, that they intentionally put poisons in our vaccines, and that they dump toxins out of airplanes to control the minds of their citizens, the proposition that the government doesn't engage in conspiracies would, if true, compromise all three of these beliefs. Accepting this proposition would be like demolishing one of Conspiracy City's main boulevards.[5] By contrast, the proposition that the government doesn't intentionally put poison in our vaccines would, if true, only challenge *this* belief; it would only require demolishing Wakefield Way and maybe putting up a (cognitive dissonance) detour.

The Patterns of Conspiracy Theories

Widespread belief in conspiracy theories isn't going away—an unfortunate prognosis given that such thinking provides fertile ground upon which bad thinking flourishes. On top of conspiracy thinking being monological, we're also burdened by the cognitive dissonance associated with conspiratorial beliefs. If our options are (1) a given conspiracy theory is true or (2) our thinking is delusional, most of us will choose the former. The reason is equally as obvious: it's easier to believe a conspiracy exists than believing there's something deeply wrong with our thinking. Fortunately, our dilemma is a false choice (chapter 4). If conspiracy thinking is the product of a combination of certain dispositional factors (Bost, 2015), then there's hope that conspiratorial beliefs can be undermined by using evidence and reason.

Since our brains are preprogrammed to see intentionality (i.e., patternicity and agenticity) (Shermer, 2011a, p. 209), it shouldn't surprise us that conspiracy theories are as common as they are. In small doses and when processed through CT filters, pattern detection can be invaluable. Conspiracies can jeopardize our well-being—a likely reason why conspiracy thinking is so alluring—and place our lives or livelihoods in danger. Thus, we err by conflating conspiracy thinking with uncritical thinking. However, when, unbeknownst to us, our brains detect malevolent hands lurking behind every scene, we're likely to encounter more faulty conclusions than true facts about the world.

Does history reveal a centuries-long Jewish conspiracy to conquer the world and subjugate the goyim (non-Jews) (Garaudy, 2000; Steele, 2005)? Are these Jewish conspirators reptilian shapeshifters masquerading as humans (Icke, 2001)? Perhaps, but when we break down those claims and carefully examine them, there are usually better ways to explain them (Shermer & Grobman, 2009). We like separating complicated systems into more simplistic models, which is what conspiracy theorists do when pigeonholing centuries of Jewish history into a simple conspiracy theory, when something like Jewish history and the Jews' relationship with the world is far more complicated (Blaskiewicz, 2014; Chomsky, 1999; Sand, 2010).

As we've seen, the brain is constantly seeking patterns. Like our vending machine that can't tell a counterfeit bill from legal tender (chapter 1), even dubious patterns can be extremely convincing. We also know that we don't like to admit we're wrong, we believe we're smarter than we are, we attach ourselves emotionally to our beliefs and conclusions, we unwittingly engage in confirmation bias, we think we cannot be fooled, we're cognitive misers, System 1 processes incoming information and patterns veridically, System 2 is lazy, and conflicting beliefs and ideas make us emotionally and cognitively uncomfortable. Conspiracy thinking could be a consequence of this smorgasbord of mental shortcomings and cognitive biases. Or maybe a nefarious group of super-intelligent entities programmed humans for believing wild conspiracy theories as a ruse to create doubt and suspicion about those who get close to the truth. Whatever it *may* be, conspiracy thinking is certainly a pitfall that is best avoided. On the other hand, if "they" really are putting nanobots in the "chemtrails," it's too late for us anyway!

Conspiracy Theories as Rumors

Rumors develop in times of uncertainty. When people believe a rumor is relevant and credible, it provides relief from the anxiety that's associated with not knowing (Alcock, 2018, pp. 203–4). In prison, rumors can (and do) get people killed. A prisoner's reputation is sometimes all he has; thus a rumor that potentially calls his reputation into question can have dire consequences. It can be especially hard for prisoners to learn whether a rumor is true, and due to the nature of prison, once a rumor is started, the die cannot be uncast. Unfortunately, prison culture is such that the only respected way to undo the damage is through violence, lest the rumor's target be seen as weak, which itself is viewed as evidence of guilt. So whether it's rumors of conspiracy theories or rumors about who did what to whom on the prison yard, learning how to properly evaluate information is an important survival strategy.

We can think of conspiracy theories as a specific class of rumor.[6] Rumors and conspiracy theories have similar aims of fact-finding, relationship building, and enhancing self-esteem (Difonzo, 2019, p. 262). Both represent pieces of yet unproven information that could be true or false (Difonzo, 2019, p. 258). Like the prison rumor that cannot be disproven, conspiracy theories are similar in this regard. For example, in 2016, a man named Edgar Maddison Welch drove from North Carolina to Washington, D.C., where, armed with an AR-15, he opened fire inside of a pizza shop. Following rumors that had been circulating on social media, Welch believed he was investigating a conspiracy that the shop was trafficking child sex slaves by high-ranking members of the Democratic Party. Although Welch found no evidence of a child sex ring, he maintained his belief, insisting that the conspiracy could still be real (Goldman, 2016).

The Fabric of Conspiracies

Criminal law typically defines a conspirator as someone who agrees to commit a crime with someone else and either attempts to or does commit that crime. I define conspiracy a bit differently. For our purposes, a conspiracy is *a plan involving more than one person that secretly aims to accomplish a common, usually sinister goal at the expense of others.*[7] The reason a secret plot to organize a surprise birthday party isn't considered a conspiracy is the lack of malicious intent (or the violation of an established law).

Whether real or imagined, conspiracies can involve ordinary folks operating for short periods of time or powerful entities operating indefinitely. They can range from small-scale, two-person operations, such as a bank robbery, to huge, large-scale operations—called *grand* or *superconspiracies*—with practically no upper limit to the number of agents involved, although the fixed number of people on Earth does place technical limits to the number of agents that can be involved. Or does it?

> There exist a lot of conspiracy theories, misinformation, and fake news about conspiracy theories, misinformation, and fake news.
> —Joseph E. Uscinski

As we'll see, for conspiracy theories to "work," they must continually build momentum. As they snowball, conspiracy theories grow to such proportions that to successfully explain them requires invoking outside agents that have unprecedented and often unlimited powers and resources; think space aliens or supernatural beings. Thus, for a conspiracy theorist, there may never be an upper limit to how many agents are involved in any given conspiracy theory.

The Cognoscenti

Before moving forward, it's important to understand who conspiracy theorists are and what they believe. Thinking of conspiracy theorists might bring to mind your tweaker buddy that doesn't go anywhere without his police scanner, a crazy old woman wearing a tinfoil hat, or a middle-aged basement-dwelling white guy who's been stuck on the internet for years. As common as these stereotypes are, they're pretty inaccurate. Conspiracy theorists are found in all demographics, and there's no one-size-fits-all conspiracy theorist profile. However, some character traits, including mis- and distrust, cynicism, paranoia, defiance, suspicion, alienation, powerlessness, and magical thinking are fairly common among those who see the world through conspiracy lenses (Abakalina-Paap, Stephan, Craig et al., 1999; Butter & Knight, 2019, p. 37; Swami, Chamorro-Pemuzic, & Furnham, 2010; Wood & Douglas, 2019, pp. 247–48).

Although direct links between personality traits and conspiracy thinking remain evasive, research shows that conspiracy thinking is negatively correlated with skepticism and open-mindedness (Swami, Voracek, Stieger, Tran, & Furnham, 2014),[8] though positively correlated with rejecting science and accepting pseudoscience (Lewandowsky, 2019, pp. 152–53; Lewandowsky, Gignac, & Oberauer, 2013; Lewandowsky, Oberauer, & Gignac, 2013). Now, these findings could mean that people who reject science and believe pseudoscience are especially keen and particularly skilled at detecting hidden realities. However, since those who investigate by using GASPs have more success and much higher standards of proof, it seems unlikely that conspiracy theorists would be better at unearthing credible information.

Women are about as likely to believe conspiracy theories as men; Hispanic, black, and other racial minorities are slightly more prone to conspiracy theory belief than whites; Generation Xers believe more conspiracy theories than the Silent Generation, Baby Boomers, or Millennials; and conspiracy theories are more prevalent among the distrustful, downtrodden, poor, and uneducated than among the upper classes (Goertzel, 1994a; Uscinski & Parent, 2014, pp. 73–104). On that last factor, some have speculated that people with lower education may be more likely to believe conspiracy theories because better educated people "become better equipped to detect nuances across judgment domains . . . [and they] learn not to simplify the complex problems that they encounter but instead reflect analytically on them" (van Prooijen, 2019, p. 435; and see Prothero & Callahan, 2017, p. 30).

Other findings suggest that conspiracy theories are for losers (Uscinski & Parent, 2014), in that the more subjugated, marginalized, or defeated the group, the more likely they'll turn to conspiracy theories for explanations. Poor people invent conspiracy theories about the rich, not the other way around, just as political losers invent conspiracies about prevailing opponents. Like a junkie chasing a fix, once one indulges in conspiracy thinking, the craving for more conspiracy theories may never be satisfied. Indeed, political scientists and leading researchers on conspiracy theories have found that most people believe in at least one conspiracy theory (Uscinski, 2019b; Uscinski & Parent, 2014, pp. 5–6). Other research suggests that belief in one is correlated with belief in several (Bost, 2015; Goertzel, 1994a; Prothero & Callahan, 2017, p. 28).

The breadth and number of conspiracy theories are virtually limitless. One survey found that 55 percent of Americans believe at least one conspiracy theory, 27 percent believe at least two, and 12 percent believe at least three (Enders & Smallpage, 2019, p. 299; Oliver & Wood, 2014). A 2016 Chapman University survey reveals a culture brimming with conspiracy theories. Huge percentages of participants said yes when asked if they think the government is concealing information about 9/11 (54.3 percent), the JFK assassination (49.6 percent), alien encounters (42.6 percent), global warming (42.1 percent), plans for a one-world government (32.9 percent), Obama's birth certificate (30.2 percent), the origin of the AIDS virus (30.2 percent), Supreme Court Justice Antonin Scalia's death (27.8 percent), and the moon landings (24.2 percent) (Poppy, 2017; Smith, 2018, pp. 32–34). Oftentimes, the media is no help. As another source put it, relentless conspiracy mongering by the media is responsible for the fact that "51% of Americans think that a conspiracy was behind Kennedy's assassination [and] only 25% agree with the demonstrated reality that Lee Harvey Oswald Acted Alone" (Prothero & Callahan, 2017, p. 29).

For all their flawed thinking, conspiracy theorists tend to harbor genuinely good intentions, if sometimes misplaced (e.g., Pizzagate). As I'll show in this chapter, the conspiracy theorist worldview is driven by the desire to satisfy psycho- and sociological needs, to question the "official" version of events, to demand satisfactory answers, and to listen to their gut when something seems amiss (which is pretty much always). Those engulfed in this mindset view conspiracism as healthy suspicion. But suspicion without skepticism (i.e., pseudoskepticism) has the potential to lead one far astray, which is why some have called conspiracy thinking "skepticism's evil twin" (Novella, 2010a). Without CT, how can conspiracy theorists begin to evaluate their claims? How effective can suspicion be if one doesn't know a good question from a bad, science from pseudoscience, or reasons from rationalizations? What good are answers if one lacks the skills that are necessary for evaluating answers?

Conspiracy Theory Demographics

One's political slant, religious conviction, or other beliefs tell us nothing about who is most likely to believe conspiracy theories (Uscinski & Parent, 2014, pp. 89–90). However, one's beliefs do predict the *kinds* of conspiracy theories one believes. As we saw, Democrats believe Republicans are conspiring against them, while Republicans believe Democrats are the conspirators. Similarly, Christians and Muslims are more likely to believe in Jewish conspiracy theories than Jews, New Agers are more likely to believe in Da Vinci Code conspiracy theories than Christians, and New England Patriots fans are less likely to believe in "Deflate-gate" conspiracy theories than fans of other football teams (Uscinski, 2019b).

Getting in Where You Fit In

In his seminal work *A Culture of Conspiracy*, political scientist Michael Barkun argues that conspiracy theorists adopt a monological belief system, viewing the universe as being controlled by an intelligent agent (or agents) (Barkun, 2013, p. 4). According to Barkun, this monological system embodies three principles: (1) nothing happens by accident, (2) nothing is as it seems, and (3) everything is connected (Barkun, 2013, pp. 3–4; Shermer, 2011a, pp. 207–27). This **schema**—that is, a worldview that organizes categories of information and the relationships between them (Hess & Pickett, 2018)[9]—I call Barkun's Script. Barkun's Script provides a system of mutually enforcing beliefs that effectively prevent contradictory information from entering the fold. The result is that "the more conspiracy theories a monological thinker believes in, the more likely he or she is to believe in any new conspiracy theory . . . which may be proposed" (Goertzel, 1994a).

Within Barkun's Script, we also find the acceptance of what Barkun calls *stigmatized knowledge*—information conspiracy theorists accept as true, even though it isn't accepted by mainstream sources of knowledge, such as universities or scientific investigators (Barkun, 2013, pp. 26–29). The domain of stigmatized

---

Barkun's Script:

- Nothing happens by accident
- Nothing is as it seems
- Everything is connected

Barkun's Script is a monological belief system that reflects how conspiracy theorists attempt to simplify how the world works.

knowledge is essentially an echo chamber where even the idea that a conspiracy theory is false gets explained away as another layer of the conspiracy theory. One problem with accepting stigmatized knowledge is its inconsistency: conspiracy theorists are perfectly willing to accept information from any source, stigmatized or not, that supports their ideas. For example, David Icke uses FBI, CIA, and media reports as evidence for his claims that the Illuminati were behind the 9/11 attacks, while simultaneously arguing that the FBI, CIA, and the media are Illuminati-run organizations that can't be trusted (Icke, 2002).[10]

From a skeptic's perspective, following Barkun's Script is problematic because conspiracy theories are virtually unfalsifiable.[11] Regardless of its truth-value or the quality of evidence used to support it, using Barkun's Script as one's driving epistemology, any piece of information can be used to reinforce a belief. Essentially, Barkun's Script builds the roadways of Street World not with concrete but with spit and mud. Moreover, any evidence that doesn't fit into a preferred narrative can be ignored or explained away (Barkun, 2013, pp. 157–58). For example, confessions by al-Qaeda about 9/11—they would take credit for something they didn't do. The release of pictures of the moon landings—those pictures were faked by NASA. Obama's birth certificate—a counterfeit that took way too long to release. Vaccines don't cause autism—they suppress information that jeopardizes profits. Because each principle of Barkun's Script mutually reinforces the other, the mere presentation of evidence does little to disrupt deeply held conspiracy beliefs: the goalpost can be inched back evermore (chapter 5), leading one to falsely conclude that a conspiracy theory is real and is logically supported by evidence.

Conspiracy theorists see themselves as evidence seekers, so it's not that they deny the usefulness of evidence. Rather, they strongly believe that *their* interpretation of the facts is the only correct one, which, unsurprisingly, is the one that supports a conspiracy theory. With that said, here, falsification poses a problem for skeptics. Immensely powerful agents should have no trouble strategically circulating false information for investigators to "discover" and make real conspiracies look like conspiracy theories. After all, the only conspiracies we know of are the ones people exposed. Huge and secretive conspiracies could be happening all around us. Unfalsifiability, then, cannot be the only indicator of a conspiracy theory. However, while things are not always as they seem, they are not *always* not as they seem.

Conspiracies can be deadly serious, so investigating whether one is true or false requires a high level of intellectual honesty. It encourages us to investigate and discover any incompatibilities with the proposition and to seek objective, outside, corroborating evidence to see where it leads. Perhaps more importantly, Thinking Straight emphasizes the necessity to amend beliefs as the evidence demands. If everything in the universe fit perfectly into Barkun's Script (as conspiracy theorists believe), then it should be no surprise to find that the world is brimming with grand conspiracies. However, better evidence points toward psychological explanations for conspiracy thinking over the existence of actual conspiracies.

How Conspiracy Theorists See the World:

- Conspirators: those who engage in grand conspiracies and have a hand in controlling world events
- Conspiracy theorists: those who've seen through the conspirators' schemes and see the world through conspiratorial lenses
- The Dupes: those who are too dumb to see that the world is roiling with conspiracies

According to conspiracy theorists, everyone in the world falls into one of these three categories.

THE WORLD DIVIDED

Embroiled in conspiracy beliefs, even unwittingly, is the idea that everyone is split into three groups: conspirators, protectors, and sheeple (Novella, 2012, pp. 148–49; Novella, 2018, p. 205). The conspirators are those immensely powerful agents, such as huge organizations, world governments, secret societies, space aliens, or omnipotent/supernatural beings. Conspiracy theorists believe that these groups are incredibly smart, are infinitely powerful, have unlimited resources, and are ill intended and completely absent of morality. Such epistemologies are not only immune from disproof (counterevidence is explained away as another layer of the conspiracy theory), but the logic behind them fails miserably. Smart, powerful, and resourceful groups without moral scruples dominate by force, not by coercion. They don't need tricks or deceptions. Slave owners didn't try to trick their slaves into believing they weren't really slaves to keep them working. Slaves kept working because rebellion was punished severely.

Additionally, while conspirators are claimed to have infinite intellectual and technological superiority, millennia of experience, and unlimited resources, they're also quite stupid, inexperienced, and unable to control everything. These shortcomings are why they make careless mistakes that reveal themselves and their sinister plans to the world. With impressive ease, conspirators can keep all evidence of the planet's deepest secrets invisible to the masses but somehow leak information about nearly every devious plan they've concocted. In other words, the same people who masterfully execute the most complex systems of conspiratorial control the world has ever seen are the same people who are incapable of keeping their most closely kept secrets off YouTube. Either conspiracy theorists have the drop on the most powerful elites on the planet or they've been seriously deluding themselves into believing they do.

The protectors are the conspiracy theorists—the morally and intellectually superior, those who stand in solidarity against the evil "others." Protectors unite around shared ideas that enhance feelings of belonging and solidarity (Alcock, 2018, p. 193).[12] Conspiracy theorists believe they possess "elite knowledge" that enables them to see through the mirage conspirators present to the world. They can identify grand conspiracies and their collaborators with ease (Robertson & Dyrendal, 2019, p. 418), placing them in tight-knit in-groups of "woke" observers who see the world for what it really is, impervious to the deceptions and malicious intentions of the evil conspirators. From this naïve realist platform (Ross & Ward, 1996), the conspiracy theorists believe they're saviors whose mission it is to save the planet from the wicked intentions of conspirators (van Prooijen & Jostmann, 2013; Wood & Douglas, 2019, p. 252).

Interestingly, research on "pathological certainty," described by the neurologist Robert Burton, looked at participants with neurological disorders—such as schizophrenia and Cotard's delusions (the delusion that one is dead or doesn't exist)—finding that the feeling that one's judgment is correct provides emotional satisfaction that "floats free of rational processes and can occasionally become wholly detached from logical or sensory evidence" (Burton, 2008; Harris, 2011, p. 127). Given that participation in a community of like-minded individuals provides myriad social benefits from belonging to social identity, Burton's theory seems to explain the lure of conspiracy theory echo chambers. If Burton

> The key flaw in conspiracy thinking is that it assumes a level of competence and secret-keeping that has never happened in the history of the world.
> —Don Prothero and Tim Callahan

is correct, an emotional attachment to certain convictions, coupled with the illusion that one has arrived at those convictions through rational processes, further explains why conspiracy theorists maintain their beliefs with unwavering surety.

Finally, if you're not a conspirator or a protector, you fall into the last category of dupes and duds who, while not part of the conspiracy theory (not yet!), cannot see what plainly exists. The premise here is simple: if you're not part of the conspiracy and you still can't see it, you're a sucker that must be cared for and guided by the majesty of those who can truly "see" what's going on. Conspiracy theorists leave no room for intellectual objections to the "evidence" of a conspiracy theory or an honest and responsible investigation that disproves one. Anyone who cannot "see" a conspiracy that's widely accepted by conspiracy theorists can only be a fool. In other words, if you've seen the evidence for a given conspiracy theory and you reject that evidence on logical or evidentiary grounds, you are (a) blind, (b) dumb, or (c) a co-conspirator.

Piercing the Psychological Veil

Recall from chapter 8 that lacking control makes us think illusory patterns are real. This concept applies especially to conspiracy theories. The world we live in is extremely complex, and strange things occur regularly that are completely out of our control. Dividing the world into three groups, or reducing it to Barkun's Script, are simplistic schemas conspiracy theorists use to turn complexity into simplicity, ever the cognitive misers who would rather avoid the mental heavy lifting of figuring things out. This schema gives believers a sense of understanding the world, allows them to gain a sense of control over their environment, and offers them a positive role in their community (Douglas, Sutton, & Cichocka, 2017). For example, conspiracy theorists have claimed that governments put fluoride into the public's water supply to brainwash and subdue the populace (because the chemtrails just aren't enough) (Uscinski, 2019c, pp. 11–12). Thinking they're being victimized with toxic levels of fluoride, conspiracy theorists take control by deciding to drink only bottled water that's been manufactured by trusted vendors.[13]

On top of this entire argument being based on the false assumption that accepted levels of water fluoridation are harmful,[14] this sort of thinking merely provides an illusion of control. For instance, how could one reliably know whether the source they get their "safe" water from can really be trusted? If the conspirators really are as powerful as claimed, wouldn't the conspiracy theorists themselves be the targets? Manufacturing a water fluoridation controversy and spreading that misinformation throughout conspiracy theorist circles would be child's play. After all, conspirators' resources and reach are virtually limitless. So wouldn't it make sense to get people—especially conspiracy theorists—to buy bottled water from select vendors? This example shows that the decision to buy bottled water merely supports a self-proclaimed illusion that could easily be the product of manipulation by "them."

The terrorist attack of September 11, 2001, was an event that caused feelings of uncertainty and powerlessness. When living in a world where people flying airliners into buildings becomes reality, the idea that our lives really are in someone else's hands can be terrifying. In chapter 1, we saw that one way of approaching an incredible claim is to ask "What else must be true for X to be true?" One version of the 9/11 conspiracy theory holds that the Twin Towers were felled by carefully planted controlled explosives (Molé, 2006) (a contention that's been flatly rejected by controlled demolition experts [Shermer, 2011a, pp.

214–18]). Here we can ask, "*If* the Twin Towers were felled by controlled demolition, what else must be true?" Well, the plan would've had to be concocted and orchestrated by many thousands of people. Then it would've had to be kept airtight. Pilots would've had to be briefed, vetted, trained, and convinced to become martyrs.[15] The explosives would've had to be requested, manufactured, purchased, transported, and expertly planted with perfect precision and without being detected. Since the charges would've had to be precisely placed, the planes had to have hit their targets spot on without disrupting the electrical wiring. Then, the Bush administration, along with all the independent journalists, investigators, detectives, demolition experts, and anyone else who had to come up with an official version of the events, would've had to conspire to destroy evidence and lie to the public. Does this sound like a level of competence that any government in history has ever demonstrated?

Researcher Jennifer Whitson surmised that belief in 9/11 conspiracy theories could have begun because "even though we were told immediately that it was al-Qaeda [who took credit for the attacks!], there was a terrible uncertainty about the future, a sense of loss of control . . . leading to the search for hidden patterns, which the 9/11 'truthers' think they found" (Shermer, 2011a, p. 81). Whitson's research has demonstrated empirically that lacking control increases belief in conspiracy theories (Whitson & Galinsky, 2008). Others suggest that conspiracy theories help make sense of complex and destabilizing events (van Prooijen, 2011). Accordingly, turning to conspiracy theories in the wake of the 9/11 attacks is exactly what we should expect (Bartlett & Miller, 2011). Interestingly, this hypothesis remains true even if it turns out not that lacking control *increases* conspiracy theory beliefs, but that having control *reduces* conspiracy theory beliefs (van Prooijen, 2019, pp. 434–35).

Identifying and exposing grand conspiracies can be especially alluring for anyone whose psychological needs aren't being met. For instance, exposing a powerful government conspiracy can make one feel like a guardian who protects their people. It can also increase one's sense of self-worth, control, and self-esteem. Of course, this doesn't mean that all conspiracy theorists lack fulfillment or even that lacking fulfillment is a requirement for belief in conspiracy theories. However, it does suggest that people lacking fulfillment may more readily adopt a conspiratorial mindset than others whose emotional and psychological needs are satisfied.

Being a part of a community of like-minded conspiracy theorists can hugely impact psychological needs. Suppose a conspiracy theorist believes that an army of evil space aliens who have subdued mankind exists, making humans slaves to further the aliens' ends, and readily explains this idea to fellow conspiracy theorists. By sharing this information, the conspiracy theorist has satisfied several intellectual, emotional, and sociocultural needs. She has increased her sense of community by relating to others with shared interests, the need to help others and to make decisions has been satisfied by virtue of sharing the information, and she has reduced any fear, either imagined or real, by exposing the conspiracy theory, thus immunizing her community from its sinister control.

Some of our basic human needs include sociocultural needs like being part of a community and having relationships with others, emotional needs like happiness and self-acceptance, and intellectual needs like helping others and making decisions. To this we could add feeling safe, relaxed, and in control, since being unsafe, anxious, and helpless is psychologically straining, even for the most mentally capable. In fact, when people feel anxious, alienated, paranoid, or helpless, they're more likely to lean toward belief in conspiracy theories (Anthony, 1973; Goertzel, 1994a). So "saving" oneself and one's community from the evil intentions of powerful and malevolent conspirators has the added effect of creating a platform for channeling feelings of powerlessness and anger.

Making Connections

Seeing conspiracy theories lurking behind every event can also be described as a form of patternicity. Like everyone else, conspiracy theorists are susceptible to System 1 errors. Their brains are constantly on the lookout for patterns, which they invariably find, both real and imagined. When someone has a tendency to believe conspiracy theories, it isn't hard for System 2 to "connect the imagined dots between disconnected events and see an invisible hand operating behind the scenes" (Novella, 2012, p. 150). However, connecting dots alone isn't an adequate means of proving a claim, especially an incredible one. If an incredible claim has any chance of garnering support, each dot must be independently supported by credible evidence, and each link in the chain of inference must stand on its own by following the rules of evidence and logic (chapter 4).

System 1 oversees pattern detection, while System 2 oversees reality testing, and as we've seen, System 1 accepts everything veridically, while System 2 prefers it that way. A huge conflict of interest is thus created: both Systems want everything to be true, both will do as little work as possible, yet to determine *if* something is true, work must be done by the same Systems that don't want to do it. Even more alarming is the fact that not everyone's pattern recognition and reality testing capabilities are equal. What may seem strange to one person may seem perfectly normal to another. In the echo chambers of conspiracy theorists, skepticism is quickly washed away by the majority view.

Once information is gathered by System 1 and subsequently fed to System 2, it's processed veridically unless that information is obviously discordant with reality or, more accurately, with the brain's internal models of reality. Just like some arguments seem logical but are in fact fallacious, some conspiracy theories are not obviously discordant: they may even appear to align with the evidence, so whenever we hear one, a healthy habit is activating System 2 for further information processing. Question the source; question the source's motivations; question the source's credentials; question the evidence, including how it was obtained, how reliable it is, and where it came from; question the claim's plausibility; compare it to alternative explanations; ask "What else must be true?" The point is to try to identify any pseudologic within the argument, biases that might have influenced the source, or violations of GASPs, to evaluate the likelihood of the claim.

What You Get Is What You See

Thankfully, researchers have shown that Thinking Straight can reduce belief in conspiracy theories (Gervais, 2015; Swami, Voracek, Stieger, Tran, & Furnham, 2014). Furthermore, history has shown that when it comes to investigating conspiracies, the less incredible an explanation is, the more likely it is to be true (recall Occam's razor from chapter 5).

As the saying goes, seeing is believing. We use this expression—itself a kind of schema reflecting how we think the world is or should be—as a rational device, believing that demanding evidence before we're willing to believe something is the most rational course of action. Oftentimes it works: we believe the things we see and most of the time we get it right—or at least not wrong (there *is* a difference). However, it's also true—and in some instances it's even *more* true—that believing is seeing, meaning that once we have a prior disposition favoring a belief, we'll see confirmation of that belief in any information regardless of what that information is (Gilovich, 1991, pp. 29–72). This motivated reasoning even turns the absence of evidence—and even negative evidence—into positive evidence for a specific conclusion, a System 1 task unique to conspiracy thinking (Keely, 1999).

Recall the first principle of Barkun's Script, that nothing happens by accident. On this view, everything happens by design, usually by a controlling agent with malicious intentions. Situational factors that better explain the data are completely ignored. Psychologists call this the fundamental attribution error, which we first met in chapter 7. We see news footage of disaster refugees laughing with their children running around playing with other refugee children without blinking an eye. Indeed, sharing a hearty laugh with friends or family after a tragic event or the death of a close friend isn't uncommon at all. Just minutes after being sentenced to thirty to forty-five years in prison, I recall laughing at some stupid joke a buddy told me just after I returned. Through tears, I said, "You know that's funny if *I'm* laughing right now." Human behavior is unpredictable and complex, yet conspiracy theorists would have none of this. Rather than think that someone might have simply responded to a situation wrongly or inappropriately, conspiracy theorists refuse to acknowledge this possibility, instead attributing the response to deliberate intent.

President George W. Bush learned the news about the 9/11 attacks while he was reading to a class of kindergartners. Some say his reaction to the news showed he wasn't surprised, as if he was expecting it. The video is touted as evidence that 9/11 was a **false flag** operation, with proponents claiming that Bush knew about the attacks in advance. Brainless blunders, such as the following quote, only fueled the fire. During an interview, Bush was quoted saying, "Our enemies are innovative and resourceful; and so are we. They never stop thinking about new ways to harm our country and our people; and neither do we."[16] Combining cherry-picked information such as the kindergarten video with the enemies quote is poor thinking and conspiracy theory–mongering.

Another favorite video of conspiracy theorists shows Robbie Parker, a father of one of the children murdered in the Sandy Hook elementary school shooting in 2012, laughing just before going on TV and expressing heartbreak at the loss of his daughter Emilie. Conspiracy theorists claim that the Sandy Hook massacre was a false flag operation meant to advance gun restriction laws and that Parker was a **crisis actor**.[17] What conspiracy theorists are doing is drawing conclusions based on how they believe Parker *should* have reacted or behaved because of how they believe they would've acted or behaved in that situation. However, reality is far more nuanced and complicated. No one can know to any degree of certainty how they would've responded. What's more, psychologists have successfully demonstrated that people are terrible at making those sorts of judgments. Bush's reaction could just as easily be described as shock or confusion as lack of surprise, and we can't know what Parker was responding to just before going on camera. Perhaps Parker was riding an emotional roller coaster that day—not unlikely, given that his whole life and the small city he lived in had just been rocked—and was being polite to someone by offering a laugh in response to some consoling words. If a cameraman were recording me laughing moments after receiving a thirty- to forty-five-year prison sentence, would conspiracy theorists claim I was a crisis actor hired to further a "get tough on crime" agenda?

> The very lack of evidence is thus treated as evidence; the absence of smoke proves that the fire is very carefully hidden.
> —C. S. Lewis

Weighing the Evidence

One of the most important factors pertaining to conspiracy theories is weighing the evidence, an important instruction for a subculture in which the standards of evidence are absent. Evidence for conspiracy theories is also complicated by the built-in objection to the lack of evidence supporting the conspiracy theory. When asked for good evidence for a given conspiracy theory, conspiracy theorists will state, "Of course there's no evidence. The conspiracy is *that* powerful!" (Dennett, 1995, p. 349) By extension, this same framework explains why beliefs in conspiracy theories are unpopular: those in power "control the minds and memories of those who 'know too much'" (Barkun, 2013, p. 134). On the other hand, when conspiracy theorists have what they believe is good evidence, they proclaim, "Of course they did it. *Look at all the evidence!*" Paradoxically, conspiracy theorists disregard any evidence that doesn't fit into their conspiratorial schema. Any evidence unearthed from within conspiracy theorist echo chambers is ironclad, but any evidence unearthed by skeptics is disregarded. By now, this sort of immunity to contradictory evidence should sound familiar. This double-standard of evidence ultimately renders conspiracy theories a matter of faith.

Thus, the conspiracy theorists' (un)reasoning insulates any conspiracy theory from honest refutation, effectively severing it from objective reality (Delplante, 2011). Asking questions such as "Where does this evidence come from?" "Is the evidence reliable?" "Is there other supporting evidence?" "Are there better alternative explanations?" "Who is making the claim?" "What do skeptics think?" is an exercise of intellectual honesty, and these are all basic questions that anyone interested in the truth should be asking.

The history of conspiracy theories tells a vastly different story than conspiracy theorists would have us believe. When governments engage in conspiracies, the public eventually finds out. The media love an exposé of government wrongdoing because the public eats it up and the exposure boosts sales and ratings, in turn igniting a media frenzy to "get to the bottom" of the story. Even so, conspiracy theorists will claim to have mountains of evidence—even of "deep state" conspiracy theories—that the press refuses to release, reasoning that the press is either part of the conspiracy theory or afraid to report it, a strange position, given that the press has covered every confirmed conspiracy in history. Moreover, the government has an extremely poor track record of keeping secrets. For instance, keeping the Manhattan Project—one of the biggest conspiracies in U.S. history—a secret was of the utmost importance, yet the government still couldn't keep a lid on it; the project was nearly compromised by a felled balloon. The Watergate scandal was foiled by a security guard. A similar story can be told for other government conspiracies, such as the Bay of Pigs disaster, WikiLeaks, and the Edward Snowden disclosures.

Like those mentioned earlier, some conspiracies are real. However, due to the nature of conspiracies, the evidence proving them can be extremely hard to get ahold of—that is, unless you're a conspiracy theorist that's been endowed with magical conspiracy theory spectacles that read through deception and lies. The idea that conspiracy theorists have nearly unfettered access to top secret evidence that's unavailable to or ignored by everyone else leads to one of the biggest problems with conspiracy theories: the obsession with proving a conspiracy theory renders it immune to disconfirming facts, creating a hermetically sealed and mutually reinforced conspiracy theory schema. It's a convenient construct, where any evidence that doesn't support the theory is explained away as being part of a conspiracy to suppress it and any lacking evidence becomes part of the cover-up.

One grand conspiracy theory holds that hostile, space-alien reptilian shape-shifters have taken over our planet and reduced human DNA from twenty-four strands to two strands, to enslave us for nefarious purposes. These reptilians, as they're called, are so powerful that they can orchestrate and execute any event on massive scales. According to the conspiracy theory, reptilians are responsible for everything from child sex trafficking rings to terrorist attacks.

Popular conspiracy theorists who perpetuate this reptilian shapeshifter conspiracy theory (yes, I'm talking about you, David Icke) remain steadfast in exposing the reptilians and "waking up" the rest of us. Icke, for example, has been trying to expose the reptilians for years, though with minimal success and even less CT from his followers. For instance, how does Icke know reptilians manipulated our DNA? He isn't a biologist and has zero scientific credentials or training. Biologists have stated that even among animals we *know* were genetically engineered, there are "no sure marks" between natural and artificial selection (Dennett, 1995, pp. 315–18), and so they'd have a hard time proving or disproving this alien-engineering hypothesis. Does this make it a reasonable hypothesis? Absolutely not. As philosopher Daniel Dennett put it, "Prehistoric fiddling by intergalactic visitors with the DNA of earthly species cannot be ruled out, except on the grounds that it is an entirely gratuitous fantasy" (Dennett, 1995, p. 318). So what makes anyone think Icke is more capable than trained biologists?

You would think that reptilians capable of traveling millions of light-years across the universe and engineering the DNA of an unknown species to enslave said species would just vaporize Icke with ray guns. However, Icke, understandably, has bodyguards that protect him from these malicious and omnipotent shapeshifters. And why can't omnipotent, space-faring, shapeshifting reptilian alien slave-makers get past Icke's bodyguards? Well, if Icke really is *that* powerful, he must be an even more omnipotent, space-faring, shapeshifting alien slave-maker than the reptilians! Or he and his followers are immensely deluded, sucked into the black hole of anti-thinking.

The Grandest of Grand

Perhaps the biggest flaw in conspiracy thinking is what's referred to as cascade logic, or "widening the conspiracy." For conspiracy theories to "work," they must get bigger and bigger, expanding until they involve highly unlikely entities such as supernatural agents or omnipotent space aliens. One of the problems with widening a conspiracy theory is the more operators are involved, the bigger the conspiracy theory gets, the harder it is to cover it up and keep others silent. To demonstrate this phenomenon, we'll watch as our next conspiracy theory grows to the point of absurdity.

A substantial fraction of the U.S. populace believes that the U.S. government has been hiding evidence of space alien encounters for decades (Prothero & Callahan, 2017). In 2019, movements like #theycantstopusall and #stormarea51 tell us that the belief that our government is hiding evidence of alien contact isn't going away anytime soon. This conspiracy theory, then, must be cross-generational, spanning at least from before the time of the Lost Generation (1883–1900) up until the time of the Millennials (1985–2004). Thousands of people had to have knowledge about the conspiracy theory and have kept it a secret for upward of one hundred years, never telling their friends or families or being caught talking about it among themselves by bystanders.

If we assume aliens would have no reason to think the United States is special, it follows that other countries would likely have seen alien aircraft in their own skies. So aliens would probably have visited other countries as well as the United States and, like the United States, their governments, along with their thousands of cross-generational cohorts, have also been keeping this a secret for years. Are all the governments of the world harboring information about aliens that even the best spies haven't been able to "unearth" in the last hundred years? How likely is it that world governments are in collusion on the alien issue? Governments tend to disagree on many issues and have frequently resorted to blackmailing each other, threatening to release sensitive information. The late Stephen Hawking once observed that "if aliens are here, I suspect the newspapers would be full of the story. And if governments are involved in a cover up, they're doing a much better job at it than they seem to do at anything else" (Hawking & Tongue, 2010).

Our cover-up could go on and on, up until the point where we reach the aliens' boss and even creator of an alien matrix system that we've all been born (or programmed?) into. What can be invoked to "explain" the conspiracy theory is virtually limitless, since entities this powerful wouldn't be restricted by technology or resources. The problem is widening a conspiracy theory does nothing to prove its veracity; it simply renders the conspiracy theory immune to disproof. Widening a conspiracy theory is an example of trying to explain something that's complicated or difficult to explain by invoking something that's even more complicated and difficult to explain.

Manufactroversies

A *manufactroversy* is a certain type of conspiracy that's usually driven for political, financial, or ideological reasons. The term "manufactroversy" is a combination of "manufacture" and "controversy" and is used to describe a contrived disagreement where no disagreement among experts exists. Manufactroversies are typically perpetuated—or at least begun—by interested parties who rely on laypeople to keep the manufactroversy alive. For the most part, the purpose of a manufactroversy is to generate doubt and confusion in the public sphere, while creating the impression that some topic that is well settled is actually unsettled.

Like the slot machine that pays off enough to keep us pulling the lever, conspiracies happen often enough to keep people hooked. The terrorist attacks of September 11, 2001, and the London bombings of July 7, 2005, are good examples of conspiracies brought to fruition. Our prior beliefs can heavily influence what facts we accept about such big events, which is one reason skepticism is so important: beliefs don't change facts and, for some, facts only sometimes change beliefs. And as we'll see, when it comes to manufactroversies, a surprising number of conspiracy theorists completely miss the actual conspiracy and contrive some nonsensical conspiracy in its place (as just one example, Andrew Wakefield conspired to publish a fraudulent research paper linking the MMR vaccine to autism, yet conspiracy theorists missed this conspiracy and invented outlandish narratives instead).

> Those claiming harm where none exists are perpetuating harm where it does exist. Those standing up and proclaiming themselves to be champions of people are the ones killing them.
> —Kevin M. Folta

Claims made by conspiracy theorists cannot be dismissed as dumb or ill-intended: conspiracy theorists usually believe they have information about some critical truth, believing it would be unethical *not* to share that information. However, conspiracy theorists tend to get sucked in by thinking traps such as pseudojargon, cognitive biases, and logical fallacies. All too often, fallacious thinking leads to an inability to tell a conspiracy from a conspiracy theory, which, ironically, can result in the creation of whole new conspiracy theories and manufactroversies.[18] As the political scientists and conspiracy researchers Joseph E. Uscinski and Joseph M. Parent point out, Richard Nixon "was concerned with Jews, the intellectual elite, the media, and the anti-war movement," who he believed were conspiring against the United States. In response, Nixon conspired to tap the headquarters of the Democratic National Committee (Uscinski & Parent, 2014, pp. 14–15).

CONSPIRACY THEORY–INSPIRED BELIEFS

A modern manufactroversy with intuitive appeal (Blancke, Van Breusegem, De Jaeger, Braeckman, & Van Montagu, 2015) is the conspiracy theory that producers of genetically engineered crops (GEs) are involved in a massive conspiracy with governments and scientists to poison the populace using GEs (Lewandowsky, Gignac, & Oberauer, 2013; Specter, 2014). Research shows that conspiracy theorists typically reject scientific facts regarding GEs (Lewandowsky, Gignac, & Oberauer, 2013), instead believing that "the establishment" hides the evidence of their harms (Räikkä, 2009). Given the tens of thousands of food manufacturers that profit from unhealthy foods, this conspiracy theory seems particularly ill-informed.

Humans have genetically modified foods since the domestication of plants and animals nearly thirteen thousand years ago (Diamond, 1997), a process known as artificial selection. At a molecular level, scientists can't even tell the difference between an artificially selected gene and a naturally selected gene (Dennett, 1995, pp. 315–18). Given that "products derived from [field corn, canola, sugar beets, and soybean] end up in about 80 percent of grocery store products in North America" (Folta, 2019, p. 103), chances are every one of us eats at least one product made from GE crops every day. To use one example, broccoli, cauliflower, kale, Brussels sprouts, collard greens, and all the different vegetables we call cabbage have all been genetically modified: not a single one of them existed before humans modified the genes of their ancestor, the wild cabbage (Dawkins, 2009, p. 27). And the argument that so-called organic foods are better fares no better. Organic foods—that is, foods that are tasty, nutritious, do minimal damage to the environment, and contain no unnecessary contaminants—are usually hauled over long distances using vastly more fossil fuels in the process, making it more costly and less fresh than local produce. They oftentimes fail to use land efficiently, and, lacking synthetic pesticides, "organic" plants may produce more natural pesticides for self-defense that could be harmful (Hall, 2009a).

Interestingly, GEs are not only the most widely studied food in history, there is broad scientific consensus that GEs are (1) better for the environment (Hollingworth et al., 2003), (2) safe for human consumption (Domingo & Bordonaba, 2011; Fahlgren, Bart, Herrera-Estrella, Rellán-Álvarez, Chitwood, & Dinneny, 2016; Nicolia, Manzo, Veronesi, & Rosellini, 2013; Ronald, 2011), and (3) no more dangerous than non-GEs (Nye, 2015, pp. 297–321). In fact, of all the studies conducted on GEs, not a single one has ever found something harmful to humans or animals linked to its genome (Nicolia, Manzo, Veronesi, & Rosellini, 2013; Panchin & Tuzhikov, 2016; Van Eenennaam & Young, 2014). Thus, labeling foods "non-GMO" is like labeling medicine "untested for safety and efficacy." If GEs were responsible for the deterioration of human health, we should expect human lifespans to plummet beginning at the time GEs began being widely produced. In fact, the opposite is true. People live longer now than they have in any time in the past (Pinker, 2018, pp. 53–61). Simply put, there's no evidence-based reason to think GEs are less safe than non-GEs or that Big Agra is conspiring to poison people using GEs.

In point of concern, it is GE conspiracy theories, rather than GEs themselves, that are often responsible for the needless starvation, malnourishment, and even death of impoverished people. Countries around the world have placed bans and restrictions on GEs (Sifferlin, 2015; Whitty et al., 2013), even though what we know about GEs is far more than we know about other foodstuffs, and table 13.1 (Folta, 2019, p. 111) shows just how off the mark those who deny the science behind GEs are.

VACCINES CAUSE AUTISM

Our next manufactroversy is one in which proponents believe that vaccines cause autism. It began with a conspiracy involving shady scientists and lawyers, and continues in spite of overwhelming evidence to the contrary. Proponents of this manufactroversy have proposed three explanations for how vaccines allegedly cause autism: one, that the MMR vaccine damages one's intestines, allowing harmful proteins to enter the bloodstream and damage brain cells; two, that a mercury-based preservative called thimerosal is toxic to the central nervous system; and three, that receiving multiple vaccines in a short period overwhelms or weakens the immune system. Experts in the field have demonstrated that more than twenty studies have failed to support any of these claims (Gerber & Offit, 2009).

Conspiracy theorists have largely ignored the truth behind this manufactroversy, demonstrating the *disconfirmation bias*—a bias in which people set impossible standards of evidence for claims that contradict their beliefs (Edwards & Smith, 1996). Now, while opposition to vaccines has been around since the first vaccine was invented in 1796 (Offit, 2015, p. 109), Andrew Wakefield's fraudulent paper (chapter 3) is predominantly responsible for the vaccines-cause-autism manufactroversy (Burgess, Burgess, & Leasak, 2006; Hall, 2009b). This multilayered conspiracy has been thoroughly exposed by the investigative journalist Brian Deer (Deer, 2004; Deer, 2006; Deer, 2011). Through careful

Table 13.3. Table of Genetic Diversity

	Traditional Breeding	Hybrids	Polyploids	Mutation Breeding	Transgenic (GMO)	Gene Editing
Examples	Many	Field corn, some tomatoes	Strawberries, sugar cane, bananas	Pears, apples, barley, rice, mint	Corn, canola, soy, sugar beets, papaya	Many coming!
Number of genes affected	30,000->50,000	30,000->50,000	30,000->50,000 (times number of genomes)	No way to assess	1-3	1
Occurs in nature?	Yes	Yes	Yes	Yes	Some examples	No
Knowledge of genes affected	No	No	No	No	Yes	Yes
Environmental assessment?	No	No	No	No	Yes	Undetermined
Tested before marketed?	No	No	No	No	Extensively	Undetermined
Organic acceptable?	Yes	Yes	Yes	Yes	No	Undetermined, probably no.
Label wanted?	No	No	No	No	Yes	Undetermined
Adverse effects?	Yes	Yes	?	?	No	Not marketed
Time to new variety?	5–50 years	5–30 years	>5 years	>5 years	<5 years	<5 years

investigations, Deer showed that Wakefield, lawyers who funded his study, and parents who hired those same lawyers to sue the makers of the MMR vaccine began this manufactroversy because they believed, without evidence, that the vaccine caused autism. Even though almost no manufactroversy has been more thoroughly debunked than this one (see Offit, 2015, pp. 92–93, and cases cited at 238–39, 240–41), it persists nonetheless.

To be sure, vaccines have saved, and continue to save, hundreds of millions of lives (Pinker, 2018, p. 64; van Panhuis et al., 2013), and although no vaccine is *perfectly* safe, and vaccine safety is continually monitored (Chen et al., 1994), adverse effects are "extremely rare" (Maglione et al., 2014). Without vaccines, people would be dying by the millions of completely preventable diseases. Before their respective vaccines were introduced to the United States, every year measles infected up to four million children and killed at least 500; pertussis (whooping cough) infected about 300,000 and killed around 7,000, almost all young children; hepatitis B infected around 200,000 and killed around 5,000; rotavirus infected nearly a million and killed about 60 (in the developing world, rotavirus kills about 2,000 children every day); chicken pox hospitalized around 10,000 and killed about 70; influenza hospitalized up to 200,000 and killed 100; lastly, it is estimated that around 500,000,000 people died from smallpox, a disease that has been eradicated by a vaccine (Offit, 2015, pp. ix, xviii, 64, 71–72, 75, 76, 106).

The beliefs of others can have serious impacts on our health and can even threaten the lives of our children. Because of the vaccines-cause-autism manufactroversy, parents began refusing to vaccinate their children out of fear of putting them at risk for autism. When a significant percentage of a population is immunized from a disease, it creates what researchers call *herd immunity*. Herd immunity acts like a protective shield, guarding populations from diseases. You can think of herd immunity like an elephant herd that surrounds a calf to protect it from predators: the more members of the herd that can surround the calf, the better the calf is protected from danger. For some—such as children too young for vaccinations, the medically exempt, transplant recipients, the immunocompromised, chemotherapy patients, or those for whom vaccinations didn't work—vaccines offer no protection. Just as the number of elephants needed to protect a calf varies depending on the level of threat—one matriarch is sufficient to ward off a single lion, whereas an entire herd would be needed against the threat of an entire pride—the percentage of a human population that must be immunized to protect everyone depends on how contagious the disease is (Offit, 2015, p. xxiii). For example, measles is highly contagious, so herd immunity must be at 95 percent or better. Mumps is less contagious and so requires 85 percent immunization or better.

The more people who refuse to vaccinate their children, the less effective a community's herd immunity, the more people are put at risk of contracting and dying from preventable diseases. Such is the situation with reported cases of measles in England and Wales, for example, where immunization rates dropped from 93 percent to 75 percent and diagnoses have increased from 56 in 1998, to 1,348 in 2008, killing two children in the process (Hall, 2009b; Mele, 2017; Randi, 2017b; Sun, 2017). The number of children's deaths due to diseases that were once nearly eradicated is also on the rise. In 2015, a measles outbreak that began at Disneyland in California was held responsible for infecting 147 people, 48 percent of whom were unvaccinated and 38 percent of whom didn't know about their vaccination status (Centers for Disease Control and Prevention, 2015a; Centers for Disease Control and Prevention, 2015b). Rather than protecting their children from autism, parents who refuse to vaccinate their kids are increasing their children's risk of dying from completely preventable diseases (Offit, 2014, p. 134).

HUMAN-CAUSED GLOBAL CLIMATE CHANGE

Another deadly serious manufactroversy is the conspiracy, spearheaded by prominent scientists Fred Seitz and Fred Singer (the Freds), that anthropogenic (human-caused) climate change isn't real. Believers in this conspiracy theory believe that "Big Oil" traffics in information suppression campaigns, suppressing information that could jeopardize their profits, such as "free energy" or water-to-gas technology. Energy conglomerates are indeed highly motivated to conceal information that could jeopardize their profits, but they aren't concerned with unscientific ideas like free energy or perpetual motion machines. The biggest threat to their bottom line is the truth behind climate change.

According to climatologist and geophysicist Michael Mann, this manufactroversy "constitutes an even greater crime against humanity than the tobacco industry's campaign to deny the health effects of tobacco" (Mann & Toles, 2018, p. 40; Union of Concerned Scientists, 2007). Indeed, their tactics were strikingly similar and even involved some of the same conspirators. While our planet is getting warmer, the overwhelming majority of experts agree that such warming cannot be explained by the normal cycles of planetary warming and cooling (Cook, 2010). In fact, the evidence for anthropogenic climate change is so overwhelming (IPCC, 2021) that even once ardent deniers now admit it. And, despite the efforts of the Freds, overwhelming scientific consensus shows that humans are responsible (Cook et al., 2016; Oreskes, 2004). Nonetheless, this manufactroversy has become immensely political (Mann & Toles, 2018,

p. 53; Parsons, 2019).[19] One survey showed that 73 percent of Democrats, 49 percent of independents, and 28 percent of Republicans agreed that greenhouse gas emissions cause global warming (Gronewold & Marshall, 2009). Other research (Alcock, 2018, p. 179) supports the "cultural cognition" hypothesis, which states that people's views on climate change will tend to reflect that of their social groups, regardless of scientific education. The politicization of climate change has likely done more harm than good (Nisbet, 2020).

The climate change manufactroversy has proceeded in several stages (Mann & Toles, 2018, pp. 53–67), each one eventually folding under the weight of the evidence. It began with the conspirators pushing denial and doubt, as a street dealer pushes meth and heroin. When those arguments no longer worked, the argument shifted to claiming that, while the planet is heating, the heating is natural and not caused by humans. After those arguments were destroyed, deniers claimed that the planet would fix itself. From there, deniers tried arguing that humans would adapt to the problem—and that climate change might even be good for us! The latest strategy is to claim that even though the planet is warming and humans are causing it, fixing the problem is too complicated and too expensive, a claim that's been widely rejected by experts in the field (Mann & Toles, 2018, pp. 64–67; Pinker, 2018, pp. 136–54).

During the peak of the climate change manufactroversy, the Freds conspired to put pressure on the media to cover the manufactroversy as if there were two equally legitimate "sides"; they exploited loopholes in the law, misrepresented the facts, and conspired with lobbyists, think tanks, and public relations firms to fight the facts of climate change (Kenner & Robledo, 2014; Oreskes & Conway, 2010a; Oreskes & Conway, 2010b). At one time, the Freds were prominent scientists, though neither had any credentials in environmental or health sciences (which would qualify their opinions on climate change).

Worse than having no relevant credentials, the Freds (and others) supported other manufactroversies where big business interests were threatened by science. They argued that chlorofluorocarbons (CFCs) were not damaging the ozone layer, that smoking doesn't cause cancer, that secondhand smoke doesn't cause cancer,[20] that asbestos is not harmful, and that the insecticide DDT wasn't harming the environment. The Freds derailed the facts on nearly any issue where corporate interests conflicted with scientific data. As American historians of science Naomi Oreskes and Erik M. Conway put it, "In case after case, [the Freds and their conspirators] steadfastly denied the existence of scientific agreement, even though they, themselves, were pretty much the only ones who disagreed" (Oreskes & Conway, 2010b, p. 7).

Wrapping Up

Even though conspiracies exist without violating scientific facts or the rules of evidence or logic, conspiracy theorists regularly make claims that do violate evidence and logic to make a conspiracy theory fit their preferred narratives. That conspiracies violate the rules of logic and proof renders them incredible claims that should be approached skeptically. Believing conspiracy theories and belonging to groups that form around conspiracy theories satisfy many of our psychological needs. This fulfillment combines with a myriad of cognitive biases and fallacious thinking in a witches' brew of flaw and fantasy that renders conspiracy thinking impossibly immune to any countervailing evidence.

Ultimately, conspiracy theories are about challenging an "official" version of events. Throughout history, challenging authority has been one of the most effective drivers of progress, but uncritical challenges can backfire, as when we challenge accepted facts as if those facts belong to "the establishment." The idea that scientific facts are untrustworthy is currently widespread, and because of how dangerous conspiracy thinking can be, it's our responsibility to Think Straight and help remove others from the cognitive trenches they've dug themselves into.

Acknowledgments

As a first-time author, realizing the number of people who contribute to writing a book has been eye-opening. Writing one from prison, where resources are scarce and even some research material is considered "contraband" by the designated authorities, especially increases the number of people involved. During much of the time I spent on this project, I was earning less than $50.00 a month and could only access the internet and use e-mail vicariously. It is for this reason that I owe my sincere gratitude to the many people who helped me navigate through the difficulties of making this project a reality. In acknowledgment sections, authors often make the claim that the book could not have been brought to fruition without the help of their supporters. In this instance, the sentiment could not be truer.

Without so much as a single complaint, hesitation, or moment of procrastination, David Reid bought me so many books, sent so many e-mails on my behalf, relayed so many messages, looked up so much information on the internet, and helped battle the prison administration when necessary that this project would've been dead in the water without his unfaltering help and friendship. Similarly, one of my most loyal friends, Janice Rice, was also instrumental in helping me get books and other research materials. The same goes for my good friends Matt Summers and Matthew Shaw, and for my wonderful mother, Janet Marshall, who all helped with the constant supply of books; research materials; e-mails; internet searches; finding, printing, and mailing scholarly papers; and the encouragement I needed to keep at it.

I also owe a huge thank you to Jake Bonar, Jonathan Kurtz, Nicole Carty, and everyone else at Prometheus Books for taking interest in this project, tirelessly fixing my mistakes, and working with me on deadlines to give it the final polishing touches it desperately needed. I've been over the manuscript so many times that any errors or omissions herein are strictly mine.

Professor Robert Blaskewicz of Stockton University was the first academic who accepted my invitation to look over this project. After committing to reading one chapter and providing some feedback, he ended up going over several drafts of the manuscript. Professor Blaskewicz is also responsible for much of why this book even exists. His style of pedagogy, breadth of knowledge, willingness to contribute, and critical examination of my work were huge contributing factors to making the book possible. From the start, Professor Blaskewicz provided extremely valuable feedback, and his ability to kindly offer criticism and tell me when I'm dead wrong is as impressive as it is appreciated.

Another big help was the late Richard Ford, professor emeritus of the University of Texas–El Paso, who read the entire manuscript twice over and select chapters several times over. Professor Ford's mastery-level understanding of English, grammar, and linguistics, along with his countless tips, pointers, and much needed feedback and criticism, was instrumental in the development of the manuscript. His patience, feedback, and willingness to provide in-depth explanations on grammatical and linguistic questions really improved the quality of both my writing as well as this book.

In no particular order, I would also like to thank Professor Bob Baugher of Highline Community College, Professor Emeritus Christopher French of the University of London, Dr. Elizabeth Loftus of the University of California–Irvine, Professor Clyde Herreid of the University of Buffalo, Professor Emeritus Ted Goertzel of Rutgers University, Professor Robert Colter of the University of Wyoming, Professor Kevin Folta of the University of Florida, and Professor James Caldwell of the University of Wyoming for reading draft versions of the entire manuscript or select chapters and keeping me motivated with their feedback and encouragement.

Perhaps a bit surprisingly, and a little begrudgingly, I also owe the Wyoming Department of Corrections a shout-out for having more research tools available for the prison population than many prisons around the country offer.

Finally, I want to thank Professor Steven Novella of Yale University School of Medicine. While not directly involved in this project, Professor Novella's lecture series *Your Deceptive Mind: A Scientific Guide to Critical Thinking Skills* was the impetus behind getting this book off the ground, and if I had never watched that lecture series, this book would most assuredly not have been written.

Notes

FOREWORD

1. Lane Cuthbert and Alexander Theodoridis, "Do Republicans Really Believe Trump Won the 2020 Election? Our Research Suggests That They Do," *Washington Post*, January 7, 2022, https://www.washingtonpost.com/politics/2022/01/07/republicans-big-lie-trump/.

2. "COVID-19 Vaccinations in the United States," Centers for Disease Control and Prevention, https://covid.cdc.gov/covid-data-tracker/#vaccinations_vacc-total-admin-rate-total.

3. "Short-Term Energy Outlook," US Energy Information Administration, https://www.eia.gov/outlooks/steo/report/global_oil.php.

CHAPTER 1. MEET SKEPTICISM

1. It may be the case that teaching children CT has a number of positive effects on cognition and behavior (Trickey & Topping, 2004). Trickey and Topping found that philosophy classes for children can improve several aspects of their academic and social lives, including improving thinking, reasoning, listening, and language skills; raising self-esteem; reducing anger; increasing group interactions; and decreasing negative comments and interactions.

2. Shermer defines "weird things" as "(1) a claim that is unaccepted by most people in a particular field of study, (2) a claim that is either logically impossible or highly unlikely, or (3) a claim for which the evidence is largely anecdotal and uncorroborated" (Shermer, 2013, p. 6).

3. The popular saying "Extraordinary claims require extraordinary evidence" is normally attributed to the science popularizer Carl Sagan, although Sagan merely popularized the phrase, first in his 1979 book, *Broca's Brain: Reflections on the Romance of Science*, and later in his 1980 hit TV series, *Cosmos*. While the saying was introduced by Marcello Truzzi—also known as the skeptic's skeptic—who used the phrase "Extraordinary claims require extraordinary proof" in a 1975 *Parapsychology Review* article, the origins of the phrase trace back to the Scottish philosopher David Hume in his work *On Miracles* (Deming, 2016; Pigliucci, 2005).

4. See Harrison (2012, pp. 28–29) for more scientific ideas that were once thought absurd.

5. Epictetus's real name is lost to history. Epictetus means "acquired," conveying that he was a slave (Pigliucci, 2017a).

6. Petty et al. did, however, find that participants who were instructed to mitigate their biases were able to do so, supporting my contention that CT is imperative for accurate decision making. Thus, realizing our biases motivates us to think more deeply about our stance by searching for accurate information.

7. Edzard Ernst notes that the concept of evidence-based medicine is a new arrival, being only about twenty years old (Ernst, 2018, p. 18). In his 1975 book, *Against Method*, philosopher of science Paul Feyerabend doesn't even recognize medicine as a legitimate scientific discipline.

8. Scientists still don't know how much of each vitamin we need for optimal health, how much we need to survive, or what the consequences are for mild vitamin deficiencies. Given the lack of scientific knowledge, it makes little sense to believe that an unregulated industry with massive financial incentives knows better (Hall, 2022).

9. In the 1970s, the CIA actually conducted research on psi phenomena in an operation known as Project Stargate. Project Stargate cost the American taxpayers over $20 million, and the only result was evidence that psi doesn't exist.

10. Logical reasoning is a skill to be practiced; it isn't something the brain does on its own, like remembering or processing information. Stimulating the temporal lobe with an electrode doesn't produce thought, although it can cause one to remember something like a name or a house one used to live in. (See Burton, 2008, p. 233, n. 3.) No one is born a professional violinist or basketball player. Those are learned skills that require know-how and practice, just as Straight Thinking is a learned skill that requires know-how and practice.

11. Epicurus. (2019, February/March). Selections from the letter to Menoeceus. *Free Inquiry, 39*(2), 47–48.

12. As Shermer puts it, "Smart people believe [incredible claims] because they are skilled at defending beliefs they arrived at for nonsmart reasons" (Shermer, 2011, pp. 35–36).

CHAPTER 2. A PHILOSOPHICAL DETOUR

1. This hypothetical study has actually been carried out and widely criticized for suffering from methodological weakness and even fraud (see Hood, 2009, p. 165).

2. The British philosopher Julian Baggini defines faith as "the holding of a belief which cannot be rationally justified" (Baggini, 2003, p. 212). For our purposes, we use Boghossian's definition, although Baggini's works just as well.

3. A personal favorite: "This next sentence is true. The previous sentence is false."

CHAPTER 3. PIERCING THE SCIENTIFIC VEIL

1. Senator Inhofe is the same man who, despite the majority of prisoners at Abu Ghraib having been arrested arbitrarily or for minor crimes, said they deserved to be tortured because "they're murderers, they're terrorists, they're insurgents. Many of them probably have American blood on their hands." Quoted in Tavris & Aronson, 2019, p. 264. Tavris and Aronson also note that a number of military intelligence officers told

the International Committee of the Red Cross that some 70 to 90 percent of Iraqi prisoners had been mistakenly arrested.

2. Popper observes that Einstein's theory of the speed of light, as well as Heisenberg's uncertainty principle, are both falsifiable. Just because the theories forbid faster-than-light travel or knowing a particle's momentum and velocity simultaneously doesn't forbid us from seeking evidence that would falsify either theory (Popper, 1959/2014, pp. 249–50).

3. Negative research findings receive more criticism and are less likely to be recommended for publication.

4. Throughout the coronavirus pandemic, Dr. Ioannidis himself began publishing nonsense research and personal opinions that others have painstakingly debunked (Gorski, 2021a; Gorski, 2022).

5. A dropout rate refers to a specific sampling bias that, if greater than 10 to 20 percent, can seriously compromise the internal validity of a study.

6. Participant demand is closely related to another bias known as the *Hawthorne effect*. The Hawthorne effect explains why people behave differently when they know they're being observed (for example, you work harder when your boss is watching, you make your presence a little more known when someone attractive is watching).

7. The late physicist Victor Stenger rejects the p-value of 0.05, arguing that in physics, where researchers aren't under pressure to publish findings that might save someone's life, nothing short of $p = 0.0001$ [1 in 10,000] suffices (Stenger, 2012, p. 242).

8. See also the list of studies at Offit, 2015, pp. 238–39, 240–41.

CHAPTER 4. THE LANGUAGE OF LOGIC

1. Federal Rules of Evidence, Rule 102.

2. Strictly speaking, a theoretical explanation, that is, a scientific explanation, *is* meant to show that something is or is not the case, rather than why it is or is not. The theory of heliocentrism, for example, shows *that* the planets in our solar system orbit the sun, thus demonstrating *that* geocentrism is wrong.

3. I am, of course, only using capital punishment as an example on how an argument is structured. For a concise treatment of the relationship between capital punishment and murder rates, see Goertzel & Goertzel, 2008.

4. Bacon believed so strongly in the power of observation that he died of pneumonia while studying how freezing can preserve meat (Stenger, 2012, p. 80).

5. He made this argument in *A Treatise of Human Nature* (1739), book 1, part iii, section 6, as well as *An Enquiry Concerning Human Understanding* (1748), section 4.

6. http://www.cbsnews.com/news/once-a-criminal-always-a-criminal.

7. After detailing the history of inductive/deductive argument, Shermer narrowly misses the opportunity to bring this issue to modern thinking (Shermer, 2011a, pp. 295–96).

CHAPTER 5. BUILDING YOUR SKEPTICAL TOOL KIT

1. Dobelli used a man driving across a bridge.

2. Each of these examples can also be understood as cognitive dissonance reducers.

3. The word "astronomy" comes from the Greek words *astron*, meaning star, and *nomos*, for law. Thus astronomy is "the law of the stars."

4. Source: http://www.quora.com/if-a-neuron-were-a-galaxy-how-would-the-num ber-of-atoms-in-it-stack-up-to-the-number-of-stars (accessed November 14, 2018).

CHAPTER 6. SELF-DECEPTION, PART 1—MEMORY

1. Rolf Dobelli uses a similar metaphor borrowing from George Orwell's classic novel *1984*. In Dobelli's version, Winston Smith, the novel's protagonist, is tirelessly working in our heads, revising the past and constructing our beliefs so that they agree with new developments. And he does it all outside of our awareness so that as old mistaken beliefs vanish, this "falsification of history" allows us to live under the illusion that our beliefs were right all along. See Dobelli, 2014, pp. 233–35.

2. So-called criminal profilers also exacerbated the situation and led to the delayed arrest of the killers. Several profilers wrongly guessed the killers' age and race and even how many gunmen there were. For a detailed analysis of the myth of the effectiveness of criminal profiling, see Fox & Farrington, 2018; Snook, Eastwood, Gendreau, Goggin, & Cullen, 2007.

3. Even the oft-repeated phrase "false memory *syndrome*" can contaminate the judicial process by giving false memories the imprimatur of an official medical disorder. Wiktionary defines "false memory syndrome" as "a condition in which a person's identity and relationships are affected by memories which are factually incorrect but strongly believed." Despite the popularity of the term, false memory syndrome isn't a real clinical diagnosis. No memory experts recognize false memory syndrome as a diagnostic label. False memories are the rule, not a condition.

4. In a particularly appalling example, after being accused of molesting his daughter, Paul Ingram started believing he was guilty of any and every absurd accusation leveled against him, including forcing his son and daughter to have sexual intercourse. For a detailed description of the Ingram case, see Loftus & Ketcham, 1994, pp. 226–63.

5. And see http://www.ojp.usdoj.gov/nij/pubs-sum/178240.htm

6. Here we can push our Street World metaphor a little further. While the content will invariably differ, a street in Boston looks pretty much the same as a street in Cleveland. Similarly, one memory to the next will differ in content, but physiologically, false memories might be nearly indistinguishable from real memories (Loftus & Ketcham, 1994, pp. 165–66).

7. Schacter notes that the brain regions that store episodic memories aren't fully developed until around the age of five (Schacter, 2002, pp. 126–27). The experimental psychologist Bruce Hood, however, cites multiple studies suggesting that long-term memory begins before birth (Hood, 2009, pp. 90–94).

8. The Harvard professor of psychology Richard McNally calls the repressed memory movement a "pernicious bit of psychiatric folklore" (McNally, 2004).

9. In response to the wave of people being exonerated by DNA evidence, rather than admit that their beliefs about their suspects were wrong, the time and resources spent obtaining the conviction was spent sending an innocent person to prison, or that they had been incapable of finding the guilty party, prosecutors fought the evidence as if they

couldn't have made these mistakes. As Tavris and Aronson have shown, a prosecutor's reluctance to admit fault results from attempting to reduce cognitive dissonance (Tavris & Aronson, 2019, p. 193).

10. Tavris and Aronson note that the National Registry's list shows around 13 to 15 percent of prisoners who were "unequivocally exonerated" had confessed to crimes they didn't commit (Tavris & Aronson, 2019, pp. 182–83).

11. Of course, brains are not *fully developed* until we reach our twenties.

12. One study of actual police interrogations showed that at least one in every six questions used by police was suggestive (Fisher, 1995; Schacter, 2002, p. 115).

CHAPTER 7. SELF-DECEPTION, PART 2—BELIEF

1. That beliefs can be reduced to electrochemical reactions invites the philosophical question, "Does a belief exist if one is not thinking about it?"

2. Strictly speaking, this approach to morality is known in philosophical circles as *consequentialism*, a branch of utilitarianism that was founded by philosopher Jeremy Bentham (1748–1832) and later refined by John Stuart Mill (1806–1873). Utilitarianism's priority is maximizing happiness for the greatest number of people. That is, an action is moral so long as it maximizes the happiness of more people than not taking the action would. The problem here, however, is that ends don't always justify means. Despite Robin Hood's valor, stealing from the rich and giving to the poor is still stealing. Deontologists hold that means are just as important as ends. It's important to protect your family and help provide the necessities of living, but it'd be highly immoral to protect your son by stomping out the fourth-grade bully on your way to robbing a furniture store for his bed.

Consequentialism, however, isn't the only route to a consequentialist morality. In his book *The Science of Good and Evil*, Michael Shermer (2004) lays out what he calls "provisional ethics," which encompasses three overarching principles: (1) the ask first principle: to find out whether an action is right or wrong, ask first; (2) the happiness principle: it is a higher moral principle to always seek happiness with someone else's happiness in mind and never seek happiness when it leads to someone else's unhappiness; and (3) the liberty principle: it is a higher moral principle to always seek liberty with someone else's liberty in mind and never seek liberty when it leads to someone else's loss of liberty.

3. Whitson and Galinsky define illusory patterns as "the identification of a coherent and meaningful interrelationship among a set of random or unrelated stimuli (such as the tendency to perceive false correlations, see imaginary figures, form superstitions, rituals, and embrace conspiracy beliefs, among others)." Their paper also points to research indicating that a strong desire for control is associated with distorting reality.

CHAPTER 8. SELF-DECEPTION, PART 3— PERCEPTION AND PATTERN RECOGNITION

1. A demonstration of the color phi phenomenon can be seen at www.yorku.ca/eye /colorphi.htm

2. https://www.testreligion.com/solarphotos.html

3. By "everyday illusion," Chabris and Simons mean illusions people experience daily, yet know little or nothing about.

4. According to philosopher of science Karl Popper, "There is no such thing as a logical method of having new ideas, or a logical reconstruction of this process" (Popper, 1959/2014, p. 32). Kant used the word "subjective" to denote the feeling of certainty (p. 44).

5. Some research has suggested that we are in fact born with size-distance constancy (see e.g., Slater, Mattock, & Brown, 1990).

6. Burton compares mishearings to a speech recognition program. If you say something into the program, such as "He's a wolf in cheap clothing," the program transcribes it as "He's a wolf in sheep's clothing" based on what's stored in its database; in this case, the saying "wolf in sheep's clothing." People, however, home in on environmental cues, such as body language or inflection, to help us decipher everyday language. As it applies to this chapter, it isn't difficult to see—pun intended—how easily we can misinterpret audio stimuli and later "remember" that we heard something we didn't.

7. Experiments conducted on humans have shown results eerily similar to those of the pigeon studies (Shermer, 2011a, pp. 65–67; Vyse, 2014, pp. 87–90). Researchers have found similar pattern-seeking behavior in other animals as well (Vyse, 2014, pp. 83–84).

CHAPTER 9. INNUMERACY AND THE DECEPTION OF COINCIDENCE

1. We'd be equally shocked if the day after our dream, a relative, classmate, coworker, acquaintance, teammate, or anyone else we knew died in a plane crash.

2. Colloquially, the phrase "Everything happens for a reason" usually implies some cosmic meaning behind otherwise random events. Indeed, the phrase is true in a strict sense: everything *does*, in fact, happen for some reason, and, as the popular T-shirt says, that reason is usually physics. Thus, the true reason is not the one we are tempted to assume, given our tendencies to assign meaning and agency to random events. Psychologists call the tendency to think everything happens for a reason *global coherence*.

3. Evidence suggests that new learned behavior spreads throughout monkey colonies more quickly when it is introduced by a dominant animal (Cialdini, 2007, p. 290, n.4).

4. I watched many of these tricks firsthand while myself and a team of guerilla skeptics investigated the self-proclaimed psychic medium Thomas John (Gerbic, 2020).

5. A failed cold reading by self-proclaimed medium James Van Praagh is available at http://youtu.be/to7UzTcApMl

6. Apparently the healing powers of Lourdes were saved for others. That fourteen-year-old girl who supposedly received the vision from the Virgin Mary, one Bernadette Soubirous, was herself immune to the miracles; she died at thirty-five from tuberculosis. Nevertheless, in 1933, Pope Pius XI officially declared Soubirous a saint.

7. The nineteenth-century playwright George Bernard Shaw said about Lourdes, "All those canes, braces, and crutches, and not a single glass eye, wooden leg, or toupee" (Offit, 2014, p. 33).

8. Wikipedia. (2018). *Probability*. Retrieved from https://en.wikipedia.org/wiki/Probability

CHAPTER 10. PRACTICALLY THINKING

1. Feyerabend argued that the process of scientific discovery is ultimately anarchic. Throughout the history of science, scientists have used both sound and unsound methods to pursue their preferred theories, including many unexpected means of "discovery" such as biases, ignorance, and, in Feyerabend's words, "pigheadedness."

2. The metaphor of rewinding the tape of history comes from Gould, S. (2007). *Wonderful life: The Burgess Shale and the nature of history.* New York: W. W. Norton.

3. The idea of looking at a paradigm as a map here is from Beckner, 2019. See also Schick & Vaughn, 2005, pp. 105–6.

4. Wikipedia. (2018). *Planck's principle.* Retrieved from https://en.wikipedia.org/wiki/Planck%27s_principle

5. Avery, McCarty, and MacLeod were, of course, following the research of Gregor Mendel (1822–1884), whose research on genetic inheritance was recognized posthumously. Likewise, Francis Crick and James D. Watson have become household names for their paper describing the double helix structure of DNA, although they were not the first to discover it. Their work built off that of others, most prominently the English chemist Rosalind Franklin (Mukherjee, 2016).

6. Skeptics have devised numerous methods. For instance, Jonathan Smith uses what he calls the FEDS Standard. Smith's approach asks us to look for Fraud, Error, Deception, and Sloppiness while investigating an incredible claim (Smith, 2018). Brian Dunning, host of the podcast *Skeptoid*, uses a similar approach that he calls the Three Cs: challenge, consider, and conclude. Dunning begins by challenging the claim being asserted. Does the claim contradict what we know about the laws of nature? Does it defy logic? Do we have to make any new assumptions about what we know about the world for the claim to be true? Like the SEARCH method, Dunning then considers alternative explanations. Are there competing explanations? Does something more prosaic explain the phenomenon? Do competing explanations better explain the data? Finally, Dunning reaches a conclusion based on an assessment of the first two criteria (Clark, Dunning, & Johnson, 2017; Dunning, 2008). Another skeptic, Melanie Trecek-King, devised the FLOATER model, which stands for Falsifiability, Logic, Objectivity, Alternative Explanations, Tentative Conclusions, Evidence, and Replicability (Trecek-King, 2022b). When used properly, the FLOATER model is like a lifesaving device that can save us from drowning in a sea of misinformation.

7. For Darwin's theory to work, the geologic column must show a progression from simple organisms in the lower layers of rock (strata) to more complex organisms in the higher strata. For example, mammals first appeared in the geologic column about 225 million years ago (MYA). If a fossil of, say, an African lion were found in strata 300 MYA, Darwin's theory could not explain this fact and thus would be a much less fruitful theory.

8. The Dutch sea captain Willem de Vlamingh was the first to "discover" black swans during a rescue mission to Australia in 1697. Because of the large number of black swans he found at a particular river site, he named the river Swan River.

CHAPTER 11. THE FRINGES OF SCIENCE

1. The term "instrumental trans communication" (ITC) is used to refer to the use of electronic equipment for communicating with incorporeal entities.

2. Dallow, A. (Director). (2009, December 28). Mini-myth mayhem [Television series episode]. In S. Christiansen (Producer), *Mythbusters*. Silver Spring, MD: Discovery Channel.

3. For a comical list of spurious correlations, see https://www.tylervigen.com/spurious-correlations.

CHAPTER 12. STANDING IN SCIENCE'S WAY

1. Who the saying is credited to is a matter of dispute. Barnum's biographer, who finds that no evidence that the term "sucker" was used as a derogatory term in Barnum's day, credits the saying to an 1880s con man named Joseph Bessimer (Hines, 2003, p. 64).

2. Meehl's work is what inspired Kahneman and Tversky's work on heuristics. Recognizing the power of popularity, Barnum is also credited with saying, "Nothing draws a crowd like a crowd."

3. https://en.wikipedia.org/wiki/Hollow_Earth?wprov=sfla1

4. Of course, Darwin was not the first to propose the idea of evolution. Darwin's contribution to the evolutionary debate was his paradigm-shifting, epoch-making theory of natural selection.

5. The reader will note that these are ad hominem attacks on Darwin's character, not legitimate attacks on his research.

6. Gregory traces the widespread rejection of Darwin's theory by intellectual scientists and theologians to the modern synthesis and Theodosius Dobzhansky's 1937 book *Genetics and the Origin of Species*.

CHAPTER 13. CONSPIRACIES—WHAT "THEY" WANT YOU TO THINK . . . MAYBE

1. I'm reminded of Brandolini's law, also known as the Bullshit Asymmetry Principle, which states that the amount of energy needed to refute bullshit is an order of magnitude bigger than that needed to produce it (Williamson, 2016).

2. Using the word "theory" or its variants in this chapter doesn't comport with my definition of theory from chapter 1. However, here I use the conventional phrase "conspiracy theory" as defined in the vocabulary builder.

3. An attempt to establish criteria for implausible conspiracy theories has been made (Brotherton, 2013). Brotherton's criteria state that an implausible conspiracy theory: (1) is unverified; (2) is less plausible than mainstream explanations; (3) is sensational; (4) assumes everything is intended; (5) assumes evil intent; (6) has low standards of evidence; and (7) is self-insulating.

4. It's worth noting that this study has been criticized for assuming that a *suspicion* of a conspiracy is the same thing as a *settled belief* that a conspiracy happened (Räikkä & Basham, 2019, p. 186, n. 16). However, the questions in the Wood et al. study were not framed in such a way to assume the participants were merely suspicious; e.g., the questions were framed as "Princess Diana *faked* her own death," rather than "*I believe* Princess Diana faked her own death."

5. Of course, if Conspiracy City was built *entirely* upon the premise that the U.S. government is engaging in totalitarian or New World Order conspiracy theories, the premise would, if accepted, effectively deconstruct Conspiracy City.

6. Others have suggested that conspiracy theories are memes (Goertzel, 2011).

7. As many have pointed out, narrowing down a definition of conspiracy isn't as easy as it might seem. For instance, siblings secretly plotting to put a "kick me" sign on Dad's back would satisfy my definition of conspiracy, yet it'd be hard to consider their plot a conspiracy as the term is generally understood. The U.S. government's plan to capture and kill Osama bin Laden satisfies my definition of conspiracy, but again, this isn't considered a conspiracy, at least not by Westerners.

8. The findings of Swami et al. align with Kahneman's model of System 1 and System 2 thinking. Given that System 1 thinking is less analytical, it should be expected that intuitive thinkers are more prone to belief in conspiracy theories, which is exactly what Swami et al. found. "Belief in conspiracy theories was significantly correlated with intuitive thinking style and need for closure. . . . Greater belief in conspiracy theories was significantly predicted by lower analytic thinking style, greater intuitive thinking, and lower open-minded thinking" (p. 576).

9. A schema for "jail" might include uniformed guards, orange scrubs, locked doors, the smell of metal, concrete and paint, etc.

10. As Uscinski observes, the numbers of diseases researchers still don't understand is huge, yet conspiracy theorists have set their sights on autism because of a single discredited research paper. Conspiracy theorists are suspicious of water fluoridation, yet haven't batted an eye to folic acid being added to breads and cereals (Uscinski, 2019a), nor have they raised an eyebrow to putting iodine into table salt to prevent goiter (Hall, 2019c). Meanwhile, real conspiracies, such as megacorporations conspiring to keep people hooked on the internet (Harrison, 2017; Pariser, 2012) or convince the world that climate change isn't real (Mann & Toles, 2018; Oreskes & Conway, 2010a), pass before the conspiracy theorists' eyes.

11. The idea that a conspiracy must be falsifiable has been challenged (Bezalel, 2019).

12. The need for belonging is extremely powerful, as evidenced by cultic groups, such as David Koresh's Branch Davidians or those who followed the Reverend Jim Jones to Jonestown.

13. Interestingly, many of the same people who take measures to drink "safe" water also believe their governments subdue and brainwash their citizenry by regularly dumping hundreds of billions of gallons of toxins out of planes. Ignoring the fact that the fluoride conspiracy theory would make more economic sense and the "chemtrail" thing would be nothing short of absolute overkill, belief in chemtrail conspiracy theories doesn't lead believers to wear surgical masks or respirators. Apparently just knowing that the government is crop dusting humans is enough to inoculate them from the effects.

Drawing a strange comparison between these two conspiracy theories, the science attesting to the safety and benefits of water fluoridation abounds (Yeung, 2008). Conspiracy theorists dismiss this science as being part of the conspiracy theory: "Of course the science says water fluoridation is safe. That's what they want you to think, so they get scientists to claim it's true." Following this logic, one would expect a large body of science supporting the claim that contrails are safe and good for our health. No such science exists.

14. Research shows that for every dollar spent on fluoridation, $38 is saved on dental care (Brody, 2012). Despite dentists' obvious financial interest in tooth decay, most strongly support water fluoridation. Hall makes a similar point, observing that around one hundred scientific organizations worldwide agree that water fluoridation is safe and effective, yet conspiracy theorists claim doctors and dentists want to keep us sick to make a profit, so why are they supporting this measure that clearly promotes dental health? As a final point of concern, the EPA has established accepted levels of water fluoridation at a maximum of 2 parts per million (ppm) to prevent enamel fluorosis and 4 ppm to prevent skeletal fluorosis. As noted in the note above, if the government wanted to poison us by putting fluoride into our water supplies, they'd manufacture scientific studies "demonstrating" that fluoride doesn't cause enamel or skeletal fluorosis and "allow" (read "put") many more ppm. The levels of fluoridation wouldn't be accepted if they were harmful. Rather, it's the very safety of water fluoridation that makes it acceptable.

15. Unless, of course, there were no planes that ran into the buildings and what the world saw were holograms casted by the powerful conspirators.

16. https://en.wikipedia.org/wiki/Bushism

17. The term "crisis actor" can be misleading. A crisis actor may be someone who's been hired to portray the role of a victim during emergency or disaster drills. Crisis actors are used in legitimate training exercises to train emergency response teams to better respond to crises by working with first responders or other emergency service personnel. Conspiracy theorists have hijacked the term. I've adopted the conspiracy theorists' usage of the term.

18. Some evidence suggests that conspiracy theorists are themselves more willing to conspire than those less predisposed to conspiracy thinking (Douglas & Sutton, 2011).

19. Guy P. Harrison has pointed out that public surveys repeatedly show that "political slant strongly influences how [Americans] think about global warming" (Harrison, 2012, pp. 162–63). As it applies to conspiracy theories, Mann and Toles observe that, "when it comes to climate change, Fox News has constructed an alternative universe where the laws of physics no longer apply, where the greenhouse effect is a myth, and where climate change is a hoax, the product of a massive conspiracy among scientists, who somehow have gotten the polar bears, glaciers, sea levels, superstorms, and megadroughts to play along" (Mann & Toles, 2018, p. 106).

20. These particularly egregious claims flew in the face of more than nine thousand studies (Oreskes & Conway, 2010b, pp. 21–23).

Works Cited

Aaen-Stockdale, C. (2012). Neuroscience for the soul. *Psychologist, 25*(7), 520–23.

Abakalina-Paap, M., Stephan, W. G., Craig, T., et al. (1999). Beliefs in conspiracies. *Political Psychology, 203*, 637–47.

Abell, G. O., & Singer, B. (Eds.). (1983). *Science and the paranormal: Probing the existence of the supernatural.* New York: Scribner's.

Alcock, J. E. (2018). *Belief: What it means to believe and why our beliefs are so compelling.* Amherst, NY: Prometheus.

Alcock, J. E., & Otis, L. (1980). Critical thinking and belief in the paranormal. *Psychological Reports, 46*, 479–82.

Alvarez, L. W. (1965). Letter to the editors. *Science, 148*(3677), 1541. Retrieved July 7, 2020, from http://science.sciencemag.org/content/148/3677/1541.1

American Philosophical Association. (1990). Critical thinking: A statement of expert consensus for purposes of educational assessment and instruction. *California Academic Press, 2.*

Ames, A. H. (2001). The polygraphs controversy. *Skeptical Inquirer, 25*(6), 72–73.

Amicus curiae. (1986). *Brief of seventy-two nobel laureates, seventeen state academies of science, and seven other scientific organizations, in support of appellees.* Submitted to the Supreme Court of the United States, October Term, 1986, *Edwin W. Edwards, in His Official Capacity as Governor of Louisiana, et al., Appellants v. Don Aguillard et al., Appellees.*

Angell, M., & Kassirer, J. P. (1998). Alternative medicine: The risks of untested and unregulated remedies. *New England Journal of Medicine, 339*, 839.

Anthony, S. (1973). Anxiety and rumor. *Journal of Personality and Social Psychology, 89*, 91–98.

Aronson, E., & Mills, J. (1959). The effect of severity of initiation on liking for a group. *Journal of Abnormal and Social Psychology, 59*, 177–81.

Arrigo, B. A., & Bullock, J. L. (2007). The psychological effects of solitary confinement on prisoners in supermax units: Reviewing what we know and recommending what should change. *Journal of Offender Therapy and Comparative Criminology, 52*(6), 622–40. doi:10.1177/0306624X07309720

Asimov, I. (1989a). The never-ending fight. *Humanist, 49*(2), 7–8, 30.

Asimov, I. (1989b). The relativity of wrong. *Skeptical Inquirer, 14*, 35–44.

Asimov, I. (1997). *The roving mind* (New ed.). Amherst, NY: Prometheus.

Atwood, K. C. (2004). Naturopathy, pseudoscience, and medicine: Myths and fallacies vs truth. *MedGenMed: Medscape General Medicine, 6*(1), 33.

Azevedo, F. A., Carvalho, L. R., Grinberg, L. T., Farfel, J. M., Feretti, R. E., Leite, R. E., & Herculano-Houzel, S. (2009). Equal number of neuronal and nonneuronal cells make the human brain an isometrically scaled-up primate brain. *Journal of Comparative Neurology, 513*(5), 532–41. Retrieved July 7, 2020, from http://www.ncbi.nlm.gov.pubmed?Db=pubmed&Cmd=ShowDetailView&TersToSearch=19226510

Bacon, F. T. (1979). Credibility of repeated statements: Memory for trivia. *Journal of Experimental Psychology: Human Learning and Memory, 5,* 241–52.

Baggini, J. (2003). *Making sense: Philosophy behind the headlines.* New York: Oxford University Press.

Baggini, J. (2006). *The pig that wants to be eaten: 100 philosophical experiments for the armchair philosopher.* New York: Plume.

Baggini, J. (2017). *The edge of reason: A rational skeptic in an irrational world.* New Haven, CT: Yale University Press.

Balcetis, E., Dunning, D., & Miller, R. L. (2008). Do collectivists know themselves better than individualists? Cross-cultural studies of the holier than thou phenomenon. *Journal of Personality and Social Psychology, 95,* 1252–67.

Ball, D. W. (2019). Cold fusion: Thirty years later. *Skeptical Inquirer, 43*(1), 36–41.

Bangerter, A., & Heath, C. (2004). The Mozart Effect: Tracking the evolution of a scientific legend. *British Journal of Social Psychology, 43,* 605–23.

Barch, D. M. (2018). Schizophrenia spectrum disorder. In R. Biswas-Diener & E. Diener (Eds.), *Noba Textbook Series: Psychology* (pp. 420–38). Chamnpaign, IL: DEF Publishers.

Barkun, M. (2013). *A culture of conspiracy: Apocalyptic visions in contemporary America* (Second ed.). Berkeley: University of California Press.

Barlow, D. H., & Ellard, K. K. (2018). Anxiety and related disorders. In R. Biswas-Diener & E. Diener (Eds.), *Discover psychology 2.0: A brief introductory text* (pp. 382–99). Champaign, IL: DEF Publishers.

Barnett, A. G., van der Pols, J. C., & Dobson, A. J. (2005). Regression to the mean: What it is and how to deal with it. *International Journal of Epidemiology, 34*(1), 215–20.

Bartlett, J., & Miller, C. (2011). A bestiary of the 9/11 truth movement: Notes from the front line. *Skeptical Inquirer, 35*(4), 43–46.

Baruš, I. (2001). Failure to replicate Electronic Voice Phenomenon. *Journal of Scientific Expiration, 15*(3), 355–67.

Basterfield, K., & Bartholomew, R. (1988). Abductions: The fantasy-prone personality hypothesis. *International UFO Review, 13*(3), 9–11.

Bazerman, M. H. (2018). Judgment and decision making. In R. Biswas-Diener & E. Diener (Eds.), *Discover psychology 2.0: A brief introductory text* (pp. 243–55). Champaign, IL: DEF Publishers.

Beall, J. (2019). Scientific soundness and the problem of predatory journals. In A. B. Kaufman & J. C. Kaufman (Eds.), *Pseudoscience: The conspiracy against science* (pp. 283–99). Cambridge, MA: MIT Press.

Beckner, S. (2019). Straw man on a slippery slope. *Skeptic, 24*(1), 42–49.

Begg, I. M., Robertson, R. K., Gruppuso, V., Anas, A., & Needham, D. R. (1996). The illusory-knowledge effect. *Journal of Memory and Language, 35*, 410–33.

Begley, C. G., & Ioannidis, J. P. (2015). Reproducibility in science: Improving the standard for basic and preclinical research. *Circulation Research, 116*(1), 116–26.

Bell, D. (2016, June 24). *David Stephan gets jail time, Collet Stephan gets house arrest in son's meningitis death.* Retrieved July 7, 2020, from CBS News: http://www.cbc.ca/news/canada/calgary/lethbridge-meningitis-trial-sentence-parents-toddler-died-1.3650653

Benos, D. J., Kirk, K. L., & Hall, J. E. (2003). How to review a paper. *Advances in Physiology Education, 27*(2), 47–52.

Benson, V. (2019). The remedies of National Geographic. *Skeptical Inquirer, 43*(5), 30–36.

Berinsky, A. (2012, July 11). *The birthers are (still) back.* Retrieved July 7, 2020, from YouGov: What the World Thinks: http://today.yougov.com/news/2012/07/11/birthers-are-still-back

Beyerstein, B. D. (1997). Why bogus therapies seem to work. *Skeptical Inquirer, 21*(5), 29–34.

Bezalel, G. Y. (2019). Conspiracy theories and religion: Reframing conspiracy theories as bilks. *Episteme*, 1–19.

Biddle, W. (1986, March). The deception of detection. *Discover*, 24–26, 28–31, 33.

Biswas-Diener, R. (2018). The brain and nervous system. In R. Biswas-Diener & E. Diener (Eds.), *Discover psychology 2.0: A brief introductory text* (pp. 47–61). Champaign, IL: DEF Publishers.

Blackmore, S. (1990). The lure of the paranormal. *New Scientist, 22*, 62–65.

Blackmore, S., & Moore, R. (1994). Seeing things: Visual recognition and belief in the paranormal. *European Journal of Parapsychology, 10*, 91–103.

Blackmore, S., & Troscianko, T. (1985). Belief in the paranormal: Probability, judgments, illusory control, and the "chance baseline shift." *British Journal of Psychology, 76*, 459–68.

Blancke, S., Van Breusegem, F., De Jaeger, G., Braeckman, J., & Van Montagu, M. (2015). Fatal attraction: The intuitive appeal of GMO opposition. *Trends in Plant Science, 20*(7), 414–18.

Blanco, F., Barberia, I., & Matute, H. (2014). The lack of side effects of an ineffective treatment facilitates the development of a belief in its effectiveness. *PLOS ONE, 9*(1), e84084. doi:10.1371/journal.pone.0084084

Blanco, F., Barberia, I., & Matute, H. (2015). Individuals who believe in the paranormal expose themselves to biased information and develop more causal illusions than nonbelievers in the laboratory. *PLOS ONE, 10*(7), e0131378. doi:10.1371/journal.pone.0131378

Blanco, F., & Matute, H. (2019). The illusion of causality: A cognitive bias underlying pseudoscience. In A. B. Kaufman & J. C. Kaufman (Eds.), *Pseudoscience: The conspiracy against science* (pp. 45–75). Cambridge, MA: MIT Press.

Blaskiewicz, R. (2014). *Jews and reptilians.* Retrieved July 7, 2020, from CSICOP: http://www.csicop.org/specialarticles/show/jews_and_reptilians

Blaskiewicz, R., & Jarsulic, M. (2018). Arthur J. Cramp: The quackbuster who professionalized American medicine. *Skeptical Inquirer, 42*(6), 45–50.

Bleske-Rechek, A., Paulich, K., & Jorgensen, K. (2019). Therapeutic touch redux: Twenty years after the "Emily event," energy therapies live on through bad science. *Skeptic, 24*(2), 24–31.

Boghossian, P. (2006). Socratic pedagogy, critical thinking, and inmate education. *Journal of Correctional Education, 57*(1), 42–63.

Boghossian, P. (2013). *A manual for creating atheists.* Durham, NC: Pitchstone.

Bohannon, J. (2015). Many psychology papers fail replication test. *Science, 349*(6251), 910–11.

Bohle, R. H. (1986). Negativism as news selection predictor. *Journalism Quarterly, 63*, 789–96.

Bortolotti, L. (2010). *Delusions and other irrational beliefs.* Oxford: Oxford University Press.

Bost, R. (2015). Crazy beliefs, sane believers: Toward a cognitive psychology of conspiracy ideation. *Skeptical Inquirer, 39*(1), 44–49.

Bouvet, R., & Bonnefon, J. F. (2015). Non-reflective thinkers are predisposed to attribute supernatural causation to uncanny experiences. *Personality and Social Psychology Bulletin, 41*(7), 955–61. doi:10.1177/0146167215585728

Boyd, R. (2008). Do people only use 10 percent of their brains? *Scientific American.* Retrieved from http://www.scientificamerican.com/article/do-people-only-use-10-percent-of-their-brains/

Brainerd, C. J., & Reyna, V. F. (2002). Fuzzy-trace theory and false memory. *Current Directions in Psychological Science, 11*, 164–69.

Brenner, S. N., & Molander, E. A. (1977, January/February). Is the ethics of business changing? *Harvard Business Review*, 57–71.

Brett, A., Phillips, M., & Beary, J. (1986). Predictive power of the polygraph: Can the "lie detector" really detect liars? *Lancet, 1*, 544–47.

Brewer, W. F., & Treyens, J. C. (1981). Role of schemata in memory for places. *Cognitive Psychology, 13*, 207–30.

Brody, J. E. (2012, January 24). Dental exam went well? Thank fluoride. *New York Times*, D7.

Brotherton, R. (2013). Towards a definition of "conspiracy theory." *PsyPag Quarterly, 88*(3), 9–14.

Brown, J. D. (2012). Understanding the better than average effect: Motives (still) matter. *Personality and Social Psychology Bulletin, 38*(2), 209–19. doi:10.1177/0146167211432763

Brown, R., & Kulik, J. (1977). Flashbulb memories. *Cognition, 5*(1), 73–99.

Bruck, M. S., & Hembrooke, H. (1997). Children's reports of pleasant and unpleasant events. In D. Read & S. Lindsay (Eds.), *Recollections of trauma: Scientific research and clinical practice* (pp. 199–219). New York: Plenum Press.

Brugger, P., Landis, T., & Regard, M. (1990). A "sheep-goat effect" in repetition avoidance: Extra-sensory perception as an effect of subjective probability? *British Journal of Psychology, 81*, 455–68.

Bruner, J. S., & Postman, L. (1949). On the perception of incongruity: A paradigm. *Journal of Personality, 18*, 206–23.

Buckhout, R. (1975). Nearly 2000 witnesses can be wrong. *Social Action and the Law, 2,* 7.

Buckner, J. E., V, & Buckner, R. A. (2012). Talking to the dead, listening to yourself: An empirical study on the psychological aspects of interpreting Electronic Voice Phenomena. *Skeptic, 17*(2), 44–49.

Budson, A. E., & Price, B. (2005). Memory dysfunction. *New England Journal of Medicine, 352*(7).

Bunge, M. (2020). The scientist's skepticism. *Skeptical Inquirer, 44*(6), 57–59.

Burgess, D., Burgess, M., & Leasak, J. (2006). The MMR vaccination and autism controversy in the United Kingdom 1998–2005: Inevitable community outrage or a failure of risk communication? *Vaccine, 24*(18), 3921–28.

Burton, R. A. (2008). *On being certain: Believing you are right even when you're not.* New York: St. Martin's Griffin.

Butter, M., & Knight, P. (2019). The history of conspiracy theory research: A review and commentary. In J. E. Uscinski (Ed.), *Conspiracy theories and the people who believe them* (pp. 33–46). New York: Oxford University Press.

Byrne, R. (Producer), & Heriot, D. (Director). (2006). *The Secret* [Motion picture]. Melbourne, Austrailia: Prime Time Productions.

Cama, T. (2015, March 13). *EPA confronts "chemtrails" conspiracy talk.* Retrieved July 7, 2020, from The Hill: https://thehill.com/policy/energy-environment/235632-epa-confronts-chemtrails-conspiracy-theory

Carlson, S. (1985). A double-blind test of astrology. *Nature, 318,* 419–25.

Carroll, A. E. (2014, October 6). *The placebo effect doesn't apply just to pills.* Retrieved July 7, 2020, from *New York Times*: http://www.nytimes.com/2014/10/07/upshot/the-placebo-effect-doesnt-apply-just-to-pills.html?_=0&abt=0002&abg=0

Carroll, R. T. (2005). *Anecdotal (testimonial) evidence.* Retrieved from the Skeptic's Dictionary, http://www.skepdic.com/testimon.html

Centers for Disease Control and Prevention. (2015a). *Measles cases and outbreaks.* Retrieved July 7, 2020, from Centers for Disease Control and Prevention: http://www.cdc.gov/measles/cases-outbreaks.html

Centers for Disease Control and Prevention. (2015b). *Year in review: Measles linked to Disneyland.* Retrieved July 7, 2020, from Centers for Disease Control and Prevention: http://blogs-origin.cdc.gov/publichealthmatters/files/2015/12/Banner_blog-banner_balloons.jpg

Centers for Disease Control and Prevention. (2017). *Data and statistics: Short sleep duration among US Adults.* Retrieved March 12, 2022, from Centers for Disease Control and Prevention: https://www.cdc.gov/sleep/data_statistics.html

Cha, A., Hecht, B. R., Nelson, K., & Hopkins, M. P. (2004). Resident physician attire: Does it make a difference to our patients? *American Journal of Obstetrics and Gynecology, 190,* 1484–88.

Chabris, C. F. (1999). Prelude or requiem for the "Mozart Effect"? *Nature, 400,* 826–27.

Chabris, C. F., & Simons, D. J. (2010). *The invisible gorilla: And other ways our intuitions deceive us.* New York: Crown.

Chaplin, C., & Shaw, J. (2015). Confidently wrong: Police endorsement of psycho-legal misconceptions. *Journal of Police and Criminal Psychology,* 1–9.

Charlson, C., & Apsell, P. S. (Producers). (1993). *Secrets of the Psychics* [Motion picture]. Arlington, VA: PBS Distribution. Retrieved from https://www.youtube.com/watch?v=41hJ6DY8xLI

Chen, R. T., Rastogi, S. C., Mullen, J. R., Hayes, S. W., Cochi, S. L., Donlon, J. A., & Wassilak, S. G. (1994). The Vaccine Adverse Event Reporting System (VAERS). *Vaccine, 12*(6), 542–50.

Cherkin, D. C., Sherman, K. J., Avins, A. L., Erro, J. H., Ichikawa, L., Barlow, W. E., . . . Deyo, R. A. (2009). A randomized trial comparing acupuncture, simulated acupuncture, and usual care for chronic low back pain. *Archives of Internal Medicine, 169*(9), 858–66.

Chiropractic. (2013). Retrieved July 7, 2020, from Science-Based Medicine: https://sciencebasedmedicine.org/reference/chiropractic/

Chomsky, N. (1999). *Fateful triangle: The United States, Israel, and the Palestinians.* New York: South End Press.

Chugh, D. (2004). Societal and managerial implications of social cognition: Why milliseconds matter. *Social Justice Research, 17*(2), 203–22.

Cialdini, R. (2007). *Influence: The psychology of persuasion* (First Collins Business Essentials ed.). New York: Harper Collins.

Clancy, S. A., McNally, R. J., Schacter, D. L., Lenzenweger, M. F., & Pitman, R. (2002). Memory distortion in people reporting abduction by aliens. *Journal of Abnormal Psychology, 111*(3), 455–61.

Clark, D. (Producer), Dunning, B. (Writer), & Johnson, J. C. (Director). (2017). *Principles of Curiosity* [Motion picture]. Bend, OR: Skeptoid Media.

Clark, S. E., & Loftus, E. F. (1996). The construction of space alien abduction memories. *Psychological Inquiry, 7*, 140–43.

Clarke, D. (1995). Experience and other reasons given for belief and disbelief in paranormal and religious phenomena. *Journal of the Society for Psychical Research, 60*, 371–84.

Clegg, D. O., Reda, D. J., Harris, C. L., Klein, M. A., O'Dell, J. R., Hooper, M. M., . . . Williams, H. J. (2006). Glucosamine, chondroitin sulfate, and the two in combination for painful knee osteoarthritis. *New England Journal of Medicine, 354*(8), 798–808. doi:10.1056/NEJMoa052771

Clement-Jones, V., McLoughlin, L., Tomlin, S., et al. (1980). Increased beta-endorphin but not met-enkephalin levels in human cerebrospinal fluid after acupuncture for recurrent pain. *Lancet, 2*, 946–49.

Clements, A. (Producer), & Dawkins, R. (Writer & director). (2006). *Root of All Evil?* [Motion picture]. London: Channel Four.

Cobb, L. A., Thomas, G. I., Dillard, D. H., et al. (1959). An evaluation of internal-mammary-artery ligation by a double-blind technic. *New England Journal of Medicine, 260*, 1115–18.

Coe, M. R., Jr. (1975). Fire-walking and related behaviors. *The Psychological Record, 7*(2), 101–10.

Cohen, G. (2003). Party over policy: The dominating impact of group influence on political beliefs. *Journal of Personality and Social Psychology, 85*(5), 808–82. doi:10.1037/0022-3514.85.5.808

Combs, B., & Slovic, P. (1979). Newspaper coverage of causes of death. *Journalism Quarterly, 56*, 837–43, 849.

Cook, J. (2010, December). *The scientific guide to global warming skepticism.* Retrieved July 7, 2020, from Skeptical Science: http://www.skepticalscience.com/docs/Guide_to_Skepticism.pdf

Cook, J., Oreskes, N., Doran, P. T., Anderegg, W. R., Verheggen, B., Maibach, E. W., . . . Green, S. A. (2016). Consensus on consensus: A synthesis of consensus estimates on human-caused global warming. *Environmental Research Letters, 11*(4), 1–8.

Cooper, C. (2010, June 30). *Belief in the paranormal remains deep.* Retrieved July 7, 2020, from CBS News: http://www.cbsnews.com/news/belief-in-the-paranormal-ramins-deep/

Cooper, C., Li, R., Lyketsos, C., & Livingston, G. (2013). Treatment for mild cognitive impairment: Systematic review. *British Journal of Psychiatry, 203*(3), 255–64.

Cooper, J. M. (2000). *The trial and death of Socrates* (Third ed.) (G. M. Grube, Trans.). Indianapolis, IN: Hackett.

Cooper, M. W. (1991). *Behold a pale horse.* Flagstaff, AZ: Light Technology Publications.

Craft, L., Ruloff, W., Sullivan, J. (Producers), Miller, K., Stein, B., Ruloff, W. (Writers), & Frankowski, N. (Director). (2008). *Expelled: No intelligence allowed* [Motion picture]. Salt Lake City, UT: Rocky Mountain Pictures.

Crandall, C. S., & Eshelman, A. (2003). A justification-suppression model of the expression and experience of prejudice. *Psychological Bulletin, 129*, 425–26.

Crist, C. (2021, March 25). *"Disinformation dozen" driving anti-vaccine content.* Retrieved from WebMD: https://webmd.com/children/vaccine/news/20210325/disinformation-driving-anti-vaccine-content

Crombag, H. F., Wagenaar, W. A., & Van Koppen, J. (1996). Crashing memories and the problem of "source monitoring." *Applied Cognitive Psychology, 10*(2), 95–104.

Cunningham, M. R., Shamblen, S. R., Barbee, A. P., & Ault, L. K. (2005). Social allergies in romantic relationships: Behavioral repetition, emotional sensitization, and dissatisfaction in dating couples. *Personal Relationships, 12*, 273–95.

Cuno, S. (2019). Jesus vs. Santa: The evidence speaks. *Free Inquiry, 39*(1), 46–47.

Cuntz, M. (2020). 100 authors against Einstein: A look in the rearview mirror. *Skeptical Inquirer, 44*(6), 48–51.

Cupp, M. J. (1999). Herbal remedies: Adverse effects and drug interactions. *American Family Physician, 59*, 1239–44.

Cutler, B. L., & Penrod, S. D. (1988). Improving the reliability of eyewitness identification: Lineup construction and presentation. *Journal of Applied Psychology, 73*, 281–99.

Damasio, A. (2002). Remembering when. *Scientific American*, 66.

Davis, M. H., & Stephan, W. G. (1980). Attributions for exam performance. *Journal of Applied Social Psychology, 10*(3), 235–48.

Dawkins, R. (2000). *Unweaving the rainbow: Science, delusion, and the appetite for wonder.* New York: Mariner Books.

Dawkins, R. (2002, February). *Militant atheism.* Retrieved March 20, 2022, from TED: https://www.youtube.com/watch?v=VxGMqKCcN6A

Dawkins, R. (2005). *The ancestor's tale: A pilgrimage to the dawn of life.* London: Phoenix.

Dawkins, R. (2006). *The God delusion.* New York: First Mariner Books.

Dawkins, R. (2009). *The greatest show on Earth: The evidence for evolution.* New York: Free Press.

Dawkins, R. (2012). *The magic of reality: How we know what's really true.* New York: Free Press.

De Craen, S., Twisk, D., Hagenzieker, M., Elffers, H., & Brookhuis, K. (2011). Do young novice drivers overestimate their driving skills more than experienced drivers? *Accident Analysis and Prevention, 43,* 1660–65.

De Martino, B., et al. (2006). Frames, biases, and rational decision-making in the human brain. *Science, 313,* 684–87.

Dean, G. (2016). Does astrology need to be true? A thirty-year update. *Skeptical Inquirer, 40,* 38–45.

DeBakcsy, D. (2019). Hot and wild sufficiency: Epicurus, the mehness of death, and the pleasures of enough. *Free Inquiry, 39*(2), 45–47.

Deer, B. (2004, February 22). MMR: The truth behind the crisis. *Sunday Times (London).*

Deer, B. (2006, December 31). MMR doctor given legal aid thousands. *Sunday Times (London).*

Deer, B. (2011). How the case against the MMR vaccine was fixed. *British Medical Journal, 342,* c5347. doi:10.1136/bmj.c5347

Delplante, K. (2011). *Critical thinking about conspiracies: An argument for default skepticism.* Retrieved July 7, 2020, from Critical Thinker Academy: http://www.critical thinkeracademy.com/017-default-skepticism-about-conspiracies.html

Deming, D. (2016). Do extraordinary claims require extraordinary evidence? *Philosophia, 44,* 1319–31.

Dennett, D. (1995). *Darwin's dangerous idea: Evolution and the meanings of life.* New York: Simon & Schuster.

Dennett, D. (1996). *Kinds of minds: Toward an understanding of consciousness.* New York: Basic Books.

Deutsch, D. (2008). Auditory illusions. In E. B. Goldstein (Ed.), *Encyclopedia of perception* (Vol. 1, pp. 160–64). Newbury Park, CA: Sage.

Deutsch, D. T., Henthorn, T., & Lapidis, R. (2011). Illusory transformation from speech to song. *Journal of the Acoustical Society of America, 129,* 2245–52.

Diamond, J. (1997). *Guns, germs, and steel: The fates of human societies.* New York: W. W. Norton.

Dickenson, D. H., & Kelly, I. W. (1985). The "Barnum effect" in personality assessment: A review of the literature. *Psychological Reports, 57,* 367–82.

Difonzo, N. (2019). Conspiracy theory rumor psychology. In J. E. Uscinski (Ed.), *Conspiracy theories and the people who believe them* (pp. 257–68). New York: Oxford University Press.

Dobelli, R. (2014). *The art of thinking clearly.* New York: Harper Collins.

Domingo, J., & Bordonaba, J. G. (2011). A literature review on the safety assessment of genetically modified plants. *Environmental International, 37*(4), 734–42.

Douglas, K. M., & Sutton, R. M. (2011). Does it take one to know one? Endorsement of conspiracy theories is influenced by personal willingness to conspire. *British Journal of Social Psychology, 20*(3).

Douglas, K. M., Sutton, R. M., & Cichocka, A. (2017). The psychology of conspiracy theories. *Current Directions in Psychological Science, 26*(6), 538–42. doi:http://doi.org/10.1177/0963721417718261

Douglas, K. M., Sutton, R. M., Jolley, D., & Wood, M. (2015). The social, political, environmental, and health-related consequences of conspiracy theories. In M. Bilewicz, A. Cichocka, & W. Soral (Eds.), *The psychology of conspiracy* (pp. 183–96). Hove, UK: Routledge.

Duguid, S., Hawkey, C., & Pawson, R. (1996, June). Using recidivism to evaluate effectiveness in prison education programs. *Journal of Correctional Education, 47*(2), 74–85.

Dunbar, R. (1997). *Grooming, gossip, and the evolution of language.* London: Faber and Faber.

Duncker, K. (1939). The influence of past experience upon perceptual properties. *American Journal of Psychology, 52*, 255–65.

Dunn, E. W., Aknin, L. B., & Norton, M. I. (2008). Spending money on others promotes happiness. *Science, 319*(5870), 1687–88. doi:10.1126/science.1150952

Dunning, B. (Producer & writer). (2008). *Here be dragons: An introduction to critical thinking* [Motion picture]. Bend, OR: Skeptoid Media.

Dunning, B. (2014, July 8). *Lie detection.* Retrieved July 7, 2020, from Skeptoid Podcast: http://skeptiod.com/episodes/4422

Dunning, B. (2019). The life and times of the moon hoax. *The Skeptic, 39*(2), 28–35.

Dybing, E. (2002). Development and implementation of the IPCS conceptual framework for evaluating mode of action of chemical carcinogens. *Toxicology, 181–82*, 121–25.

Ebert, R. J., & Griffin, R. W. (2013). *Business essentials* (Ninth ed.). New York: Pearson.

Ecker, U. K., Lewandowsky, S., Swire, B., & Chang, D. (2011). Correcting false information in memory: Manipulating the strength of misinformation encoding and its retraction. *Psychonomic Bulletin Review, 18*, 570–78. doi:10.3758/s13423-011-0065-1

Editors of the Lancet Retraction. (2010). Ileal lymphoid nodular hyperplasia, non-specific colitis, and pervasive developmental disorder in children. *Lancet, 375*(9713), 445.

Edwards, K., & Smith, E. E. (1996). A disconfirmation bias in the evaluation of arguments. *Journal of Personality and Social Psychology, 71*, 5–24.

Ehrilinger, J., & Gilovich, T. (2005). Peering into the bias blind spot: People's assessment of bias in themselves and in others. *Personality and Social Psychology Bulletin, 31*(5), 680–91. doi:10.1177/0146167204271570

Ehrlinger, J., Johnson, K., Banner, M., Dunning, D., & Kruger, J. (2008). Why the unskilled are unaware: Further explorations of (absent) self-insight among the incompetent. *Organizational Behavior and Human Decision Processes, 105*(1), 98–121. doi:10.1016/j.obhdp.2007.05.002

Ehrsson, H. E. (2007). The experimental induction of out-of-body experiences. *Science, 317*, 1048.

Eisenberg, D. M., Kessler, R., Foster, C., Norlock, F., Calkins, D., & Delbanco, T. (1993). Unconventional medicine in the United States. *New England Journal of Medicine, 328*(4), 246–52.

Elgat, G. (2019). "Prove that I am wrong!": What QAnon, Descartes, and brains in vats have in common. *Skeptic, 24*(4), 46–47.

Enders, A. M., & Smallpage, S. M. (2019). Polls, plots, and party politics: Conspiracy theories in contemporary America. In J. E. Uscinski (Ed.), *Conspiracy theories and the peole who believe them* (pp. 298–318). New York: Oxford University Press.

Engelhard, I. M., McNally, R. J., & van Schie, K. (2019). Retrieving and modifying traumatic memories: Recent research relevant to three controversies. *Current Directions in Psychological Science, 28*(1), 91–96.

Epley, N., & Dunning, D. (2000). Feeling "holier than thou": Are self-serving assessments produced by errors in self- or social prediction? *Journal of Personality and Social Psychology, 37*, 861–75.

Epstein, S., Pacini, R., Denes-Raj, V., & Heier, H. (1996). Individual differences in intuitive-experiential and analytical-rational thinking styles. *Journal of Personality and Social Psychology, 71*(2), 390–405.

Ernst, E. (2002). A systematic review of systematic reviews of homeopathy. *British Journal of Pharmacology, 54*(6), 577–82.

Ernst, E. (2007). Adverse effects of spinal manipulation: A systematic review. *Journal of the Royal Society of Medicine, 100*(7), 330–38.

Ernst, E. (2010a). Deaths after chiropractic: A review of published cases. *International Journal of Clinical Practice, 64*(8), 1162–65.

Ernst, E. (2010b). Vascular accidents after neck manipulation: Cause or coincidence? *International Journal of Clinical Practice, 64*(6), 673–77.

Ernst, E. (2018). *SCAM: So-called alternative medicine.* La Vergn, TN: Ingram.

Ernst, E., & Abbot, N. C. (1999). I shall please: The mysterious power of placebos. In S. Della Sala (Ed.), *Mind myths: Exploring popular assumptions about the mind and brain* (pp. 209–13). Chichester, UK: John Wiley & Sons.

Ernst, E., Lee, M. S., & Choi, T. Y. (2011). Acupuncture: Does it alleviate pain and are there serious risks? A review of reviews. *Pain, 152*(4), 755–64. doi:10.1016/j.pain .2010.11.004

Evans, J. S. (2003). In two minds: Dual process accounts of reasoning. *Trends in Cognitive Science, 7*(10), 454–59.

Fahlgren, N., Bart, R., Herrera-Estrella, L., Rellán-Álvarez, R., Chitwood, D. H., & Dinneny, J. R. (2016). Plant scientists: GM technology is safe. *Science, 351*(6275), 824.

Fazio, L. K., Brashier, N. M., Payne, B. K., & Marsh, E. J. (2015). Knowledge does not protect against illusory truth. *Journal of Experimental Psychology: General, 144*, 993–1002.

Festinger, L., & Carlsmith, J. M. (1959). Cognitive consequences of forced compliance. *Journal of Abnormal and Social Psychology, 58*, 203–10.

Feyerabend, P. (1975/2010). *Against method* (Fourth ed.). New York: Verso.

Feynman, R. P. (1974). Cargo cult science. Retrieved March 6, 2022, from https:// calteches.library.caltech.edu/51/2/CargoCult.htm?fbclid=IwAR07ozLS02Ayn4x6432 oqbh176RuUW3EutqG6sEeWvIwBk0H4hSocJ4P3-E

Fincke, D. (2011, August). *How faith poisons religion.* Retrieved July 7, 2020, from Patheos: http://www.patheos.com/blogs/camelswithhammers/2011/08/disambiguat ing-faith-how-faith-poisons-religion/

Fisher, R. P. (1995). Interviewing victims and witnesses of crime. *Psychology, Public Policy, and Law, 1*, 732–64.

Folta, M. (2019). Food-o-science pseudoscience: The weapons and tactics in the war on crop biotechnology. In A. B. Kaufman & J. C. Kaufman (Eds.), *Pseudoscience: The conspiracy against science* (pp. 103–35). Cambridge, MA: MIT Press.

Fontanarosa, B., & Lundberg, G. D. (1998). Alternative medicine meets science. *Journal of the American Medical Association, 280*, 1618–19.

Forer, B. R. (1949). The fallacy of personal validation: A classroom demonstration of gullibility. *Journal of Abnormal and Social Psychology, 44*(1), 118–23.

Forrest, B. (2017). Methodological naturalism and philosophical naturalism: Clarifying the connection. *Free Inquiry, 37*(5), 34–41.

Foster, C. A. (2018). Flat-Earth anxieties reflect misplaced priorities. *Skeptical Inquirer, 42*(3), 10–11.

Foster, C. A., & Ortiz, S. M. (2017). Vaccines, autism, and the promotion of irrelevant research: A science-pseudoscience analysis. *Skeptical Inquirer, 41*(3), 44–48.

Foster, R. G., & Roenneberg, T. (2008). Human responses to the geophysical daily, annual, and lunar cycles. *Current Biology, 18*(17), R784–94.

Foucault, M. (1995). *Discipline and punish: The birth of the prison* (Second Vintage Books ed.). New York: Vintage Books.

Fox, B., & Farrington, D. P. (2018). What have we learned from offender profiling? *Psychological Bulletin, 144*(12), 1247–74.

Franks, B., Bangerter, A., Bauer, M. W., Hall, M., & Noort, M. C. (2017). Beyond "monologicality"? Exploring conspiracist worldviews. *Frontiers in Psychology, 8*, 861.

French, C. C., Haque, U., Bunton-Stasyshyn, R., & Davis, R. (2009). The "Haunt" Project: An attempt to build a "haunted" room by manipulating complex electromagnetic fields and infrasound. *Cortex, 45*, 619–29. doi:10.1016/j.cortex.2007.10.011

Frenda, S. J., Patihis, L., Loftus, E. F., Lewis, H. C., & Fenn, K. M. (2014). Sleep deprivation and false memories. *Psychological Science, 25*(9), 1674–81.

Galtung, J., & Ruge, M. H. (1965). The structure of foreign news. *Journal of Peace Research, 2*(1), 64–91.

Garaudy, R. (2000). *The founding myths of modern Israel.* Fountain Valley, CA: Institute for Historical Review.

Gardner, M. (1988–1989). Water with memory? The dilution affair. *Skeptical Inquirer, 13*, 132–41.

Gavura, S. (2021, April 1). *The "disinformation dozen" spreading anti-vaccine messaging on social media.* Retrieved from sciencebasedmedicine: https://sciencebasedmedicine.org/the-disinformation-dozen-spreading-vaccine-messaging-on-social-media

Gawande, A. (2009, March 23). Hellhole. *The New Yorker*, 36–45.

Gell-Mann, M. (2019). Reality is out there . . . and it's beautiful. *Skeptical Inquirer, 43*(5), 11.

Gelman, S. A. (2004). Psychological essentialism in children. *Trends in Cognitive Sciences, 8*(9), 404–9.

Gerber, J. S., & Offit, P. A. (2009). Vaccines and autism: A tale of shifting hypotheses. *Vaccines, 48*, 456–61. doi:10.1086/596476

Gerbic, S. (2020). Right turns only! Circling back to Seatbelt Psychic. Retrieved March 13, 2022, from https://skepticalinquirer.org/exclusive/right-turns-only-circling-back-to-seatbelt-psychic/

Gervais, W. M. (2015). Overriding the controversy: Analytic thinking predicts endorsement of evolution. *Cognition, 142,* 312–21.

Gettier, E. (1963). Is justified true belief knowledge? *Analysis, 23,* 966.

Gilbert, D. (1991). How mental systems believe. *American Psychologist, 46*(2), 107–19.

Gilovich, T. (1991). *How we know what isn't so: The fallibility of human reason in everyday life.* New York: Free Press.

Gilovich, T., Vallone, R., & Tversky, A. (1985). The hot hand in basketball: On the misperception of random sequences. *Cognitive Psychology, 17,* 295–314.

Glaser, E. M. (1941). *An experiment in the development of critical thinking.* New York: Bureau of Publications, Teachers College, Columbia University.

Glick, P., Gottesman, D., & Jolton, J. (1989). The fault is not in the stars: Susceptibility of skeptics and believers in astrology to the Barnum effect. *Personality and Social Psychology Bulletin, 15,* 572–83.

Gnatta, J. R., Kurebayashi, L. F., & Paes da Silva, M. J. (2013). Atypical mycobacterias associated to acupuncuture: An integrative review. *Revista Latino-Americana de Enfermagem, 21*(1), 450–58. doi:10.1590/s0104-11692013000100022

Goertzel, T. (1994a). Belief in conspiracy theories: A pilot study. *Political Psychology, 15*(4), 731–42.

Goertzel, T. (1994b). Measuring the prevalence of false memories: A new interpretation of a "UFO abduction survey." *Skeptical Inquirer, 18*(3), 266–72.

Goertzel, T. (2011). The conspiracy meme: Why conspiracy theories appeal and persist. *Skeptical Inquirer, 35*(1), 28–37.

Goertzel, T., & Goertzel, B. (2008). Capital punishment and homicide rates: Sociological realities and econometric distortions. *Critical Sociology, 34*(2), 239–54. doi:10.1177/0896920507085519

Goldman, A. (2016, December 7). The Comet Ping Pong gunman answers our reporter's questions. *New York Times.*

Gorski, D. H. (2018). Integrative medicine: Integrating quackery with science-based medicine. In A. B. Kaufman & J. C. Kaufman (Eds.), *Pseudoscience: The conspiracy against science* (pp. 309–29). Cambridge, MA: MIT Press.

Gorski, D. (2021a). *What the heck happened to John Ioannidis?* Retrieved March 10, 2022, from Science-Based Medicine: https://sciencebasedmedicine.org/what-the-heck-happened-to-john-ioannidis/

Gorski, D. (2021b). *What the heck happened to the BMJ?* Retrieved March 16, 2022, from Science-Based Medicine: https://sciencebasedmedicine.org/what-the-heck-happened-to-the-bmj/

Gorski, D. (2022). *John Ioannidis uses the Kardashian Index to attack critics of the Great Barrington Declaration.* Retrieved March 10, 2022, from Science-Based Medicine: https://sciencebasedmedicine.org/ioannidis-uses-kardashian-index-to-attack-gbd-critics/

Gould, S. J. (1983). *Hen's teeth and horse's toes: Further reflections in natural history.* New York: W. W. Norton.

Grams, N. (2019). The un-diluted truth. *The Skeptic, 39*(1), 28–32.

Granqvist, P., Fredrikson, M., Unge, P., Hagenfeldt, A., Valind, S., Larhammar, D., & Larsson, M. (2005). Sensed presence and mystical experiences are predicted by suggestibility, not by the application of transcranial weak complex magnetic fields. *Neuroscience Letters, 379*(1), 1–6. doi:10.1016/j.neulet.2004.10.057

Gray, S. B. (2018). Humbling humanity: Reality need not diminish our concept of our place in the cosmos. *Skeptic, 23*(2), 42–44.

Greeley, A. (1987, January/February). Mysticism goes mainstream. *American Health,* 47–49.

Greene, B. (2003). *The elegant universe: Superstrings, hidden dimensions, and the quest for the ultimate theory.* New York: W. W. Norton.

Greenfield, S. (Presenter). (2000). *Brain Story* [Motion picture]. London: BBC Productions.

Gregory, F. (2008). *Darwinian revolution.* Chantilly, VA: Teaching Co.

Gronewold, N., & Marshall, C. (2009, December 3). *Rising partisanship sharply erodes U.S. public's belief in global warming.* Retrieved July 7, 2020, from New York Times: http://www.nytimes.com/cwire/2009/12/03/03climatewire-rising-partisanship-sharply -erodes-us-public-47381.html

Gudjonsson, G. H. (1991). Suggestibility and compliance among alleged false confessors and resistors in criminal trials. *Medicine, Science and the Law, 31*(2), 147–51.

Gudjonsson, G. H., & MacKeith, J. A. (1990). A proven case of false confession: Psychological aspects of the coerced-compliant type. *Medicine, Science and the Law, 30,* 329–35.

Guiley, R. E. (2008). *The encyclopedia of witches, witchcraft and wicca* (3rd edition). New York: Facts on File.

Hacking, I. (2012). Introductory essay. In T. Kuhn, *The structure of scientific revolution* (Fourth ed., p. xi). Chicago: Chicago University Press.

Hadjikhani, N., Kveraga, K., Naik, P., & Ahlfors, S. (2009). Early (M170) activation of face-specific cortex by face-like objects. *Neuroreport, 20,* 403–7.

Hagen, K. (2016). The state of tumor-town: The cancer-care industry's marketing is among the most deceptive on the consumer landscape. *Skeptic, 21*(4), 43–49.

Hagen, L. K. (2018). The last one forgotten: Bruce Perkins and another terrible tragedy of the recovered memory movement. *Skeptic, 23*(2), 20–25.

Halford, S. (2020, March 12–14). *From trustworthy to dubious: Conspiracy movements, their discursive strategies, and the process of symbolic recording in the anti-vaccination movement.* Paper prepared for presentation at the "Conference on Conspiracy Theories" (pp. 1–30). University of Miami.

Hall, H. (2009a). *Top ten things you should know about alternative medicine.* Retrieved July 7, 2020, from skeptic.com: www.skeptic.com

Hall, H. (2009b, June 3). *Vaccines and autism: A deadly manufactroversy.* Retrieved July 7, 2020, from eSkeptic: www.skeptic.com/eskeptic/09-06-03

Hall, H. (2010). Power balance technology: Pseudoscientific silliness suckers card-carrying surfers. *Skeptical Inquirer, 34*(3), 47–49.

Hall, H. (2012). Energy medicine and fantasy physics: How real physics has been kidnapped by alternative medicine practitioners. *Skeptic, 17*(2), 4–5.

Hall, H. (2014). Faith healing: Religious freedom vs. child protection. *Skeptical Inquirer, 38*(4), 42–46.

Hall, H. (2017). *Magnets provide amusement, but not health benefits.* Retrieved March 16, 2022, from Science-Based Medicine: https://sciencebasedmedicine.org/magnets -provide-amusement-but-not-health-benefits/

Hall, H. (2019a). Is low-dose radiation good for you? The questionable claims for hormesis. *Skeptic, 24*(1), 4–5.

Hall, H. (2019b). Science envy in alternative medicine. *Skeptical Inquirer, 43*(4), 21–23.

Hall, H. (2019c). Water fluoridation: Public health, not poison. *Skeptic, 24*(4), 4–5.

Hall, H. (2019d). Whither chiropractic? *Skeptical Inquirer, 43*(6), 24–26.

Hall, H. (2021). *Energy medicine pain relief patches are laughable quackery.* Retrieved March 16, 2022, from Science-Based Medicine: https://sciencebasedmedicine.org /energy-medicine-pain-relief-patches-are-laughable-quackery/

Hall, H. (2022). Misconceptions about vitamins. *Skeptical Inquirer, 46*(3), 19–21.

Hare, W. (2009). What open-mindedness requires. *Skeptical Inquirer, 33*, 36–39.

Harland-Logan, S., & Morin, G. P. (n.d.). *Innocence Canada.* Retrieved July 7, 2020, from Innocence Canada: http://www.aidwyc.org/cases/historical/guy-paul-morin

Harris, I. (2016). *Ian Harris at Reason Rally 2016.* Retrieved July 7, 2020, from You-Tube: http://holesinthefoam.us/ian-harris-at-reason-rally-2016

Harris, S. (2010, February). *Science can answer moral questions.* Retrieved July 7, 2020, from TED: www.ted.com/talks

Harris, S. (2011). *The moral landscape: How science can determine human values.* New York: Free Press.

Harrison, G. P. (2012). *50 popular beliefs that people think are true.* New York: Prometheus.

Harrison, G. P. (2013). *Think: Why you should question everything.* Amherst, NY: Prometheus.

Harrison, G. P. (2015). *Good thinking: What you need to know to be smarter, safer, wealthier, and wiser.* Amherst, NY: Prometheus.

Harrison, G. P. (2017). *Think before you like: Social media's effect on the brain and the tools you need to navigate your newsfeed.* New York: Prometheus.

Hasher, L., Goldstein, D., & Toppino, T. (1977). Frequency and the conference of referential validity. *Journal of Verbal Learning and Verbal Behavior, 16*(1), 107–12.

Hastie, R., Schkade, D. A., & Payne, J. W. (1999). Juror judgment in civil cases: Hindsight effects on judgments of liability for punitive damages. *Law and Human Behavior, 23*, 597–614.

Hawking, S. (1998). *A brief history of time* (Updated and expanded tenth anniversary ed.). New York: Bantam.

Hawking, S., & Tongue, S. (Director). (2010). *Into the universe* [TV miniseries]. Silver Spring, MD: Discovery Channel.

Hawkins, J. (2021, April 15). How your brain understands the world and why it sometimes goes wrong. *Skeptical Inquirer Presents.* (L. Lord, Interviewer). Retrieved April 19, 2021, from http://centerforinquiry.org/video/how-your-brain-understands-the -world-and-why-it-sometimes-goes-wrong/

Hawkins, S. A., & Hastie, R. (1990). Hindsight: Biased judgments of past events after the outcomes are known. *Psychological Bulletin, 107*, 311–27.

Hayes, D. P. (2007). Nutritional hormesis. *European Journal of Clinical Nutrition, 61*, 147–59.

Hecht, D. K. (2019). Pseudoscience and the pursuit of truth. In A. B. Kaufman & J. C. Kaufman (Eds.), *Pseudoscience: The conspiracy against science* (pp. 3–20). Cambridge, MA: MIT Press.

Helfand, D. J. (2017). Surviving the misinformation age. *Skeptical Inquirer, 41*(3), 34–39.

Herculano-Houzel, S. (2009). The human brain in numbers: A linearly scaled-up primate brain. *Frontiers of Human Neuroscience, 3*, 31.

Hergovich, A. (2004). The effect of pseudo-psychic demonstrations as dependent on belief in paranormal phenomena and suggestibility. *Personality and Individual Differences, 36*(2), 365–80.

Hergovich, A., & Arendasy, M. (2005). Critical thinking ability and belief in the paranormal. *Personality and Individual Differences, 38*(8), 1805–12. doi:10.1016/j.paid.2004.11.008

Hermes, B. M. (2019). An inside look at naturopathic medicine: A whistleblower's deconstruction of its core principles. In A. B. Kaufman & J. C. Kaufman (Eds.), *Pseudoscience: The conspiracy against science* (pp. 137–69). Cambridge, MA: MIT Press.

Hermes, B. M. (2020). Beware the naturopathic cancer quack. *Skeptical Inquirer, 44*(2), 38–44.

Hespel, P., Maughan, R. J., & Greenhaff, L. (2006). Dietary supplements for football. *Journal of Sports Sciences, 24*(7), 749–61.

Hess, Y. D., & Pickett, C. L. (2018). Social cognition and attitudes. In R. Biswas-Diener & E. Diener (Eds.), *Discover psychology 2.0: A brief introductory text* (pp. 485–502). Champaign, IL: DEF Publishers.

Hill, K. (2013). A million poisoning planes. *Skeptical Inquirer.* Retrieved from https://www.csicop.org/specialarticles/show/a_million_poisoning_planes

Hines, T. (2003). *Pseudoscience and the paranormal* (Second ed.). Amherst, NY: Prometheus.

Hirst, W., & Phelps, E. (2016). Flashbulb memories. *Current Directions in Psychological Science, 25*(1), 36–41.

Hirst, W., Phelps, E., et al. (2015). A ten-year follow-up of a study of memory for the attack of September 11, 2001: Flashbulb memories and memories for flashbulb events. *Journal of Experimental Psychology: General, 44*(3), 604–23.

Hofling, C. K., Brotzman, E., Dalrymple, S., Graves, N., & Pierce, C. M. (1966). An experimental study in nurse-physician relationships. *Journal of Nervous and Mental Disease, 143*(2), 171–80.

Hofmann, S. G., Lehman, C. L., & Barlow, D. H. (1997). How specific are specific phobias? *Journal of Behavioral Therapy and Experimental Psychiatry,, 28*(3), 233–40.

Hollingworth, R. M., Bjeldanes, L. F., Bolger, M., Kimber, I., Meade, B. J., Taylor, S. L., & Wallace, K. B. (2003). The safety of genetically modified foods produced through biotechnology. *Toxicological Sciences, 71*(1), 2–8.

Honda, H., Shimizu, Y., & Rutter, M. (2005). No effect of MMR withdrawal on the incidences of autism: A total population study. *Journal of Child Psychology and Psychiatry, 46*(6), 572–79. doi:doi.org/10.1111/j.1469-7610.2005.01425.x

Hood, B. (2009). *Supersense: From superstition to religion—the brain science of belief.* London: Constable.

Howard, J., & Reiss, D. R. (2019). The anti-vaccine movement: A litany of fallacy and errors. In A. B. Kaufman & J. C. Kaufman (Eds.), *Pseudoscience: The conspiracy against science* (pp. 195–219). Cambridge, MA: MIT Press.

Hróbjartsson, A., & Gøtzsche, C. (2001). Is the placebo powerless? An analysis of clinical trials comparing placebo with no treatment. *New England Journal of Medicine, 344*(21), 1594–1602.

Hurley, J. (2012). *A concise introduction to logic* (Eleventh ed.). Boston, MA: Clark Baxter.

Hyman, I. E., Husband, T. H., & Billings, F. J. (1995). False memories of childhood experiences. *Applied Cognitive Psychology, 9*(3), 181–97.

Hyman, R. (1977, Spring/Summer). Cold reading: How to convince strangers that you know all about them. *The Zetetic,* 18–37.

Hyman, R. (2001). *How people are fooled by ideomotor action.* Retrieved July 7, 2020, from Quackwatch: http://www.quackwatch.org/01QuackeryRelatedTopics/ideo motor.html

Iacono, W. G. (2001). Forensic "lie detection": Procedures without scientific basis. *Journal of Forensic Psychology Practice, 1*(1), 75–86.

Iacono, W. G., & Lykken, D. T. (1997). The validity of the lie detector: Two surveys of scientific opinion. *Journal of Applied Psychology, 82*(3), 426–33.

Icke, D. (2001). *Children of the Matrix: How an interdimensional race has controlled the world for thousands of years—and still does.* Isle of Wight: David Icke Books.

Icke, D. (2002). *Alice in Wonderland and the World Trade Center disaster: Why the official story of 9/11 is a monumental lie.* Hoboken, NJ: Bridge of Love Publications.

IEP staff. (n.d.). *Deductive and inductive arguments.* Retrieved July 7, 2020, from Internet Encyclopedia of Philosophy: http://www.iep.utm.edu/ded-ind/

Innes, M. (2020). Techniques of disinformation: Constructing and communicating "soft facts" after terrorism. *British Journal of Sociology,* 1–16. doi:https://doi .org/10.1111.1468-4446.12735

Innocence Project. (2009). *Eyewitness misidentification.* Retrieved July 7, 2020, from Innocent Project: http://innocenceproject.org/understand/Eyewitness-Misidentifi cation.php

Interlandi, J. (2016). The acupuncture myth. *Scientific American, 315*(2), 24–25.

Ioannidis, J. (2004). Contradicted and initially stronger effects in highly cited clinical research. *Journal of the American Medical Association, 294,* 218–28.

Ioannidis, J. P. (2005). Why most published research findings are false. *PLoS Medicine, 2*(8), e124. doi:10.1371/journal.pmed.0020124

IPCC (2021). *Climate change 2021: The physical science basis.* Contribution of Working Group I to the Sixth Assessment Report of the Intergovernmental Panel on Climate Change (Masson-Delmotte, V., P. Zhai, A. Pirani, S. L. Connors, C. Péan, S. Berger, N. Caud, Y. Chen, L. Goldfarb, M. I. Gomis, M. Huang, K. Leitzell, E. Lonnoy, J. B. R. Matthews, T. K. Maycock, T. Waterfield, O. Yelekçi, R. Yu, and B. Zhou [Eds.]). Cambridge: Cambridge University Press.

Irwin, H. J., Dagnall, N., & Drinkwater, K. (2012). Paranormal belief and biases in reasoning underlying the formation of delusions. *Australian Journal of Parapsychology, 12*(1), 7–21.

Jacobson, J. W., Mulick, J. A., & Schwartz, A. A. (1995). A history of facilitated communication: Science, pseudoscience, and antiscience. *American Psychologist, 50,* 750–65.

Jain, A., Marshall, J., Buikema, A., Bancroft, T., Kelly, J. P., & Newschaffer, C. J. (2015). Autism occurrence by MMR vaccine status among US children with older siblings with and without autism. *Journal of the American Medical Association, 313*(15), 1534. doi:doi.org/10.1001/jama.2015.3077

Jaspers, K. (1913/1997). *General psychopathology* (Vol. 1 and 2). (J. Hoenig & M. W. Hamilton, Trans.). Baltimore, MD: Johns Hopkins University Press.

Jenkins, H. M., & Ward, W. C. (1965). Judgment of contingency between responses and outcomes. *Psychological Monographs: General and Applied, 79*(1), 1–17

Johnson, D. D., & Fowler, J. H. (2011). The evolution of overconfidence. *Nature, 477*(7364), 317–20.

Johnson, D. K. (2018). Countless counterfeits. In R. Arp, S. Barbone, & M. Bruce (Eds.), *Bad arguments: 100 of the most important fallacies in Western philosophy* (pp. 140–44). Hoboken, NJ: Wiley-Blackwell.

Johnson, D. K. (2020). Countless counterfeits: A new logical fallacy? *Skeptic, 25*(1), 52–53.

Johnson, J. T., Cain, L. M., Falke, T. L., Hayman, J., & Perillo, E. (1985). The "Barnum effect" revisited: Cognitive and motivational factors in the acceptance of personality descriptions. *Journal of Personality and Social Psychology, 49*(5), 1378–91.

Johnson, M., & Ghuman, P. (2020). *Blindsight: The (mostly) hidden ways marketing reshapes our brains.* Dallas, TX: BenBella.

Jolley, D. (2013). Are conspiracy theories just harmless fun? *The Psychologist, 26*(1), 60–62.

Judd, C. M., & Park, B. (1993). Definition and assessment of accuracy in social stereotypes. *Psychological Review, 100*(1), 109–28.

Judd, C. M., Park, B., Ryan, C. S., et al. (1995). Stereotypes and ethnocentrism: Diverging interethnic perceptions of African American and white American youth. *Journal of Personality and Social Psychology, 69,* 460–81.

Kahneman, D. (2013). *Thinking, fast and slow.* New York: Farrar, Straus, and Giroux.

Kalichman, S. C. (2018). "HIV does not cause AIDS": A journey into AIDS denialism. In A. B. Kaufman & J. C. Kaufman (Eds.), *Pseudoscience: The conspiracy against science* (pp. 419–39). Cambridge, MA: MIT Press.

Kanowski, S., Hermann, W. M., Stephan, K., Wierich, W., & Horr, R. (1996). Proof of efficacy of the ginkgo biloba special extract EGb 761 in outpatients suffering from mild to moderate primary degenerative dementia of the Alzheimer type or multi-infarct dementia. *Pharmacopsychiatry, 29,* 47–56.

Kassan, P. (2016). I am not living in a computer simulation, and neither are you. *Skeptic, 21*(4), 37–39.

Kassin, S. (2005). On the psychology of confessions: Does innocence put innocents at risk? *American Psychologist, 60*(3), 215–28. doi:10.1037/0003-066X.60.3.215

Kassin, S. M., Ellsworth, C., & Smith, V. L. (1989). The "general acceptance" of psychological research on eyewitness testimony: A survey of the experts. *American Psychologist, 44,* 1089–98.

Kassin, S. M., & Kiechel, K. L. (1996). The social psychology of false confessions: Compliance, internalization, and confabulation. *Psychological Science, 7*(3), 125–28.

Keely, B. (1999). Of conspiracy theories. *Journal of Philosophy, 96*(3), 109–26.

Keinan, G. (1994). Effects of stress and tolerance of ambiguity on magical thinking. *Journal of Personality and Social Psychology, 67,* 48–55.

Keinan, G. (2002). The effects of stress and desire for control on superstitious behavior. *Personality and Social Psychology Bulletin, 28,* 102–8.

Kelly, I. W. (1997). Modern astrology: A critique. *Psychological Reports, 81,* 931–62.

Kelly, I. W. (1998). Why astrology doesn't work. *Psychological Reports, 82,* 527–46.

Kenner, R., Robledo, M. (Producers), Kenner, R., Roberts, K. (Writers), & Kenner, R. (Director). (2014). *Merchants of Doubt* [Motion picture]. Los Angeles, CA: Participant Media.

Kienle, G. S., & Kiene, H. (1997). The powerful placebo: Fact or fiction? *Journal of Clinical Epidemiology, 50*(12), 1311–18.

King, L. A., Burton, C. M., & Hicks, J. A. (2007). Ghosts, UFOs and magic: Positive affect and the experiential system. *Journal of Personality and Social Psychology, 92*(5), 905–19. doi:10.1037/0022-3514.92.5.905

Kiran, C., & Chaudhury, S. (2009). Understanding delusions. *Industrial Psychiatry Journal, 18,* 3–18. doi:10.4103/0972-6748.57851

Kloor, K. (2019). UFOs won't go away. *The Skeptic, 39*(2), 46–56.

Knox, R. A. (1993, October 14). Mozart makes you smarter, California researchers suggest. *The Boston Globe.*

Kolers, P., & von Grünau, M. (1976). Shape and color in apparent motion. *Vision Research, 16,* 329–35.

Kompf, M., & Bond, R. (2001). The craft of teaching adults. *Critical Reflection in Adult Education,* 21–38.

Korva, N., Porter, S., O'Connor, B. P., Shaw, J., & ten Brinke, L. (2013). Dangerous decisions: Influence of juror attitudes and defendant appearance on legal decision-making. *Psychiatry, Psychology and Law, 20*(3), 384–98.

Krauss, L. (2013). *A universe from nothing: Why there is something rather than nothing.* New York: Atria.

Krauss, L. (2017). *The greatest story ever told . . . so far: Why are we here?* London: Simon & Schuster.

Kruger, J., & Dunning, D. (1999). Unskilled and unaware of it: How difficulties in recognizing one's own incompetence lead to inflated self-assessments. *Journal of Personality and Social Psychology, 77*(6), 1121–34.

Kuhn, D., Weinstock, M., & Flaton, R. (1994). How well do jurors reason? Competence dimensions of individual variation in a juror reasoning task. *Psychological Science, 5,* 289–96.

Kuhn, T. (1962/2012). *The structure of scientific revolutions* (Fourth ed.). Chicago: University of Chicago Press.

Kunda, Z. (1990). The case for motivated reasoning. *Psychological Bulletin, 108*(3), 480–98.

Ladendorf, B., & Ladendorf, B. (2018). Wildlife apocolypse: How myths and superstitions are driving animal extinctions. *Skeptical Inquirer, 42*(4).

Landau, M. J., Kay, A. C., & Whitson, J. A. (2015). Compensatory control and the appeal of a structured world. *Psychological Bulletin, 141*(3), 694.

Lander, U. (n.d.). *Introduction to logic: Deductive and inductive arguments.* Retrieved July 7, 2020, from Philosophy: http://philosophy.lander.edu/logic/ded_ind.html

Laney, C., & Loftus, E. F. (2018). Eyewitness testimony and memory bias. In R. Biswas-Diener & D. Diener (Eds.), *Discover psychology 2.0: A brief introductory text* (pp. 230–41). Champaign, IL: DEF Publishers.

Langer, E. J. (1975). The illusion of control. *Journal of Personality and Social Psychology, 32,* 311 28. doi10.1037/0022-3514 32 2 311

Langer, E. J., & Roth, J. (1975). Heads I win, tails it's chance: The illusion of control as a function of the sequence of outcomes in a purely chance task. *Journal of Personality and Social Psychology, 32*(6), 951–55.

Larson, E. J. (2002). *The theory of evolution: A history of controversy.* Chantilly, VA: Great Courses.

Laudan, L. (1981). A confutation of convergent realism. *Philosophy of Science, 48*(1), 19–49.

Law, S. (2007). *Philosophy, Eyewitness Companion Guides.* New York: DK Publishing.

Law, S. (2011). *Believing bullshit: How not to get sucked into an intellectual black hole.* Amherst, NY: Prometheus.

Law, S. (2017). Must humanists be naturalists? *Free Inquiry, 37*(6), 44–46.

Laws, K. R., Sweetnam, H., & Kondel, T. K. (2012). Is ginkgo biloba a cognitive enhancer in healthy individuals? A meta-analysis. *Human Psychopharmacology, 27*(6), 527–33.

Lease, D. (1999). Th-that's all folks! It's the end of the world . . . again. *Skeptic, 7*(3), 52–53.

Lee, B. (2016, October 10). *Model Katie May's death raises more questions about chiropractors.* Retrieved July 7, 2020, from *Forbes*: http://forbes.com/sites/brucelee/2016/10/23/model-katie-mays-death-raises-more-questions-about-chiropractors/#3276c87212aa

Leeper, R. (1935). A study of a neglected portion of the field of learning—The development of sensory organization. *Journal of Genetics and Psychology, 46,* 41–75.

Lefevre, J. (2000). Research on the development of academic skills: Introduction to the special issue on early literacy and early numeracy. *Canadian Journal of Experimental Psychology, 54*(2), 57–60. doi:10.1037/h0088185

Le Guin, U. K. (2000). *The telling.* San Diego, CA: Harcourt.

Leikind, B., & McCarthy, W. (1985–1986). An investigation of firewalking. *Skeptical Inquirer, 10,* 23–24.

Lenggenhager, B., Tadi, T., Metzinger, T., & Blanke, O. (2007). Video ergo sum: Manipulating bodily self-consciousness. *Science, 317*(5841), 1096–99.

Lent, R., Azevedo, F. A., Andrade-Moraes, C. H., & Pinto, A. V. (2012). How many neurons do you have? Some dogmas of quantitative neuroscience under revision. *European Journal of Neuroscience, 35,* 1–9.

Levin, A. L. (1968). *Problems and materials on trial advocacy.* Mineola, NY: Foundation Press.

Levin, D. T., & Angelone, B. L. (2008). The visual metacognition questionnaire: A measure of intuitions about vision. *American Journal of Psychology, 121,* 451–472.

Levine, J. D., Gordon, N. C., & Fields, H. L. (1978). The mechanism of placebo analgesia. *Lancet, 312,* 654–57.

Levine, R. V. (2018). Persuasion: So easily fooled. In R. Biswas-Diener & E. Diener (Eds.), *Discover psychology 2.0: A brief introductory text* (pp. 516–33). Champaign, IL: DEF Publishers.

Lewandowsky, S. (2019). In whose hands the future? In J. E. Uscinski (Ed.), *Conspiracy theories and the people who believe them* (pp. 149–77). New York: Oxford University Press.

Lewandowsky, S., Gignac, G. E., & Oberauer, K. (2013). The role of conspiracist ideation and worldviews in predicting rejection of science. *PLoS One, 8*(10), e75637.

Lewandowsky, S., Oberauer, K., & Gignac, G. E. (2013). NASA faked the moon landings—therefore, (climate) science is a hoax: An anatomy of the motivated rejection of ccience. *Psychological Science, 24,* 622–33.

Lichtenstein, S., & Fischhoff, B. (1977). Do those who know more also know more about how much they know? *Organizational Behavior and Human Performance, 20,* 159–83.

Lichtenstein, S., Slovic, P., Fischhoff, B., Layman, M., & Combs, B. (1978). Judged frequency of lethal events. *Journal of Experimental Psychology: Human Learning and Memory, 4*(6), 551–78. doi:10.1097/0278-7393.4.6.551

Lie, E., & Newcombe, N. S. (1999). Elementary school children's explicit and implicit memory for faces of preschool classmates. *Developmental Psychology, 35*(1), 102.

Lilienfeld, S. O. (2005). Scientifically unsupported and supported interventions for childhood psychopathy: A summary. *Pediatrics, 115*(3), 761–64.

Lilienfeld, S. O. (2007). Psychological treatments that cause harm. *Perspectives on Psychological Science, 2,* 53–70.

Lilienfeld, S. O. (2017). Teaching skepticism: How early should we begin? *Skeptical Inquirer, 41*(5), 30–31.

Lilienfeld, S. O., Lynn, S. J., & Lohr, J. M. (2014). Science and pseudoscience in clinical psychology: Initial thoughts, reflections, and considerations. In S. O. Lilienfeld, S. J. Lynn, & J. M. Lohr (Eds.), *Science and pseudoscience in clinical psychology* (Second ed., pp. 1–18). New York: Guilford.

Lillquist, O., & Lindeman, M. (1998). Belief in astrology as a strategy for self-verification and coping with negative life events. *European Psychologist, 3*(3), 302.

Lindeman, M. (2011). Biases in intuitive reasoning and belief in complementary and alternative medicine. *Psychology and Health, 26*(3), 371–82.

Lindman, M., & Aarnio, K. (2007). Superstitious, magical, and paranormal beliefs: An integrative model. *Journal of Research in Personality, 41*(4), 731–44.

Lindner, I., et al. (2010). Observation inflation: Your actions become mine. *Psychological Science, 21*(9), 1291–99.

Lindsay, R. A. (2017). Why skepticism? Sasquatch, broken windows, and public policy. *Skeptical Inquirer, 41*(2), 46–50.

Linson, A., Chaffin, C., Bell, R. G. (Producers), & Fincher, D. (Director). (1999). *Fight Club* [Motion picture]. Los Angeles, CA: 20th Century Fox.

Little, N. J. (2019). Suing for science. *Skeptical Inquirer, 43*(5), 60–63.

Lobato, E., Mendoza, J., Sims, V., & Chin, M. (2014). Examining the relationship between conspiracy theories, paranormal beliefs, and pseudoscience acceptance among a university population. *Applied Cognitive Psychology, 28*(5), 617–25. doi:10.1002/acp.3042

Lobato, E. J., & Zimmerman, C. (2018). The psychology of (pseudo) science: Cognitive, social, and cultural factors. In A. B. Kaufman & J. C. Kaufman (Eds.), *Pseudoscience: The conspiracy against science* (pp. 21–43). Cambridge, MA: MIT Press.

Loftus, E. (1975). Leading questions and the eyewitness report. *Cognitive Psychology, 562*, 562. Retrieved from https://webfiles.uci.edu/eloftus/CognitivePsychology75.pdf

Loftus, E. (1993). The reality of repressed memories. *American Psychologist, 48*(5), 524.

Loftus, E. (2005). Planting misinformation in the human mind: A 30-year investigation of the malleability of memory. *Learning and Memory, 12*(4), 361–66.

Loftus, E. (2018, June). *How reliable is your memory?* Retrieved July 7, 2020, from TED: http://www.ted.com/talks/elizabeth_loftus_the_fiction_of_memory

Loftus, E., & Hoffman, H. G. (1989). Misinformation and memory: The creation of new memories. *Journal of Experimental Psychology, 188*(1), 100–4.

Loftus, E., & Ketcham, K. (1994). *The myth of repressed memory: False memories and the allegations of sexual abuse.* New York: St. Martin's.

Loftus, E., & Palmer, J. (1974). Reconstruction of automobile destruction: An example of the interaction between language and memory. *Journal of Verbal Learning and Verbal Behavior, 13*(5), 585–89.

Loftus, E. F. (1995). The formation of false memories. *Psychiatric Annals, 25*, 720–25.

Loftus, E. F. (1997). Creating false memories. *Scientific American, 277*(3), 70–75.

Loftus, E. F., Miller, D. G., & Burns, H. J. (1978). Semantic integration of verbal information into a visual memory. *Journal of Experimental Psychology: Human Learning and Memory, 4*, 19–31.

Lord, C. G., Ross, L., & Lepper, M. R. (1979). Biased assimilation and attitude polarization: The effects of prior theories on subsequently considered evidence. *Journal of Personality and Social Psychology, 37*(11), 2098.

Loxton, D. (2017). Terrifying! Impossible! Chemtrails! *Skeptic, 22*(2), 64–73.

Loxton, D. (2018a). Secrets of the Ouija board. *Skeptic, 23*(4), 64–73.

Loxton, D. (2018b). The perpetual quest for perpetual motion. *Skeptic, 23*(2), 64–73.

Loxton, D. (2018c). The startling truth behind claims of astral projection. *Skeptic, 23*(3).

Loxton, D. (2019). Understanding flat Earthers. *Skeptic, 24*(4), 10–23.

Lunsford, A. A., RuszKiewicz, J. J., & Walters, K. (2013). *Everything's an argument: With readings.* Boston, MA: Bedford.

Lusty, N. (2006). Teaching quantitative reasoning. *APS Observer, 19*, 35–36.

Lynn, S. J., Gautam, A., Ellenberg, S., & Lilienfeld, S. O. (2018). Hypnosis: Science, pseudoscience, and nonsense. In A. B. Kaufman & J. C. Kaufman (Eds.), *Pseudoscience: The conspiracy against science* (pp. 331–49). Cambridge, MA: MIT Press.

Macdonald, T. (2014, February 14). *How do we really make decisions?* Retrieved July 7, 2020, from BBC News: https://www.bbc.com/news/science-environment-26258662

Macrae, C. N., & Bodenhausen, G. V. (2000). Social cognition: Thinking categorically about others. *Annual Review of Psychology, 51*, 93–120.

Macrae, C. N., Milne, A. B., & Bodenhausen, G. V. (1994). Stereotypes as energy-saving devices: A peek inside the cognitive toolbox. *Journal of Personality and Social Psychology, 66*(1), 37–47.

Maddox, J., Randi, J., & Stewart, W. (1998). "High-dilution" experiments a delusion. *Nature, 344*(6180), 287–90.

Madsen, M. V., Gøtzsche, C., & Hróbjartsson, A. (2009). Acupuncture treatment for pain: Systematic review of randomized clinical trials with acupuncture, placebo acupuncture, and no acupuncture groups. *British Medical Journal, 3115*. Retrieved from http://bmj.com/cgi/reprint/338/jan27_2/a3115?maxtoshow=&HITS=10&hits=10& RESULTFORMAT=&fulltext=sham+acupuncture&searchid=1&FIRSTINDEX=0 &resourcetype=HWCIT

Maglione, M. A., Das, L., Raaen, L., Smith, A., Chari, R., Newberry, S., & Gidengil, C. (2014). Safety of vaccines used for routine immunization of US children: A systematic review. *Pediatrics, 134*(2), 325–37. doi:10.1542/peds.2014-1079

Mahoney, M. J. (1977). Publication prejudices: An experimental study of confirmation bias in the peer review system. *Cognitive Therapy and Research, 1*(2), 161–75.

Mann, M. E., & Toles, T. (2018). *The madhouse effect: How climate change denial is threatening our planet, destroying our politics, and driving us crazy.* New York: Columbia University Press.

Matute, H., Yarritu, I., & Vadillo, M. (2011). Illusions of causality at the heart of pseudoscience. *British Journal of Psychology, 102*, 392–405. doi:10.1348/000712610X532210

Mauer, M. (2005). Thinking about prison and its impact in the twenty-first century. *Ohio State Journal of Criminal Law, 2*, 607–18.

Mazzoni, G. A., & Loftus, E. F. (1998). Dream interpretation can change beliefs about the past. *Psychotherapy, 35*, 177–87.

McCarthy, A. (2005). Literature review. *Journal of Pediatric Health Care, 19*, 337–38.

McDermott, K. B. (2018). Memory (encoding, storage, retrieval). In R. Biswas-Diener & E. Diener (Eds.), *Discover psychology 2.0: A brief introductory text* (pp. 210–29). Champaign, IL: DEF Publishers.

McKee, R. D., & Squire, L. R. (1992). Equivalent forgetting rates in long-term memory for diencephalic and medial temporal lobe amnesia. *Journal of Neuroscience, 12*, 3765–72.

McKenzie-Mcharg, A. (2019). Conspiracy theory: The nineteenth-century prehistory of a twentieth-century concept. In J. E. Uscinski (Ed.), *Conspiracy theories and the people who believe them* (pp. 62–81). New York: Oxford University Press.

McKinstry, B., & Wang, J. (1991). Putting on the style: What patients think of the way their doctor dresses. *British Journal of General Practice, 412*, 275–78.

McNally, R. (2003). *Remembering trauma.* Cambridge, MA: Harvard University Press.

McNally, R. J. (2004). The science and folklore of traumatic amnesia. *Clinical Psychology: Science and Practice, 11*, 29–33.

McNally, R. J. (2012). Explaining "memories" of space alien abduction and past lives: An experimental psychopathology approach. *Journal of Experimental Psychopathology, 3*(1), 2–16. https://doi.org/10.5127/jep.017811

McNeil, B., Pauker, S. G., Sox, H. C., Jr., & Tversky, A. (1982). On the elicitation of preferences for alternative therapies. *New England Journal of Medicine, 306*, 1259–62.

Measom, T., Weinstein, J. (Producers), Weinstein, J., Measom, T., O'Toole, G. (Writers), Weinstein, J., & Measom, T. (Directors). (2014). *An Honest Liar* [Motion picture]. Salt Lake City, UT: Left Turn Films.

Meehl, E. (1956). Wanted—A good cook book. *American Psychologist, 11*(6), 262–72.

Mehl, M. R. (2018). Conducting psychology research in the real world. In R. Biswas-Diener & E. Diener (Eds.), *Discover psychology 2.0: A brief introductory text* (pp. 30–45). Champaign, IL: DEF Publishers.

Mele, C. (2017, May 5). Minnesota sees largest outbreak of measles in almost 30 years. *New York Times*.

Memmert, D. (2006). The effects of eye movements, age, and expertise on inattentional blindness. *Consciousness and Cognition, 15*, 620–27.

Michno, G. F. (2014). Ghostbusters busted at Sand Creek. *Skeptic, 19*(1), 52–56.

Mielczarek, E., & Engler, B. (2012). Measuring mythology: Startling concepts in NCCAM grants. *Skeptical Inquirer, 36*(1), 35–43.

Mikkelson, B., & Mikkelson, D. (n.d.). *Doll talk*. Retrieved July 7, 2020, from Snopes: http://graphics1.snopes.com/business/audio/mammycoo.mp3

Milgram, S. (1963). Behavioral study of obedience. *Journal of Abnormal and Social Psychology, 67*(4), 371–78.

Miller, R. A., & Albert, K. (2015). If it leads, it bleeds (and if it bleeds, it leads): Media coverage and fatalities in militarized interstate disputes. *Political Communication, 32*, 61–82. doi:10.1080/10584609.2014.880976

Mlowdinow, L. (2009). *The drunkard's walk: How randomness rules our lives.* New York: Vintage Books.

Mohr, C., Landis, T., & Brugger, P. (2006). Lateralized semantic priming: Modulation by levodopa, semantic distance, and participants' magical beliefs. *Neuropsychiatric Disease and Treatment, 2*(1), 71–84.

Molé, P. (1999). Celestine profits: A critical analysis of James Redfield and the Celestine Prophesy. *Skeptic, 7*(3), 76–81.

Molé, P. (2006, September 11). *9/11 Conspiracy Theories: The 9/11 Truth Movement in Perspective.* Retrieved July 7, 2020, from eSkeptic: www.skeptic.com/eskeptic/06-09-11

Moore, D. W. (2005, June 16). *Three in four Americans believe in paranormal.* Retrieved July 7, 2020, from Gallup News Service: http://gallup.com/poll/16915/Three-Four-Americans-Believe-Paranormal.aspx

Moreland, R. L., & Beach, S. R. (1992). Exposure effects in the classroom: The development of affinity among students. *Journal of Experimental Social Psychology, 28*, 255–76.

Moseley, B., O'Malley, K., Petersen, N. J., Menke, T. J., Brody, B. A., Kuykendall, D. J., . . . Wray, N. P. (2002). A controlled trial of arthroscopic surgery for osteoarthritis of the knee. *New England Journal of Medicine, 347*(2), 81–88.

Mukherjee, S. (2016). *The gene: An intimate history.* New York: Scribner.

Mullen, R., & Linscott, R. J. (2010). A comparison of delusions and overvalued ideas. *Journal of Nervous and Mental Diseases, 198*, 35–38.

Murrie, D. C., et al. (2013). Are forensic experts biased by the side that retained them? *Psychological Science, 24*, 1889–97.

354 THINK STRAIGHT

Musch, J., & Ehrenberg, K. (2002). Probability misjudgment, cognitive ability, and belief in the paranormal. *British Journal of Psychology, 93*(2), 169–77.

National Health and Medical Research Council. (2015). *NHMRC statement on homeopathy and NHMRC information paper: Evidence on the effectiveness of homeopathy for treating health conditions* (NHMRC Publication No. CAM02). Retrieved from http://www.nhmrc.gov.au/guidelines-publications/cam02 or http://www.nhmrc.gov.au/about-us/publications/homeopathy

National Research Council. (2003). *The polygraph and lie detection.* Washington, DC: National Academies Press.

Nazé, Y. (2018). A doctoral dissertation on a geocentric flat Earth: "Zetetic" astronomy at the university level. *Skeptical Inquirer, 42*(3), 12–14.

Nees, M. A., & Phillips, C. (2015). Auditory pareidolia: Effects of contextual priming on perceptions of purportedly paranormal and ambiguous auditory stimuli. *Applied Cognitive Psychology, 29*(1), 129–35.

Neisser, U., & Harsh, N. (1992). Phantom flashbulbs: False recollections of hearing the news about *Challenger.* In E. Winograd & U. Neisser (Eds.), *Affect and accuracy in recall: Studies of "flasbulb" memories* (pp. 9–31). New York: Cambridge University Press.

Nesse, R. M. (2001). The smoke detector principle. *Annals of the New York Academy of Sciences, 935*(1), 75–85. doi:10.1111/j.1749-6632.2001.tb03472.x

Nesse, R. M., & Ellsworth, P. C. (2009). Evolution, emotion, and emotional disorders. *American Psychologist, 64*, 129–39.

Newport, F., & Strausberg, M. (2001, June 8). *Americans' belief in psychic and paranormal phenomena is up over the last decade.* Retrieved July 7, 2020, from Gallup News Service: http://www.gallup.com/poll/4483/Americans-Beilef-Psychic-Paranormal-Phenomena-Over-Last-Decade.aspx

Nichols, A. L., & Maner, J. K. (2008). The good-subject effect: Investigating participant demand characteristics. *Journal of General Psychology, 135*(2), 151–65.

Nickell, J. (2019). Magic waters. *Skeptical Inquirer, 43*(5), 44–49.

Nickell, J., & Biddle, K. (2020). So you have a ghost in your photo. *Skeptical Inquirer, 44*(4), 39–43.

Nickerson, R. S. (1998). Confirmation bias: A ubiquitous phenomenon in many guises. *Review of General Psychology, 2*, 175–220.

Nicolia, A., Manzo, A., Veronesi, F., & Rosellini, D. (2013). An overview of the last 10 years of genetically engineered crop safety research. *Critical Review in Biotechnology, 34*(1), 77–88.

Nirenberg, E. (2020). *Long-term effects of COVID-19 vaccines: Should you be worried?* Retrieved March 14, 2022, from https://edwardnirenberg.medium.com/long-term-effects-of-covid-19-vaccines-should-you-be-worried-c3c3a547b565

Nisbet, M. C. (2020). Against climate change tribalism: We gamble with the future by dehumanizing our opponents. *Skeptical Inquirer, 44*(1), 26–28.

Novella, S. (2010a, December 20). *Conspiracy thinking: Skepticism's evil twin.* Retrieved July 7, 2020, from NeuroLogica Blog: http://theness.com/neurologicablog/index.php/conspiracy-thinking-skepticisms-evil-twin

Novella, S. (2010b). The poor, misunderstood placebo. *Skeptical Inquirer, 34*(6), 33–35.

Novella, S. (2012). *Your deceptive mind: A scientific guide to critical thinking skills.* Chantilly, VA: The Teaching Company.

Novella, S. (2018). *The skeptics' guide to the universe: How to know what's really real in a world increasingly full of fake.* New York: Grand Central.

Novella, S. (2019). *Black salve still thriving online.* Retrieved March 16, 2022, from Science-Based Medicine: https://sciencebasedmedicine.org/black-salve-still-thriving-online/

Nuzzo, R. (2015). Fooling ourselves. *Nature, 526*(7572), 182–85.

Nye, B. (2015). *Undeniable: Evolution and the science of creation.* New York: St. Martin's Griffin.

Nyhan, B., & Reifler, J. (2015). Does correcting myths about the flu vaccine work? An experimental evaluation of the effects of corrective information. *Vaccine, 33,* 459–64. doi:10.1016/j.vaccine.2014.11.017

Nyhan, B., & Reifler, J. (2018). The roles of information deficits and identity threat in the prevalence of misperceptions. *Journal of Elections, Public Opinion and Parties, 29*(2), 222–44.

Nyhan, B., Reifler, J., Richey, S., & Freed, G. L. (2014, June 14). Effective messages in vaccine promotion: A randomized trial. *Indian Pediatrics, 51,* 491–93. doi:10.1542/peds.20133-2365

O'Keeffe, C., & Wiseman, R. (2005). Testing alleged mediumship: Methods and results. *British Journal of Psychology, 96,* 165–79.

Offit, P. A. (2014). *Do you believe in magic? Vitamins, supplements, and all things natural: A look behind the curtain.* New York: Harper.

Offit, P. A. (2015). *Deadly choices: How the anti-vaccine movement threatens us all.* New York: Basic Books.

Offit, P. A., Quarles, J., Gerber, M. A., Hackett, C. J., Marcuse, E. K., Kollman, T. R., . . . Landry, S. (2002). Addressing parents' concerns: Do multiple vacines overwhelm or weaken the infants' immune system? *Pediatrics, 109*(1), 124–29.

Ofshe, R. J. (1989). Coerced confessions: The logic of seemingly irrational action. *Cultic Studies Journal, 6,* 1–15.

Oliver, J. (2017). Vaccines. *Last Week Tonight with John Oliver.* Retrieved March 15, 2022, from https://www.youtube.com/watch?v=7VG_s2PCH_c

Oliver, J. E., & Wood, T. J. (2014). Conspiracy theories and the paranoid style(s) of mass opinion. *American Journal of Political Science, 58*(4), 952–66.

Olmsted, K. S. (2019). Conspiracy theories in U.S. history. In J. E. Uscinski (Ed.), *Conspiracy theories and the people who believe them* (pp. 285–97). New York: Oxford University Press.

Oreskes, N. (2004). Beyond the ivory tower: The scientific consensus on climate change. *Science, 306*(5702), 1686.

Oreskes, N., & Conway, E. M. (2010a). Defeating the merchants of doubt. *Nature, 465*(10), 686–87.

Oreskes, N., & Conway, E. M. (2010b). *Merchants of doubt: How a handful of scientists obscured the truth on issues from tobacco smoke to global warming.* New York: Bloomsbury Press.

Orne, M. T. (1962). On the social psychology of the psychological experiment: With particular reference to demand characteristics and their implications. *American Psychologist, 17*(11), 776–83.

Orne, M. T., Witehouse, W. G., Orne, E. C., & Dinges, D. F. (1996). "Memories" of anomalous and traumatic autobiographical experiences: Validation and consolidation of fantasy through hypnosis. *Psychological Inquiry, 7*, 168–72.

Otani, D., & Dixon, P. (1976). Power function between duration of friendly interaction and conformity in perception and judgments. *Perceptual and Motor Skills, 43*, 975–78.

Palmer, M. A., Brewer, N., Weber, N., & Nagesh, A. (2013). The confidence-accuracy relationship for eyewitness identification decisions: Effects of exposure duration, retention interval, and divided attention. *Journal of Experimental Psychology: Applied, 19*(1), 55–71.

Panchin, A. Y., & Tuzhikov. (2016). Published GMO studies find no evidence of harm when corrected for multiple comparisons. *Critical Reviews in Biotechnology*, 1–5. doi: 10.3109/07388551.2015.1130684

Pariser, E. (2012). *The filter bubble: What the internet is hiding from you.* New York: Penguin.

Parsons, A. (2019). Reporting on climate change: The good, the bad, and the ugly. *ScienceWriters*, 14–16.

Paulos, J. A. (2001). *Innumeracy: Mathematical illiteracy and its consequences.* New York: Hill and Wang.

Pennycook, G., Cheyne, J. A., Barr, N., Koehler, D. J., & Fugelsang, J. A. (2015). On the reception and detection of pseudo-profound bullshit. *Judgment and Decision Making, 10*(6), 549.

Pennycook, G., Cheyne, J. A., Seli, P., Koehler, D. J., & Fugelsang, J. A. (2012). Analytic cognitive style predicts religious and paranormal belief. *Cognition, 123*(3), 335–46. doi:http://dx.doi.org/10.1016/j.cognition.2012.03.003

Pennycook, G., Fugelsang, J. A., & Koehler, D. J. (2015). Everyday consequences of analytic thinking. *Current Directions in Psychological Science, 24*(6), 425–32.

Perez, B. M., & Hines, T. (2011). The aura: A brief review. *Skeptical Inquirer, 35*(1), 38–40.

Petrosino, A., Turpin-Petrosino, C., & Buehler, J. (2004). "Scared straight" and other juvenile awareness programs for preventing juvenile delinquency. *Campbell Systematic Reviews 2004.2*, 1–38. doi:10.4073/csr.2004.2

Petty, R. E., Wegener, D. T., & White, P. H. (1998). Flexible correction processes in social judgment: Implications for persuasion. *Social Cognition, 16*(1), 93–113.

Pezdek, K. (2003). Event memory and autobiographical memory for the events of September 11, 2001. *Applied Cognitive Psychology, 17*(9), 1033–45.

Pigliucci, M. (2005). Do extraordinary claims require extraordinary evidence? *Skeptical Inquirer, 29*(2), 14.

Pigliucci, M. (2017a). *How to be a Stoic.* New York: Basic Books.

Pigliucci, M. (2017b). The virtuous skeptic. *Skeptical Inquirer, 41*(2), 54–57.

Pinker, S. (2018). *Enlightenment now: The case for reason, science, humanism, and progress.* New York: Penguin.

Pinker, S. (2019). What can science learn from religion? Steven Pinker on religious beliefs and rituals. (M. Shermer, Ed.) *Skeptic, 24*(2), 32–34.

Pirsig, R. (1974). *Zen and the art of motorcycle maintenance.* New York: William Morrow.

Polidoro, M. (2020a). Living on air? The crazy ideas and consequences of breatharians. *Skeptical Inquirer, 44*(1), 18.

Polidoro, M. (2020b). Stop the epidemic of lies! Thinking about COVID-19 misinformation. *Skeptical Inquirer, 44*(4), 15–16.

Popper, K. (1959/2014). *The logic of scientific discovery.* New York: Basic Books.

Poppy, C. (2017). Survey shows Americans fear ghosts, the government, and each other. *Skeptical Inquirer, 41*(1), 16–18.

Porter, S., Yuille, J. C., & Lehman, D. R. (1999). The nature of real, implanted, and fabricated memories for emotional childhood events: Implications for the recovered memory debate. *Law and Human Behavior, 23*(5), 517–37.

Posadzki, P., Alotaibi, A., & Ernst, E. (2012). Adverse effects of homeopathy: A systematic review of published case reports and case series. *International Journal of Clinical Practice, 66*(12), 1178–88.

Posner, G. P. (2000). The face behind the "face" on Mars. *Skeptical Inquirer, 24*(6), 20–26.

Pronin, E., Gilovich, T., & Ross, L. (2004). Objectivity in the eye of the beholder: Divergent perceptions of bias in self versus others. *Psychological Review, 111*, 781–99.

Pronin, E., Kruger, J., Savitsky, K., & Ross, L. (2011). You don't know me, but I know you: The illusion of asymmetric insight. *Journal of Personality and Social Psychology, 81*, 639–56.

Pronin, E., Lin, D. Y., & Ross, L. (2002). The bias blind spot: Perceptions of bias in self versus others. *Personality and Social Psychology Bulletin, 28*(3), 369–81.

Prothero, D., & Callahan, T. (2017). *UFOs, chemtrails and aliens: What science says.* Bloomington: Indiana University Press.

Radford, B. (1999). The ten-percent myth. *Skeptical Inquirer, 23*, 52–53. Retrieved from http://www.csicop.org/si/9903/ten-percent-myth.html

Radford, B. (2009a). Curious contrails: Death from the sky? *Skeptical Inquirer, 33*(2), 25.

Radford, B. (2009b). Psychic exploits horrific abduction case. *Skeptical Inquirer, 36*(6), 6–7.

Radford, B. (2010a). Ghost-hunting mistakes: Science and pseudoscience in ghost investigations. *Skeptical Inquirer, 34*(6), 44–47.

Radford, B. (2010b). The psychic and the serial killer: Examining the "best case" for psychic detectives. *Skeptical Inquirer, 34*(2), 32–37.

Radford, B. (2011). Holly Bobo still missing: Psychics hurt investigation. *Skeptical Inquirer, 35*, 9.

Radford, B. (2017). *Investigating ghosts: The scientific search for spirits.* New York: Rhombus.

Radford, B. (2018). The curious case of ghost taxonomy. *Skeptical Inquirer, 43*(2), 47–49.

Radford, B. (2020). Coronavirus crisis: Chaos, counting, and confronting our biases. *Skeptical Inquirer, 44*(4), 10–14.

Räikkä, J. (2009). The ethics of conspiracy theorizing. *Journal of Value Inquiry, 43*, 457–68.

Räikkä, J., & Basham, L. (2019). Conspiracy theory phobia. In J. E. Uscinski (Ed.), *Conspiracy theories and the people who believe them* (pp. 178–86). New York: Oxford University Press.

Randi, J. (1982–1983). Nostradamus: The prophet for all seasons. *Skeptical Inquirer, 7*(1), 30–37.

Randi, J. (1993). Secrets of the psychics [TV series episode]. *Nova.* Boston, MA: WGBH.

Randi, J. (2007, February). *Homeopathy, quackery and fraud.* Retrieved July 7, 2020, from TED: www.ted.com/talks

Randi, J. (2017a). A consistently erroneous technology. *Skeptical Inquirer, 41*(5), 16–19.

Randi, J. (2017b). The dangerous delusion about vaccines and autism. *Skeptical Inquirer, 41*(2), 29–31.

Rauscher, F. H., Robinson, K. D., & Jens, J. J. (1998). Improved maze learning through early music exposure in rats. *Neurological Research, 20,* 427–32.

Rauscher, F. H., Shaw, G. L., & Ky, K. N. (1993). Music and spatial task performance. *Nature, 365,* 611.

Rauscher, F. H., Shaw, G. L., & Ky, K. N. (1995). Listening to Mozart enhances spatial-temporal reasoning: Towards a neurophysiological basis. *Neuroscience Letters, 185,* 44–47.

Reber, S. A., & Alcock, J. E. (2019). Why parapsychological claims cannot be true. *Skeptical Inquirer, 43*(4), 8–10.

Reber, S. A., & Alcock, J. E. (2020). Searching for the impossible: Parapsychology's elusive quest. *American Psychologist, 75*(3), 391–99. doi:http://doi.org/10.1037/amp0000486

Redelmeier, D. A., & Tversky, A. (1996). On the belief that arthritis pain is related to the weather. *Proceedings of the National Academy of Sciences, 93,* 2895–96.

Redlawsk, D. P., Civettini, A. J., & Emmerson, K. M. (2010). The affective tipping point: Do motivated reasoners ever "get it?" *Political Psychology, 31*(4), 563–93.

Reed, P., Wakefield, D., Harris, J., Parry, M., Cella, M., & Tsakanikos, E. (2008). Seeing non-existent events: Effects of environmental conditions, schizotypal symptoms, and sub-clinical characteristics. *Journal of Behavior Therapy and Experimental Psychiatry, 39*(3), 276–91.

Rehman, S. U., Nietert, P. J., Cope, D. W., & Kilpatrick, A. O. (2005). What to wear today? Effect of doctor's attire on the trust and confidence of patients. *American Journal of Medicine, 118,* 1279–86.

Reiser, M., & Klyver, N. (1982). A comparison of psychics, detectives and students in the investigation of major crimes. In M. Reiser (Ed.), *Police psychology: Collected papers* (pp. 260–67). Los Angeles: Lehi.

Reiser, M., Ludwig, L., Saxe, S., & Wagner, C. (1979). Evaluation of the use of psychics in the investigation of major crimes. *Journal of Police Sciences and Administration, 7*(1), 18–25.

Renckens, C. N., & Dorlo, T. P. (2019). Quackery at the WHO: A Chinese affair. *Skeptical Inquirer, 43*(5), 39–43.

Rice, S. (2020). Creationist funhouse, episode five: God's pet bunny. *Skeptical Inquirer, 44*(6), 52–56.

Ridley, M. (2002). Crop circle confession. *Scientific American, 287*(2), 25.

Riekki, T. (2012). Paranormal and religious believers are more prone to illusory face perception than skeptics and non-believers. *Applied Cognitive Psychology, 27,* 150–55.

Robertson, D. G., & Dyrendal, A. (2019). Conspiracy theories and religion: Superstition, seekership, and salvation. In J. E. Uscinski (Ed.), *Conspiracy theories and the people who believe them* (pp. 411–21). New York: Oxford University Press.

Rokeach, M. (1960). *The open and closed mind: Investigations into the nature of belief systems and personality systems.* New York: Basic Books.

Ronald, P. (2011). Plant genetics, sustainable agriculture and global food security. *Genetics, 188*(1), 11–20.

Rosa, E. (1998). TT and me. *Skeptic, 6*(2), 97–99.

Rosa, L., Rosa, E., Sarner, L., & Barrett, S. (1998). A closer look at therapeutic touch. *Journal of the American Medical Association, 279*(13), 1005–10.

Rosenthal, R. (1979). The "file drawer problem" and tolerance for null results. *Psychological Bulletin, 86*(3), 638–41.

Ross, D. F., Ceci, S. J., Dunning, D., & Toglia, M. P. (1994). Unconscious transference and mistaken identity: When a witness misidentifies a familiar but innocent person. *Journal of Applied Psychology, 79,* 918–30.

Ross, L., Lepper, M. R., & Hubbard, M. (1975). Perseverance in self-perception and social perception: Biased attributional processes in the debriefing paradigm. *Journal of Personality and Social Psychology, 32*(5), 880–92.

Ross, L., & Ward, A. (1996). Naive realism in everyday life: Implications for social conflict and misunderstanding. In T. Brown, E. S. Reed, & E. Turiel (Eds.), *Values and knowledge* (pp. 103–35). Hillsdale, NJ: Erlbaum.

Ross, M. (1989). Relation of implicit theories to the construction of personal histories. *Psychological Review, 96,* 341–57.

Ross, R. R., Fabiano, E. A., & Ewles, C. D. (1988). Reasoning and rehabilitation. *International Journal of Offender Therapy and Comparative Criminology, 32*(1), 29–35.

Rottenberg, A. T. (1995). *Elements of argument: A text and reader.* Boston, MA: Bedford Books.

Rotton, J., & Kelly, I. W. (1985). Much ado about the full moon: A meta-analysis of lunar-lunacy research. *Psychological Bulletin, 97*(2), 286–306.

Rowland, I. (2009). The art of cold reading. *Skeptic, 14*(4), 79–89.

Russell, B. (1929). *Marriage and morals.* London: Allen and Unwin.

Russell, D., & Jones, W. (1980). When superstition fails: Reactions to disconfirmation of paranormal beliefs. *Personality and Social Psychology Bulletin, 6*(1), 83–88.

Sagan, C. (1980). *Cosmos.* New York: Random House.

Sagan, C. (1994). *Pale blue dot: A vision of the human future in space.* New York: Ballantine Books.

Sagan, C. (1996). *The demon-haunted world: Science as a candle in the dark.* New York: Ballantine.

Sagan, C. (2007). *The varieties of scientific experience: A personal view of the search for God.* New York: Penguin.

Saher, M., & Lindeman, M. (2005). Alternative medicine: A psychological perspective. *Personality and Individual Differences, 39,* 1169–78.

Sahu, K. N., Naidu, D. C., & Sankar, K. J. (2014). Radar based lie detection technique. *Global Journal of Researches in Engineering: F Electrical and Electronics Engineering, 14*(5), Version 1.

Salomons, T. V., Johnstone, T., Backonja, M., & Davidson, R. J. (2004). Perceived controllability modulates the neural response to pain. *Journal of Neuroscience, 24,* 7199–203.

Samuels, S. M., & McCabe, G. P. (1986, February 27). More lottery repeaters are on the way. *New York Times.*

Sand, S. (2010). *The invention of the Jewish people* (English ed.). (Y. Lotan, Trans.) New York: Vergo.

Sanitioso, R., Kunda, Z., & Fong, G. T. (1990). Motivated recruitment of autobiographical memories. *Journal of Personality and Social Psychology, 59,* 229–41.

Saxe, L., Dougherty, D., & Cross, T. (1985). The validity of polygraphy testing. *American Psychologist, 40,* 355–66.

Schachter, S., & Singer, J. E. (1962). Cognitive, social, and psychological determinants of emotional states. *Psychological Review, 69*(5), 379–99.

Schacter, D. (2002). *The seven sins of memory: How the mind forgets and remembers* (First Houghton Mifflin Paperback ed.). New York: Houghton Mifflin.

Schacter, D. L., Koutstaal, W., Johnson, M. K., Gross, M. S., & Angell, K. A. (1997). False recollection induced by photographs: A comparison of older and younger adults. *Psychology and Aging, 12,* 203–15.

Schick, T., Jr., & Vaughn, L. (2005). *How to think about weird things: Critical thinking for a new age* (Fourth ed.). Boston, MA: McGraw Hill.

Schlecht, L. F. (1991). Classifying fallacies logically. *Teaching Philosophy, 14*(1), 53–64.

Schlosser, R. W., Balandin, S., Hemsley, B., Iacono, T., Probst, P., & von Tetzchner, S. (2014). Facilitated communication and authorship: A systematic review. *ACC: Augmentative and Alternative Communication, 30*(4), 359–68.

Schneiderman, A. G. (2015). *Major retailers asked to halt sale of certain herbal supplements as DNA tests fail to detect plant materials listed on majority of products tested.* Retrieved July 7, 2020, from http://on.ny.gov/1BSm53a

Schwartz, N., Bless, H., Strack, F., Klumpp, G., Rittenauer-Schatka, H., & Simons, A. (1991). Ease of retrieval as information: Another look at the availability heuristic. *Journal of Personality and Social Psychology, 61*(2), 195.

Scollon, C. N. (2018). Research designs. In R. Biswas-Diener, & E. Diener (Eds.), *Discover psychology 2.0: A brief introductory text* (pp. 16–29). Champaign, IL: DEF Publishers.

Scott, A. J. (2018). Grand illusions and existential angst. *Skeptical Inquirer, 42*(6), 51–55.

Scriven, M., & Paul, R. (Summer 1987). Statement presented at the 8th Annual International Conference on Critical Thinking and Education Reform. Available at: https://www.criticalthinking.org/pages/defining-critical-thinking/766

Seegmiller, J. K., Watson, J. M., & Strayer, D. L. (2011). Individual differences in susceptibility in inattentional blindness. *Journal of Experimental Psychology: Learning, Memory, and Cognition,* 785–91.

Sergent, J., Ohta, S., & MacDonald, B. (1992). Functional neuroanatomy of face and object processing: A positron emission tomography study. *Brain, 115*(Pt. 1), 15–36.

Shaw, J. (2017). *The memory illusion: Remembering, forgetting and the science of false memory.* Toronto: Anchor Canada.

Shaw, J., & Potter, S. (2015). Constructing rich false memories of committing crime. *Psychological Science, 26*(3), 291–301.

Shearer, C., West, M., Caldeira, K., & Davis, S. J. (2016). Quantifying expert consensus against the existence of a secret, large-scale atmospheric spraying program. *Environmental Research Letters, 11*, 084011. doi:10.1088/1748-9326/11/12/129501

Shermer, M. (1997). *Why people believe weird things: Pseudoscience, superstition, and other confusions of our time.* New York: W. H. Freeman.

Shermer, M. (1999). A skeptic in the trenches: A skeptical firewalk. *Skeptic, 7*(3), 10–11.

Shermer, M. (2003a). *How we believe: Science, skepticism, and the search for God* (Second ed.). New York: Henry Holt.

Shermer, M. (2003b). Why smart people believe weird things. *Skeptic, 10*(2), 62–73.

Shermer, M. (2004). *The science of good and evil: Why people cheat, gossip, care, share and follow the golden rule.* New York: Henry Holt and Company.

Shermer, M. (2006, February). *Why people believe weird things.* Retrieved July 7, 2020, from TED: http://www.ted/talks/

Shermer, M. (2007). *Why Darwin matters: The case against intelligent design.* New York: Owl Books.

Shermer, M. (2008a). *The mind of the market: How biology and psychology shape our economic lives.* New York: Henry Holt and Company.

Shermer, M. (2008b). *Patternicity: Finding meaningful patterns in meaningless noise.* Retrieved March 17, 2022, from https://www.scientificamerican.com/article/patter nicity-finding-meaningful-patterns/

Shermer, M. (2010, February). *The pattern behind self-deception.* Retrieved July 7, 2020, from TED: www.ted/talks

Shermer, M. (2011a). *The believing brain: From ghosts and gods to politics and conspiracies—how we construct beliefs and reinforce them as truths.* New York: St. Martin's Press.

Shermer, M. (Producer). (2011b). *Bill Nye "The Science Guy" and James "The Amazing" Randi* [Motion picture]. Los Angeles, CA: Skeptics Society.

Shermer, M. (2013). *Skepticism 101: How to think like a scientist.* Chantilly, VA: The Teaching Company.

Shermer, M. (2020). Why people believe conspiracy theories. *Skeptic, 25*(1), 12–17.

Shermer, M., & Grobman, A. (2009). *Denying history: Who says the Holocaust never happened and why do they say it?* Berkeley: University of California Press.

Shermer, M., Davies, H. B. (Producers), & Shermer, M. (Director). (2007). *Geology, creationism, and evolution: The breathtaking inanity of flood geology* [Motion picture]. Los Angeles, CA: Skeptics Society.

Sherry, D. F., & Schacter, D. L. (1987). The evolution of multiple memory systems. *Psychological Review, 94*, 439–54.

Shtulman, A. (2013). Epistemic similarities between students' scientific and supernatural beliefs. *Journal of Educational Psychology, 105*(1), 199–212. doi:10.1037/a0030282

Siegler, R. (2018). Cognitive development in childhood. In R. Biswas-Diener & E. Diener (Eds.), *Discover psychology 2.0: A brief introductory text* (pp. 100–113). Champaign, IL: DEF Publishers.

Sifferlin, A. (2015, October 3). Over half of E.U. countries are opting out of GMOs. *Time.*

Sihvonen, R., Paavola, M., Malmivaara, A., Itälä, A., Joukainen, A., Nurmi, J., . . . Järvinen, T. L. (2013). Arthroscopic partial meniscetomy versus sham surgery for a degenerative meniscal tear. *New England Journal of Medicine, 369*(26), 2515–24. doi:10.1056/NEJMoal1305189

Simmons, J. A. (1979). Perception of echo phase information in bat sonar. *Science, 204*(4399), 1336–38. doi:10.1126/science.451543

Simons, D. J., & Ambinder, M. S. (2005). Change blindness. *Current Directions in Psychological Science, 14*(1), 44–48.

Simons, D. J., & Chabris, C. F. (1999). Gorillas in our midst: Sustained inattentional blindness for dynamic events. *Perception, 28*, 1059–74.

Simons, D. J., & Chabris, C. F. (2011, August 3). *What people believe about how memory works: A representative survey of the U.S. population.* Retrieved July 7, 2020, from PLoSONE: http://www.plosone.org/article/info%3Adoi%2F10.1371%2Fjournal.pone.0022757

Simons, D. J., & Levin, D. T. (1998). Failure to detect changes to people during a real-world conversation. *Psychonomic Bulletin & Review, 5*(4), 644–49.

Singer, B., & Benassi, V. A. (1980). Fooling some of the people all of the time. *Skeptical Inquirer,* 17–24.

Skeptical Raptor [Blog] (2022). VAERS once more with science—how to use it and how to abuse it. Retrieved March 12, 2022, from https://www.skepticalraptor.com/skepticalraptorblog.php/vaers-once-more-with-science-how-to-use-it-and-how-to-abuse-it/

Skinner, B. F. (1936). The verbal summator and a method for the study of latent speech. *Journal of Psychology, 2*, 71–107.

Skinner, B. F. (1948). "Superstition" in the pigeon. *Journal of Experimental Psychology, 38*, 168–72.

Skurnik, I. C., Yoon, C., Park, D. C., & Schwartz, N. (2005). How warnings about false claims become recommendations. *Journal of Consumer Research, 31*, 713–24. Retrieved from http://research.chicagobooth.edu/cdr/docs/FalseClaims_dpark.pdf

Slater, A., Mattock, A., & Brown, E. (1990). Size constancy at birth: Newborn infants' responses to retinal and real size. *Journal of Experimental Child Psychology, 49*(2), 314–22.

Smedslund, G., & Hagen, K. B. (2011). Does rain really cause pain? A systematic review of the associations between weather factors and severity of pain in people with rheumatoid arthritis. *European Journal of Pain, 15*(1), 5–10.

Smith, G. (2016). Luck and regression to the mean: One of the most fundamental sources of error in human judgment. *Skeptic, 24*(4), 50–53.

Smith, G. (2017). Torturing the data in the name of nonsense. *Skeptic, 22*(1), 20–26.

Smith, G. (2019). Duped by data mining. *Skeptic, 24*(1), 18–21.

Smith, J. C. (2018). *Critical thinking: Pseudoscience and the paranormal* (Second ed.). Hoboken: Wiley-Blackwell.

Smith, M. C., Bibi, U., & Sheard, D. E. (2003). Evidence for the differential impact of time and emotion on personal and event memories for September 11, 2001. *Applied Cognitive Psychology, 17*(9), 1047–55.

Smolin, L. (1997). *The life of the cosmos.* New York: Oxford University Press.

Snook, B., Eastwood, J., Gendreau, P., Goggin, C., & Cullen, R. M. (2007). Taking stock of criminal profiling. *Criminal Justice and Behavior, 34*(4), 437–53.

Sobel, S. (2018). Facilitated communication redux: Persistence of a discredited technique. *Skeptic, 23*(3), 6–9.

Spanos, N. P., Burgess, C. A., & Burgess, M. F. (1994). Past-life identities, UFO abductions, and satanic ritual abuse: The social construction of memories. *International Journal of Clinical and Experimental Hypnosis, 42*, 433–46.

Spanos, N. P., Cross, P. A., Dickenson, K., & DuBreuil, S. C. (1993). Close encounters: An examination of UFO experiences. *Journal of Abnormal Psychology, 102*(4), 624–32.

Specter, M. (2014, August 25). Seeds of doubt: An activist's controversial crusade against genetically modified crops. *New Yorker*.

Spector, R. (2019). Antioxidant megavitamins for brain health: Puffery vs. fact. *Skeptical Inquirer, 43*(2), 46–50.

Spence, D. P. (1996). Abduction tales as metaphors. *Psychological Inquiry, 7*, 177–79.

Sporer, S., Penrod, S., Read, D., & Cutler, B. L. (1995). Choosing, confidence, and accuracy: A meta-analysis of the confidence-accuracy relation in eyewitness identification studies. *Psychological Bulletin, 118*, 315–27.

Stanovich, K. E. (2013). *How to think straight about psychology* (Tenth ed.). New York: Pearson.

Stanovich, K. E., & West, R. F. (2000). Individual differences in reasoning: Implications for the rationality debate? *Behavioral and Brain Sciences, 23*, 645–726.

Steele, E. J. (2005). *Defensive racism: An unapologetic examination of racial differences*. Sagle, ID: ProPer Press.

Steele, K. M., Bass, K. E., & Crook, M. D. (1999). The mystery of the Mozart Effect: Failure to replicate. *Psychological Science, 10*, 366–69.

Stefanek, M., & Jordan, C. D. (2020). Integrative cancer care: Below the bar of science. *Skeptic, 25*(1), 35–39.

Stenger, V. (1995). *The unconscious quantum: Metaphysics in modern physics and cosmology*. New York: Prometheus.

Stenger, V. J. (2000). The pseudophysics of therapeutic touch. In B. Scheiber & C. Selby (Eds.), *Therapeutic touch* (pp. 302–11). Amherst, NY: Prometheus.

Stenger, V. J. (2012). *God and the folly of faith: The incompatibility of science and religion*. Amherst, NY: Prometheus.

Sternberg, R. J. (2004). Why smart people can be so foolish. *European Psychologist, 9*, 145–50.

Strange, D., Sutherland, R., & Garry, M. (2006). Event plausibility does not determine children's false memories. *Memory, 14*(8), 937–51.

Stromberg, J., & Zielinski, S. (2011, October 18). *Ten threatened and endangered species used in traditional medicine*. Retrieved July 7, 2020, from Smithsonianmag.com: http://smithsonian.com/science-nature/ten-threatened-and-endangered-species-use

Sugrue, T. (1973). *There is a river: The story of Edgar Casey*. New York: A.R.E. Press.

Sun, L. H. (2017, May 5). Anti-vaccine activists spark a state's worst measles outbreak in decades. *Washington Post*.

Susskind, L. (2006). *The cosmic landscape: String theory and the illusion of intelligent design*. New York: Back Bay Books.

Svedholm, A., Lindeman, M., & Lipsanen, J. (2010). Believing in the purpose of events—Why does it occur, and is it supernatural? *Applied Cognitive Psychology, 24*(2), 252–65. doi:10.1002/acp.1560

Swami, V., Chamorro-Pemuzic, T., & Furnham, A. (2010). Unanswered questions: A preliminary investigation of personality and individual difference predictors of 9/11 conspiracist beliefs. *Applied Cognitive Psychology, 24*, 749–61. doi:10.1002/acp.1583

Swami, V., Coles, R., Steiger, S., Pietschnig, J., Furnham, A., Rehim, S., & Voracek, M. (2011). Conspiracist ideation in Britain and Austria: Evidence of a monological belief system and associations between individual psychological differences and real-world and fictitious conspiracy theories. *British Journal of Psychology, 102*, 443–63.

Swami, V., Voracek, M., Stieger, S., Tran, U. S., & Furnham, A. (2014). Analytical thinking reduces belief in conspiracy theories. *Cognition, 133*(3), 572–85.

Talarico, J. M., LaBar, K. S., & Rubin, D. C. (2004). Emotional intensity predicts autobiographical memory experience. *Memory & Cognition, 32*(7), 1118–32.

Talarico, J. M., & Rubin, D. C. (2003). Confidence, not consistency, characterizes flashbulb memories. *Psychological Science, 14*(5), 455–61.

Tavel, M. (2018). Quackery in America: An inglorious and ongoing history. *Skeptic, 23*(4), 28–33.

Tavris, C. (2019). The persistence of memory . . . and of the memory wars. *Skeptic, 24*(3), 6–7.

Tavris, C., & Aronson, E. (2019). *Mistakes were made (but not by me)* (Revised and updated ed.). London: Pinter & Martin.

Taylor, K. (2017). *Brainwashing: The science of thought control* (Second ed.). Oxford: Oxford University Press.

Taylor, L. E., Swerdfeger, A. L., & Eslick, G. D. (2014). Vaccines are not associated with autism: An evidenced-based meta-analysis of case-control and cohort studies. *Vaccine, 32*(29), 3623–29. doi:10.1016/j.vaccine.2014.04.085

Tenney, E. R., Cleary, H. M., & Spellman, B. A. (2009). Unpacking the doubt in "beyond a reasonable doubt": Plausible alternative stories increase not guilty verdicts. *Basic and Applied Social Psychology, 31*, 1–8. doi:10.1080/01973530802659687

Thomas, D. (2008). The "chemtrail conspiracy." *Skeptical Inquirer, 18*(3).

Tietz, J. (2012, December 6). Slow-motion torture: How solitary confinement—once reserved for the most dangerous and disobedient inmates—became standard practice in America. *Rolling Stone*, 58–66.

To, K. (2002, December 6). Lie detection: The science and development of the polygraph. *Illumin, 5*(1).

Todd, J. T. (2012, July 13). The moral obligation to be empirical: Comments on Boynton's "Facilitated communication—What harm it can do: Confessions of a former facilitator." *Evidence-Based Communication Assessment and Intervention, 6*(1), 36–57.

Trecek-King, M. (2022a). Teach skills, not facts. *Skeptical Inquirer, 46*(1), 39–42.

Trecek-King, M. (2022b). A life preserver for staying afloat in a sea of misinformation. *Skeptical Inquirer, 46*(2), 44–49.

Trickey, S., & Topping, K. J. (2004). "Philosophy for children": A systematic review. *Research Papers in Education, 19*(3), 365–80.

Turnbull, C. M. (1961). Some observations regarding the experiences and behavior of the Ba Mbuti Pygmies. *American Journal of Psychology, 74*, 304–8.

Tversky, A., & Kahneman, D. (1971). Belief in the law of small numbers. *Psychological Bulletin, 76*, 105–10.

Tversky, A., & Kahneman, D. (1973). Availability: A heuristic for judging frequency and probability. *Cognitive Psychology, 4*, 207–32.

Tversky, A., & Kahneman, D. (1974). Judgment under uncertainty: Heuristics and biases. *Science, 185*(4157), 1124–31. doi:10.1126/science.185.4157.1124

Tversky, A., & Kahneman, D. (1981). The framing of decisions and the psychology of choice. *Science, 211*(4481), 453–58.

Tversky, A., & Kahneman, D. (1983). Extensional versus intuitive reasoning: The conjunction fallacy in probability judgment. *Psychological Review, 90*, 293–315.

Union of Concerned Scientists. (2007). *Smoke, mirrors & hot air: How ExxonMobil used big tobacco tactics to "manufacture uncertainty" on climate science.* Retrieved July 7, 2020, from http://www.ucsusa.org/assets/documents/global_warming/exxon_report.pdf

United States Environmental Protection Agency. (2000). *Aircraft contrails factsheet.* Retrieved July 7, 2020, from http://www.epa.gov

Upton, E. (2014). *The man who accurately estimated the circumference of the Earth over 2,000 years ago.* Retrieved July 7, 2020, from TodayIFoundOut: http://www.todayifoundout.com/index.php/2014/01/amazing-eratosthenes/

Uscinski, J. E. (2019a). Are conspiracy theories "anti-science"? In J. E. Uscinski (Ed.), *Conspiracy theories and the people who believe them* (pp. 199–200). New York: Oxford University Press.

Uscinski, J. E. (2019b). Conspiring for the common good. *Skeptical Inquirer, 43*(4), 40–43.

Uscinski, J. E. (2019c). Down the rabbit hole we go! In J. E. Uscinski (Ed.), *Conspiracy theories and the people who believe them* (pp. 1–32). New York: Oxford University Press.

Uscinski, J. E., & Parent, J. M. (2014). *American conspiracy theories.* New York: Oxford University Press.

Vallone, R. P., Griffin, D. W., Lin, S., & Ross, L. (1990). Overconfident predictions of future actions and outcomes by self and others. *Journal of Personality and Social Psychology, 58*(4), 582–92.

Van Eenennaam, A. L., & Young, A. E. (2014). Prevalence and impacts of genetically engineered feedstuffs on livestock populations. *Journal of Animal Science, 92*(10), 4255–78.

Van Lange, A. M., Taris, T. W., & Vonk, R. (1997). Dilemmas of academic practice: Perceptions of superiority among social psychologists. *European Journal of Social Psychology, 27*, 675–85.

van Panhuis, W. G., Grefenstette, J., Jung, S. Y., Chok, N. S., Cross, A., Eng, H., & Burke, D. S. (2013). Contagious diseases in the United States from 1888 to the present. *New England Journal of Medicine, 369*(22), 2152–58. doi:10/1056/NEJMms1215400

van Prooijen, J.-W. (2011). Suspicions of injustice: The sense-making function of belief in conspiracy theories. In E. Kals & J. Maes (Eds.), *Justice and conflict: Theoretical and empirical contributions* (pp. 121–32). Berlin: Springer Verlag.

van Prooijen, J.-W. (2019). Empowerment as a tool to reduce belief in conspiracy theories. In J. E. Uscinski (Ed.), *Conspiracy theories and the people who believe them* (pp. 432–42). New York: Oxford University Press.

van Prooijen, J.-W., & Jostmann, N. B. (2013). Belief in conspiracy theories: The influence of uncertainty and perceived morality. *European Journal of Social Psychology, 43*(1), 109–15.

Vaughn, L. (2019). *The power of critical thinking: Effective reasoning about ordinary and extraordinary claims* (Sixth ed.). New York: Oxford University Press.

Vilches, T. N., Moghadas, S. M., Sah, P., et al. (2022). Estimating COVID-19 infections, hospitalizations, and deaths following the U.S. vaccination campaigns during the pandemic. *JAMA Network Open, 5*(1), e2142725. doi:10.1001/jamanetworkopen.2021.42725

Vokey, J. R., & Read, J. D. (1985). Subliminal messages: Between the devil and the media. *American Psychologist, 40*(11), 1231–39.

Vyse, S. (2014). *Believing in magic: The psychology of superstition* (Updated ed.). New York: Oxford University Press.

Vyse, S. (2018a). The enduring legend of the changeling. *Skeptical Inquirer, 42*(4), 23–26.

Vyse, S. (2018b). Yes, we do need experts. *Skeptical Inquirer, 42*(1), 20–22.

Vyse, S. (2019a). An adventure in peer review. *Skeptical Inquirer, 43*(6), 27–31.

Vyse, S. (2019b). An artist with a science-based mission. *Skeptical Inquirer, 43*(2), 36–39.

Vyse, S. (2019c). Who are more biased: Liberals or conservatives? *Skeptical Inquirer, 43*(4), 24–27.

Waber, R. L., Shiv, B., Carmon, Z., & Ariely, D. (2008). Commercial features of placebo and therapeutic efficacy. *Journal of the American Medical Association, 299*(9), 1016–17.

Wade, C., & Tavris, C. (2017). *Psychology* (Twelfth ed.). New York: Pearson.

Wade, K. A., Sharman, S. J., Garry, M., et al. (2006). False claims about false memories research. *Consciousness and Cognition, 16*, 18–28.

Wagenaar, W. A., & Keren, G. (1986). The seat belt paradox: Effects of adopted roles on information seeking. *Organizational Behavior and Human Decision Processes, 38*, 1–6.

Wakefield, A. J., et al. (1998). RETRACTED Ileal lymphoid nodular hyperplasia, non-specific colitis, and pervasive developmental disorder in children. *Lancet, 351*(9), 637–41.

Walker, W. R., Hoekstra, S. J., & Vogl, R. J. (2001). Science education is no guarantee for skepticism. *Skeptic, 9*(3), 24–29.

Walton, D. (1996). The argument of the beard. *Informal Logic, 18*, 235–59.

Wegner, D. (2002). *The illusion of conscious will.* London: Bradford Books.

Wegner, D. M., Fuller, V. A., & Sparrow, B. (2008). Clever hands: Uncontrolled intelligence in facilitated communication. *Journal of Personality and Social Psychology, 85*, 5–19.

Weinberg, S. (1994). *Dreams of a final theory: The search for the fundamental laws of nature.* New York: Vintage Books.

Weiser, B. (2012, July 19). New Jersey court issues guidance for juries about reliability of eyewitnesses. *New York Times.*

Weitzenhoffer, A. M., & Hilgard, E. R. (1959). *Stanford hypnotic susceptibility scales, forms A and B.* Palo Alto, CA: Consulting Psychologists Press.

Weller, J. A., Dieckmann, N. F., Tusler, M., Mertz, C. K., Burns, W. J., & Peters, E. (2013). Development and testing of an abbreviated numeracy scale: A Rasch analysis approach. *Journal of Behavioral Decision Making, 26*(2), 198–212. doi:10.1002/bdm.1751

Wells, G. L., & Bradfield, A. L. (1998). Good, you identified the suspect: Feedback to eyewitnesses distorts their reports of the witnessing experience. *Journal of Applied Psychology, 83*, 360–76.

Wells, G. L., Malpass, R. S., Lindsay, R. C., Fisher, R. P., Turtle, J. W., & Fulero, S. M. (2000). From the lab to the police station: A successful application of eyewitness research. *American Psychologist, 55*, 581–98.

Wells, G. L., & Olson, E. A. (2002). Eyewitness identification: Information gain from incriminating and exonerating behaviors. *Journal of Experimental Psychology: Applied, 8*(3), 155–67. doi:10.1037//1079-898X.8.3.155

West, E. J. (2004). Perry's legacy: Models of epistemological development. *Journal of Adult Development, 11*(2), 61–70. doi:10.1023/B:JADE.0000024540.12150.69

West, R. F., Meserve, R. J., & Stanovich, K. E. (2012). Cognitive sophistication does not attenuate the bias blind spot. *Journal of Personality and Social Psychology, 103*(3), 506.

Westen, D., Blagov, P., Harenski, K., Kilts, C., & Hamann, S. (2006). The neural basis of motivated reasoning: An fMRI study of emotional constraints on political judgment during the U.S. presidential election of 2004. *Journal of Cognitive Neuroscience, 18*(11), 1947–58.

Whitbourne, S. K. (2012, June 12). *What psychology can tell you about your extraterrestrial beliefs.* Retrieved July 7, 2020, from *Psychology Today*: http://www.psychologytoday.com/blog/fulfillment-any-age/201206/what-psychology-can-tell-you-about-your-extraterrestrial-beliefs

Whitlock, C., & White, J. (2002, October 26). Cops slow to link Caprice to killings. Retrieved July 7, 2020, from *The Washinigton Post*: http://www.sfgate.com/news/article/Cops-slow-to-link-caprice-to-killings-They-2759421.php

Whitson, J. A., & Galinsky, A. D. (2008). Lacking control increases illusory pattern perception. *Science, 322*(5898), 115–17.

Whitty, C. J., et al. (2013). Biotechnology: Africa and Asia need a rational debate on GM crops. *Nature, 497*(7477).

Whysner, J., & Williams, G. M. (1996). Saccharin mechanistic data and risk assessment: Urine composition enhanced cell proliferation, and tumor promotion. *Pharmacology and Therapeutics, 71*(1–2), 225–52.

Wiegrebe, L. (2008). An autocorrelation model of bat sonar. *Biological Cybernetics, 98*(6), 587–95. doi:10.1007/s00422-008-0216-2

Wiley, L. (1988). Teaching critical thinking in the classroom. *Journal of Correctional Education, 39*(2), 90–92. doi:128.205.204.27

Wilkinson, L. (1999). Statistical methods in psychology journals: Guidelines and explanations. *American Psychologist, 54*, 595–604.

Williams, E. F., & Gilovich, T. (2008). Do people really believe they are above average? *Journal of Experimental Social Psychology, 44*, 1121–28. doi:10.1016/j.jesp.2008.01.002

Williamson, P. (2016). Take the time and effort to correct misinformation. *Nature, 540*, 171. doi:10.1038/540171a

Wilson, E. O. (1980). *Sociobiology* (Abridged ed.). New York: Harvard University Press.

Wilson, E. O. (1998). *Consilience: The unity of knowledge.* New York: Vintage Books.

Wilson, K., & French, C. C. (2006). The relationship between susceptibility to false memories, dissociativity, and paranormal belief and experience. *Personality and Individual Differences, 41*(8), 1493–1502.

Wilson, S. (2013). Vitalistic thinking in adults. *British Journal of Psychology, 104*(4), 512–24.

Wilson, S. C., & Barber, T. X. (1981). Vivid fantasy and hallucinatory abilities in the life histories of excellent hypnotic subjects ("somnabules"): Preliminary report with female subjects. In E. Klinger (Ed.), *Concepts, results, and applications* (pp. 133–50). Boston, MA: Springer.

Wiseman, R. (2011). *Paranormality: Why we believe the impossible.* London: Macmillan.

Wiseman, R., Greening, E., & Smith, M. (2003). Belief in the paranormal and suggestion in the seance room. *British Journal of Psychology, 94*(3), 285–97.

Wiseman, R., & Watt, C. (2006). Belief in psychic ability and the misattribution hypothesis: A qualitative review. *British Journal of Psychology, 97*(3), 323–38.

Wiseman, R., West, D., & Stemman, R. (1996a). An experimental test of psychic detection. *Journal of the Society for Psychical Research, 61*(842), 34–45.

Wiseman, R., West, D., & Stemman, R. (1996b). Psychic crime detectives: A new test for measuring their successes and failures. *Skeptical Inquirer, 20*(1), 38.

Wolfson, R. (2000). *Einstein's relativity and the quantum revolution: Modern physics for non-scientists.* Chantilly, VA: Teaching Co.

Wood, M. J., & Douglas, K. M. (2019). Conspiracy theory psychology: Individual differences, worldviews, and states of mind. In J. E. Uscinski (Ed.), *Conspiracy theories and the people who believe them* (pp. 245–56). New York: Oxford University Press.

Wood, M., Douglas, K. M., & Sutton, R. M. (2012). Dead and alive: Beliefs in contradictory conspiracy theories. *Social Psychological and Personality Science, 3*(6), 767–73.

Woodruff, F. (1998). *Secrets of a telephone psychic.* Hillsboro, OR: Beyond Words.

Wright, D. (1993). Recall of the Hillsborough disaster over time: Systematic biases of "flashbulb" memories. *Applied Cognitive Psychology, 7*, 129–38.

Wright, J. C. (1962). Consistency and complexity of response sequences as a function of schedules of noncontingent reward. *Journal of Experimental Psychology, 63*, 601–9.

Wyman, A., & Vyse, S. (2008). Science versus the stars: A double-blind test of the validity of the NEO five factor inventory and computer generated astrological natal charts. *Journal of General Psychology, 135*(3), 287–300.

Xu, S., Wang, L., Cooper, E., Zhang, M., Manheimer, E., Berman, B., & Lao, L. (2013). Adverse events of acupuncture: A systematic review of case reports. *Evidence-Based Complementary and Alternative Medicine*, 1–15. doi:10.1155/2013/581203

Yeung, C. A. (2008). A systematic review of the efficacy and safety of fluoridation. *Evidence-Based Dentistry, 9*(2), 39–43.

Zeigler, D. (2020). What's to be gained through rationality? *Skeptic, 25*(2), 41–43.

Zimbardo, P. (2008). *The Lucifer effect: Understanding how good people turn evil.* New York: Random House.

Index

Note: Page numbers in *italics* refer to figures and tables.